Foundation Vibration Analysis:
A Strength-of-Materials Approach

Foundation Vibration Analysis:
A Strength-of-Materials Approach

John P. Wolf
Andrew J. Deeks

AMSTERDAM • BOSTON • HEIDELBERG • LONDON • NEW YORK • OXFORD
PARIS • SAN DIEGO • SAN FRANCISCO • SINGAPORE • SYDNEY • TOKYO

ELSEVIER

Elsevier
Linacre House, Jordan Hill, Oxford OX2 8DP
200 Wheeler Road, Burlington, MA 01803

First published 2004

Copyright © 2004, John P. Wolf and Andrew J. Deeks. All rights reserved

The right of John P. Wolf and Andrew J. Deeks to be identified as the authors of this
work has been asserted in accordance with the Copyright, Designs
and Patents Act 1988

No part of this publication may be reproduced in any material form (including
photocopying or storing in any medium by electronic means and whether
or not transiently or incidentally to some other use of this publication) without
the written permission of the copyright holder except in accordance with the
provisions of the Copyright, Designs and Patents Act 1988 or under the terms of
a licence issued by the Copyright Licensing Agency Ltd, 90 Tottenham Court Road,
London, England W1T 4LP. Applications for the copyright holder's written
permission to reproduce any part of this publication should be addressed
to the publisher

Permissions may be sought directly from Elsevier's Science & Technology Rights
Department in Oxford, UK: phone: (+44) 1865 843830, fax: (+44) 1865 853333,
e-mail: permissions@elsevier.co.uk. You may also complete your request on-line via
the Elsevier homepage (http://www.elsevier.com), by selecting 'Customer Support'
and then 'Obtaining Permissions'

British Library Cataloguing in Publication Data
A catalogue record for this book is available from the British Library

Library of Congress Cataloguing in Publication Data
A catalogue record for this book is available from the Library of Congress

ISBN 0 7506 6164 X

For information on all Elsevier publications
visit our website at http://books.elsevier.com

Printed and bound in Great Britain by Biddles, Kings Lynn

Contents

Foreword   *ix*
Preface   *xi*
Acknowledgements   *xiv*

1 Introduction   1
    1.1 Statement of the problem   1
    1.2 Organisation of the text   7

2 Concepts of the cone model   10
    2.1 Rigorous methods   10
    2.2 Wave propagation in a truncated semi-infinite homogeneous cone   11
    2.3 Wave reflection and refraction at a material discontinuity in a cone   17
    2.4 Disk on the surface of a layered half-space   23
    2.5 Disk embedded in a layered half-space   24
    2.6 Foundation embedded in a layered half-space   26
    2.7 Features of the cone model   28

3 Initial cone with outward wave propagation   29
    3.1 Translational cones   30
    3.2 Rotational cones   35
    3.3 Interpretation of the rocking cone   40
    3.4 Incompressible or nearly-incompressible half-space   44
       3.4.1 Wave velocity   44
       3.4.2 Trapped mass   44
    3.5 Foundation on the surface of a homogeneous half-space   46
    3.6 Double cones   52

4 Wave reflection and refraction at a material discontinuity   55
    4.1 Reflection coefficient for a translational cone   55
    4.2 Reflection coefficient for a rotational cone   60
    4.3 Dynamic stiffness of a surface foundation on a layer overlying a half-space   63
    4.4 Disk embedded in a homogeneous half-space   70

	4.5	Computer implementation	72
	4.6	Termination criteria	81

5 Foundation embedded in a layered half-space — 84
5.1 Stack of embedded disks — 84
5.2 Dynamic flexibility of the free field — 86
5.3 Dynamic stiffness and effective foundation input motion — 87
5.4 Computer implementation — 93
5.5 Examples — 97

6 Evaluation of accuracy — 106
6.1 Foundation on the surface of a layered half-space — 107
6.2 Foundation embedded in a layered half-space — 108
6.3 Large number of cone segments — 112
6.4 Cutoff frequency — 113
6.5 Incompressible case — 114
6.6 Hemi-ellipsoid embedded in a homogeneous half-space — 115
6.7 Sphere embedded in a homogeneous full-space — 117

7 Engineering applications — 121
7.1 Machine foundation on the surface of a layered half-space — 121
    7.1.1 Dynamic load of single cylinder machine — 121
    7.1.2 Dynamic load of two-cylinder machine with cranks at 90° — 124
    7.1.3 Dynamic system — 125
    7.1.4 Equations of motion — 127
7.2 Seismic analysis of a structure embedded in a layered half-space — 129
7.3 Offshore wind turbine tower with a suction caisson foundation — 135

8 Concluding remarks — 142

Appendix A: Frequency-domain response analysis — 146
A.1 Alternative descriptions of harmonic motion — 146
A.2 Complex frequency response function — 148
A.3 Periodic excitation — 151
A.4 Arbitrary excitation — 152

Appendix B: Dynamic soil-structure interaction — 154
B.1 Equations of motion in total displacement — 154
B.2 Free-field response of site — 157

Appendix C: Wave propagation in a semi-infinite prismatic bar — 162

Appendix D: Historical note — 168

Appendix E: Program CONAN (CONe ANalysis) – user's guide — 170
E.1 Program overview — 170
E.2 Problem description — 171

E.3	Using CONAN	173
E.4	Further processing of results	176

Appendix F: MATLAB® Procedures for cone analysis — 177

F.1	MATLAB overview	177
F.2	Problem description	178
F.3	General functions	179
F.4	Heart of the procedure	181
F.5	Dynamic stiffness of the free field	185
F.6	Dynamic stiffness of the foundation	186
F.7	Effective foundation input motion	190
F.8	Worked example: seismic response	191

Appendix G: Analysis directly in time domain — 196

G.1	Flexibility analysis for translation	197
G.2	Interaction force-displacement relationship in the time domain for translation	201
G.3	Seismic analysis of a rigid block on the surface of a layered half-space	204
G.4	Rotation analysis	205

References   209
Dictionary   212
Index   215

Foreword

The exponential rise in computing capability over the last few decades has permitted the solution, in principle, of even the most complex of soil-structure interaction problems by means of detailed numerical analysis. However, there is still a gulf between the potential to analyse such problems and the practical ability to do so, either for a particular application or for research purposes. Instead, recourse is made to a range of simplifications in order to focus on that part of the total problem which is deemed to be critical. Thus foundation engineers may represent the superstructure by an idealised elastic block, or even as a uniform load applied to the ground, while structural engineers may represent the ground as a distributed bed of springs or as a rigid boundary.

Where the superstructure is modelled, it will generally be discretised not into three-dimensional continuum elements, but into a collection of one-dimensional (bar or beam) and two-dimensional (shell or plate) elements. Such elements represent what is referred to as the 'strength-of-materials' approach in this book. They make use of the dominant geometric axes evident in most structural components, together with simplifying assumptions that allow quantification of appropriate stiffness matrices for the elements. They may also incorporate more subtle rules of response that allow for local three-dimensional effects, such as edge buckling of a beam, even though the basic model is one or two dimensional.

Analysis of the ground response is less obviously amenable to treatment using simplified elements. Instead, continuum elements are almost universally adopted, although often the overall geometry of the problem is simplified to two dimensions, either in plane strain or axial symmetry. In static problems, advantage can be taken of the (commonly assumed) horizontal stratification of soil and rock using the finite layer techniques pioneered by Booker and Small, reducing the three-dimensional problem to two dimensions, although sophisticated software development is still necessary. In dynamic problems, a more typical simplification of the ground response has been the crude idealisation by lumped springs and dashpots, perhaps incorporating a plastic slider to represent the limit on bearing capacity. Such models have been used in the analysis of impact and vibration response of piles embedded in soil, but cannot easily be extended to deal with stratified soil, or with more complex foundation geometries.

The authors of the present book have set out to establish a set of 'strength-of-materials' idealisations for the geotechnical response in the analysis of foundation problems. The idealisations are akin to the bar and beam elements familiar in structural engineering, and are based on the truncated cone model developed by Meek in collaboration with Veletsos and Wolf. In his previous book (Foundation Vibration Analysis Using Simple Physical Models, Prentice Hall, 1994) Wolf extended the truncated cone model to a range of physical problems and different modes of dynamic excitation, and also documented various lumped parameter models. While the alternative solutions provided in that book have proved most useful to practising geotechnical engineers, the present book focuses on a self-contained development of the truncated cone models. This will allow practitioners to develop their own armoury of techniques for problems of interest, whether arising from structural-induced vibrations or from seismic events. Of particular merit is the inclusion of MATLAB software in the book, together with a full executable program, CONAN, which provides individuals with a starting point for custom-designed solutions.

The basic solutions are developed in the frequency domain, and thus are restricted to linear response of the soil. The building blocks are the vertical, horizontal, rotational and torsional response of a disk foundation resting on the surface of a homogeneous half-space. The power of the book rests in the detailed development of these primitive solutions, using the concept of reflection and refraction at layer interfaces, to address embedded foundations with quite arbitrary shapes in stratified soil deposits underlain by either a rigid base or an infinite medium. With well-chosen example applications (supplemented by segments of MATLAB code), the reader is shown how to apply the techniques to assess the response to free-field (seismic) ground motions or vibrations internally generated within the structure.

Treatment of the subject is comprehensive, with detailed appendices covering development of the basic solutions and their integration to address practical problems. The flow of the main text is left deliberately uncluttered in this way, and even historical documentation of the truncated cone solutions is summarised neatly in a separate appendix.

This is the sixth book by John Wolf and I am confident that it will prove as much a landmark as his previous books on dynamic soil-structure interaction. He is joined in the present book by a co-author, Andrew Deeks, whose positive influence can be seen clearly in the elegant MATLAB code and CONAN executable. The authors have taken pains to evaluate the accuracy of their approach against closed form and rigorous numerical solutions. As they point out, even at worst the errors are relatively minor in comparison with other uncertainties in the problem, particularly those associated with characterising the dynamic properties of the geotechnical medium. Just as in structural analysis there are limitations to conventional 'strength-of-materials' solutions, so there will be situations where the approaches described here may prove insufficient. However, this book establishes a powerful basis for a 'strength-of-materials' approach to dynamic foundation problems and will no doubt prove invaluable across the spectrum of practising engineers, researchers and teachers.

Mark F. Randolph
University of Western Australia

Preface

Most structural analysis is performed based on the strength-of-materials approach using bars and beams. Postulating the deformation behaviour ('plane sections remain plane'), the complicated exact three-dimensional elasticity is replaced by a simple approximate one-dimensional description that is adequate for design. The approach is very well developed, permitting complicated structural systems, such as curved skewed prestressed bridges with moving loads, to be modelled with one-dimensional bars and beams. This strength-of-materials theory is extensively taught in civil and mechanical engineering departments using the excellent textbooks available in this field.

In contrast, in geotechnical engineering, the other field of civil engineering where modelling is important, the strength-of-materials approach is not being used extensively. There are two main reasons for this. First, while in structural engineering the load bearing elements to be analysed tend to have a dominant direction determining the axes and cross-sectional properties of the bars and beams, in geotechnical engineering three-dimensional media, the soil and rock, are present. The choice of the axes and especially the cross-sectional properties (tributary section), which must be able to represent all essential features with the prescribed deformation behaviour, is thus more difficult in geotechnical engineering than in structural engineering. Second, up to quite recently, the state of development of the method was severely limited. Even just over ten years ago, only surface foundations on a homogeneous half-space representing the soil could be modelled with a strength-of-materials approach using conical bars and beams, which are called cones in the following. As the soil properties in an actual site will change with depth, this approach was only of academic interest.

This pioneering effort did, however, form the basis of important recent developments. Today, based on the same assumptions, reasonably complicated practical cases can be analysed. The site can exhibit any number of horizontal layers, permitting the modelling of a general variation of the properties with depth. Besides surface foundations, embedded foundations can be analysed. Seismic excitation can be processed without introducing any additional assumptions. Thus, the cone models can be used to model the foundation in a dynamic soil-structure-interaction analysis. Cone models work well for the low- and intermediate-frequency ranges important for machine vibrations and earthquakes, for the limit of very high frequencies as occurring for impact loads, and for the other limit, the static case. By simplifying the physics of the problem, conceptual clarity with physical insight results. In the cone models, the wave pattern is clearly postulated. The wave propagates outwards away from the disturbance spreading in the direction of propagation within the cross-section of the cone. When a discontinuity of the material properties corresponding to an interface of the soil layers is encountered, two new waves are generated: a reflected wave and a refracted wave, propagating in their own cones. When modelling with cones, the analyst feels at ease, as the same familiar concepts of strength of materials used daily in structural analysis are applied. This is in contrast to using rigorous methods, based on three-dimensional elastodynamics with a considerable mathematical complexity, which tend to intimidate practitioners and obscure physical insight. Due to the simplification of the physical problem, the mathematics of the cone models can be solved rigorously. The fundamental principles of wave propagation and dynamics are thus satisfied exactly for the cones. Closed-form solutions exist for these one-dimensional cases. This leads to simplicity in a practical application. The use of cone models does indeed lead to some loss of precision compared to applying the rigorous methods of elastodynamics. However, this is more than compensated by the many advantages mentioned above. It must also be remembered that the accuracy of any analysis will always be limited by significant uncertainties, such as in the material properties of the soil, which cannot be avoided. Summarising, the ease of use with physical insight especially, the sufficient generality and the good accuracy allow the cone models to be applied for foundation vibration and dynamic soil-structure-interaction analyses in everyday cases in a design office. It is fair to state that a balanced design using cone models leads to simplicity that is based on rationality, which is the ultimate sophistication!

Starting from scratch, the one-dimensional strength-of-materials theory for conical bars and beams, called cones, is developed and applied to practical foundation vibration problems. No prerequisites other than elementary notions of mechanics, which are taught in civil engineering departments of all universities, are required. In particular, concepts of structural dynamics are not needed to calculate the dynamic behaviour of a foundation. (To perform a dynamic soil-structure-interaction analysis the structure must also be modelled, which is, however, outside the scope of this book.) The elementary treatment is restricted to harmonic excitation (the frequency domain) in the main text, with a direct time domain analysis developed as an extension in an appendix. The transformation from the time domain to the frequency domain using a Fourier series is described in an appendix. The equations of motion of dynamic soil-structure interaction are also addressed in an appendix. As the transformation to modal coordinates, which is so powerful in structural dynamics, cannot be used for foundations because they are semi-infinite domains, wave propagation plays a key role. Wave motion in prismatic bars is introduced in an appendix, and wave propagation in one-dimensional cones is described in great detail throughout the book. Only two aspects of wave motion are actually needed: the outward propagation of waves in the initial cone away from the disturbance and the generation of the reflected and refracted waves at a material discontinuity corresponding to a soil layer interface. By tracking the reflection and refraction of each incident wave sequentially, the superimposed wave pattern up to a certain stage can be established. This yields a significant simplification in formulation and programming. A thorough evaluation of the accuracy for a wide range of actual sites is performed. A short computer program written in MATLAB forms an integral part of the book. It is introduced in stages in the various chapters of the book. A full understanding of all aspects of the code, which can easily be modified by the user, results. In addition, an executable computer program called CONAN (CONe ANalysis) with a detailed description of the input and output is provided, which can be used to analyse practical cases. A complicated machine foundation problem, a typical seismic soil-structure-interaction problem and an offshore wind turbine tower with a suction caisson foundation are analysed as examples. A dictionary translating the key technical expressions into various languages increases the international acceptance of the book.

Many of the ideas contained in this book developed gradually over the last ten years. Another book by the senior author, titled *Foundation Vibration Analysis Using Simple Physical Models*

(Prentice Hall, 1994), was written primarily to appeal to geotechnical consultants and contains a very complete description of simple physical models, where, besides cones, lumped-parameter models (spring-dashpot-mass models) and prescribed horizontal wave patterns are also derived. However, the completeness and thus redundancy of the book tend to irritate the reader. Also, significant advances have been made in the area of cone models since the publication of that book. This leads to the current book, which is self-contained, without any prerequisites, and concentrates on the method of cones, which is developed using the standard assumptions of the theory of strength of materials only. Very recent research by the authors, which streamlines the formulation, is incorporated. Following the suggestions of readers of the previous book over the years, a computer program for the analysis of practical cases is fully integrated and explained in detail. The new book is a state-of-the-art treatise regarding cone models, but can also be used as the basis for a first course in soil dynamics of geotechnical engineering (at the final year undergraduate or first year postgraduate level), and can be taught in a course in structural dynamics, as all structures have foundations that have to be analysed. As the students study bars and beams extensively in elementary structural engineering, the basis for the extension to dynamics is very solid. In addition, the book will be valuable to practising geotechnical engineers, who should only apply a computer program when they fully grasp the computational procedure it is based on. The computational procedure detailed in the book will be familiar to them, as the strength-of-materials approach is the same as used routinely in structural analysis.

The contribution of Matthias Preisig in his Diploma-thesis, which clearly demonstrates the potential of the streamlined formulation using cones, is noted. The creative research of Dr Jethro W. Meek, performed in an informal, enthusiastic and collegial atmosphere with the senior author in the beginning of the 1990s, which forms the basis of the strength-of-materials approach, is gratefully acknowledged. The authors are indebted to Professors Eduardo Kausel of MIT and John Tassoulas of the University of Texas at Austin who calculated on our request the results for comparison. Without this support a systematic evaluation of the accuracy would not have been possible. Provision of simulated strong ground motion for the 1989 Newcastle earthquake by Dr Nelson Lam of the University of Melbourne is also acknowledged with thanks. The authors are indebted to Professor Mark Randolph, Director of the Centre for Offshore Foundation Systems at the University of Western Australia, for writing the Foreword.

John P. Wolf
Swiss Federal Institute of Technology
Lausanne

Andrew J. Deeks
The University of Western Australia
Perth

Acknowledgements

The authors would like to thank the publishers below, who kindly allowed permission for the reproduction of their copyrighted figures and tables. The numbers given in the square brackets relate to the reference list.

Chapter 2: Figures 2.3, 2.5 and 2.7 [37] are reproduced by permission of Prentice Hall.
Chapter 3: Figure 3.13 and Table 3.1 [37] are reproduced by permission of Prentice Hall.
Chapter 6: Figures 6.1 and 6.6 [42] are reproduced by permission of John Wiley and Sons Ltd.
Chapter 7: Figure 7.2 [37] is reproduced by permission of Prentice Hall.

1

Introduction

1.1 Statement of the problem

The following preliminary remark is appropriate. To address the goal of foundation vibration analysis, certain terms such as dynamic stiffness or effective foundation input motion are introduced. At this stage of development only a sketchy qualitative description without a clear definition is possible. The reader should not become irritated. From Chapter 2 onwards the treatment is systematic, from the bottom up and rigorous.

The objective of foundation vibration analysis is illustrated in Fig. 1.1. The response of a massless cylindrical foundation of radius r_0 embedded with depth e in a layered soil half-space is to be calculated for all degrees of freedom. The vertical wall and the horizontal base of the foundation are assumed to be rigid. As a special case a circular surface foundation can be addressed, which corresponds to $e = 0$. Horizontal layering exists with constant material properties in each layer. The jth layer with thickness d_j has shear modulus G_j, Poisson's ratio ν_j, mass density ρ_j and a hysteretic damping ratio ζ_j ($j = 1, 2, \ldots, n - 1$). The underlying homogeneous half-space is denoted with the index n. The site can also be fixed at its base (rigid underlying half-space). Linear behaviour of the site is assumed, meaning that the soil is assumed to remain *linearly elastic with hysteretic material damping* during dynamic excitation. This can be justified by noting that the allowable displacements of foundations for satisfactory operation of machines are limited to fractions of a millimetre. It should also be noted that all waves propagating towards infinity decay due to geometric spreading, resulting in soil which can be regarded as linear towards infinity. Inelastic deformations are thus ruled out.

Two types of dynamic loads, which vary with time, are considered. These consist of loads acting directly on the rigid foundation at point O (Fig. 1.1), originating from rotating machinery, for example, and excitations introduced through the soil, from seismic waves, for example. For the latter excitations only vertically propagating waves are considered, with the particle motion in either the horizontal or the vertical direction. The so-called *free-field motion*, i.e. the displacements in the virgin site before excavation, is shown schematically for these horizontal and vertical earthquakes on the left-hand side of Figs 1.2a and 1.2b respectively.

As a slight extension, any axi-symmetric foundation can be examined (Fig. 1.3). The wall does not have to be vertical, but the base must remain horizontal. The wall and the base are again assumed to be rigid.

2 Foundation Vibration Analysis: A Strength-of-Materials Approach

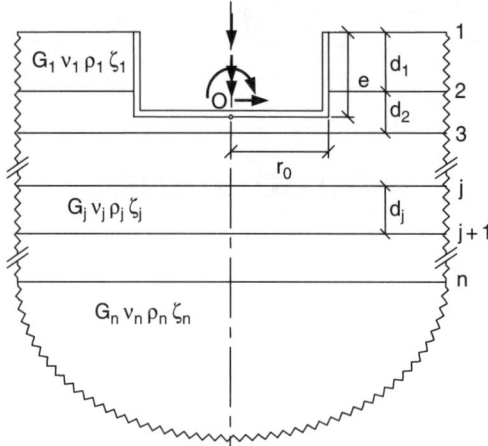

Figure 1.1 Cylindrical foundation embedded in layered soil half-space with degrees of freedom

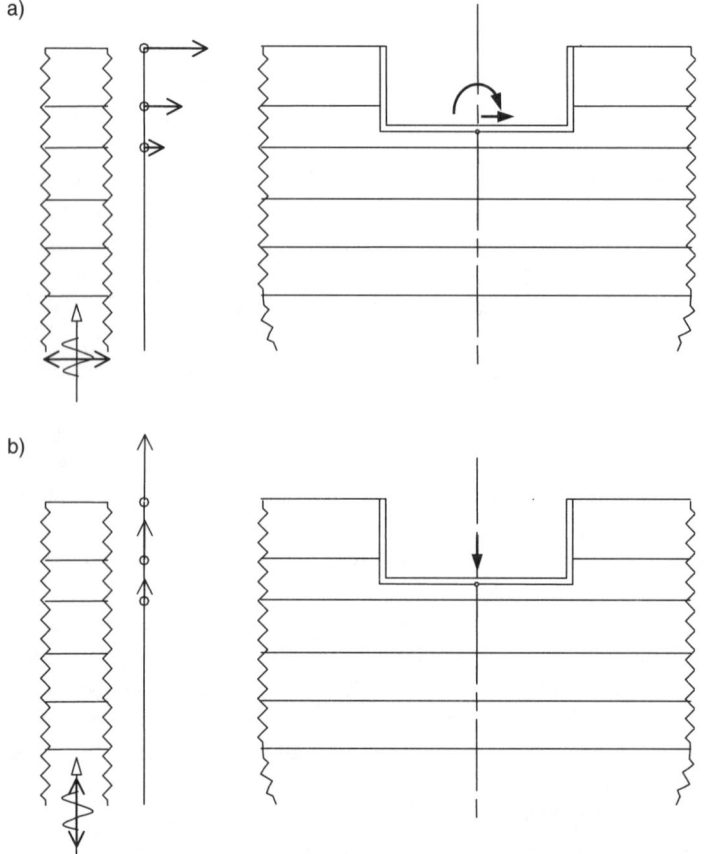

Figure 1.2 Free-field motion and effective foundation input motion for vertically propagating seismic excitation. a) Horizontal earthquake. b) Vertical earthquake

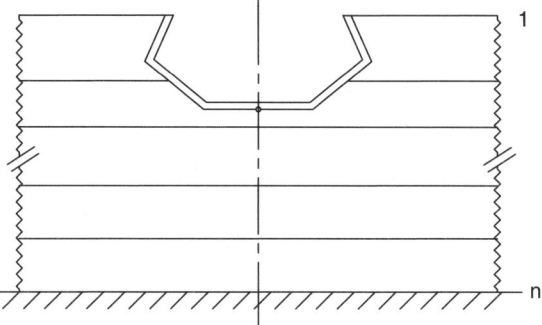

Figure 1.3 Axi-symmetric foundation embedded in soil layers fixed at base

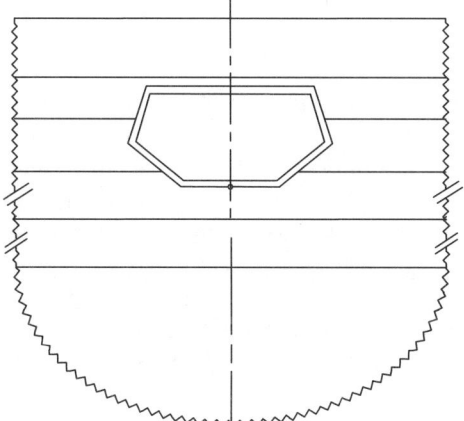

Figure 1.4 Fully-embedded foundation in layered soil half-space

Axi-symmetric inclusions can also be processed (Fig. 1.4). They are termed fully-embedded foundations, while the cases where the wall intersects the free surface are referred to as partially-embedded where a distinction is appropriate. Of course, a foundation in a full-space is always fully-embedded.

More general foundations can be transformed to axi-symmetric cases. This can be accomplished by equating a certain quantity of the general foundation to the corresponding quantity of the axi-symmetric case. For instance, when translational degrees of freedom dominate, areas in the horizontal section can be equated, while when rotational degrees of freedom dominate, moments of inertia in the horizontal section can be equated.

In many applications a structure is also present, with the structure-soil interface coinciding with the rigid wall and base of the foundation (Fig. 1.5). In this case two substructures are present, the foundation embedded in the soil and the structure. The two substructures are connected at point O forming a coupled system. This defines a *dynamic unbounded soil-structure-interaction problem*. Exterior loads can also be applied to the structure, and, as already mentioned, the dynamic excitation can be introduced through the soil (by seismic waves, for example). In such problems the responses of the structure and, to a lesser extent, of the soil are to be determined.

The coupling of the substructures enforces equilibrium and compatibility of the displacements and rotations at O. However, the dynamic behaviour of the unbounded soil, a semi-infinite domain,

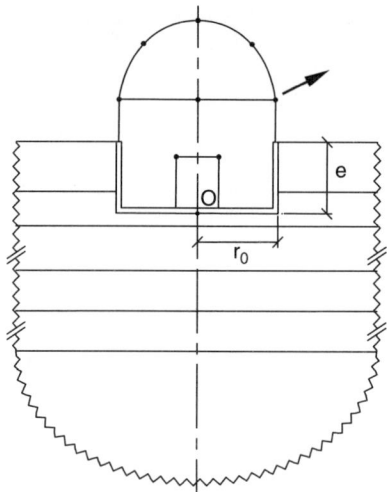

Figure 1.5 Soil-structure interaction with structure embedded in layered soil half-space

is significantly different from that of the bounded structure with finite dimensions. The motion of the structure-soil interface triggers waves propagating in all directions in the soil towards infinity. Reflections occur at the free surface, and both reflections and refractions occur at the soil layer interfaces. This complicated wave pattern radiates energy towards infinity, outside the dynamic system. The unbounded soil thus acts as an energy sink, resulting in damping (which is known as *radiation damping*) even in a linear system, in contrast to the bounded structure. The challenge in analysing the dynamic soil-structure interaction consists of modelling the soil illustrated in Fig. 1.1.

Thus, the rigid foundation embedded in a layered soil half-space, or, as it can also be described, the unbounded layered soil containing an excavation with a rigid interface, is addressed. This substructure's dynamic properties are defined on the interface with the other substructure, the structure, at point O. For seismic excitation two quantities must be determined: first, the interaction force-displacement relationship determining the contribution of the unbounded soil to the dynamic stiffness of the coupled equations; and second, the so-called effective foundation input motion arising from the seismic excitation introduced through the soil.

As the unbounded soil remains linear, the dynamic analysis can be performed in the frequency domain. As outlined in Appendices A.3 and A.4, the dynamic excitation in the time domain is expressed as the sum of a series of harmonic components (Fourier series and integral). It is thus sufficient to address a discrete harmonic excitation with a specific frequency ω, characterised by the corresponding complex amplitude, as discussed in Appendix A.1. The amplitude of the response for this harmonic excitation follows as the product of the complex frequency response function (examined in Appendix A.2) and the amplitude of the excitation.

In general, the loading applied to the structure will not be axi-symmetric. The interaction force-displacement relationship for harmonic excitation at point O (Fig. 1.1) is

$$\{P(\omega)\} = [S(\omega)]\{u(\omega)\} \tag{1.1}$$

with $\{u(\omega)\}$ denoting the amplitudes of the three displacements and three rotations at O, $\{P(\omega)\}$ the amplitudes of the three forces and three moments at O and $[S(\omega)]$ representing the *dynamic-stiffness matrix*, the complex frequency response function. As the foundation is axi-symmetric, the vertical and torsional degrees of freedom are uncoupled. Coupling does, however, exist for the horizontal and rotational (rocking) degrees of freedom, leading to an off-diagonal term in $[S(\omega)]$. In the basic equation of motion of the coupled soil-structure system derived in Appendix B.1, this dynamic-stiffness matrix is denoted as $[S_{00}^g(\omega)]$ (Eq. B.4, Fig. B.2). (The unbounded soil with the excavation is also denoted as the soil system 'ground', leading to the superscript g.)

For each coefficient $S(\omega)$ of the matrix

$$P(\omega) = S(\omega)u(\omega) \tag{1.2}$$

is formulated, omitting the indices indicating the position of the coefficient. In foundation dynamics it is appropriate to introduce the *dimensionless frequency*

$$a_0 = \frac{\omega r_0}{c_{s1}} \tag{1.3}$$

with r_0 representing a characteristic length of the foundation, for instance the radius of the cylinder, and c_{s1} denoting the shear-wave velocity of the first soil layer

$$c_{s1} = \sqrt{\frac{G_1}{\rho_1}} \tag{1.4}$$

Using a stiffness coefficient K to non-dimensionalise the dynamic-stiffness coefficient

$$S(a_0) = K[k(a_0) + ia_0 c(a_0)] \tag{1.5}$$

is formulated. (Such a decomposition is also introduced in Appendix A.2 (Eq. A.17) with very simple $k(a_0)$ and $c(a_0)$ (Eq. A.18).) The dimensionless spring coefficient $k(a_0)$ governs the force that is in phase with the displacement, and the dimensionless damping coefficient $c(a_0)$ describes the force that is 90° out of phase. The dynamic-stiffness coefficient $S(a_0)$ can thus be interpreted as a spring with the frequency-dependent coefficient $Kk(a_0)$ and a parallel dashpot with the frequency-dependent coefficient $(r_0/c_{s1})Kc(a_0)$ (since $\dot{u}(a_0) = i\omega u(a_0)$, Eq. A.9). This relationship is illustrated in Fig. 1.6.

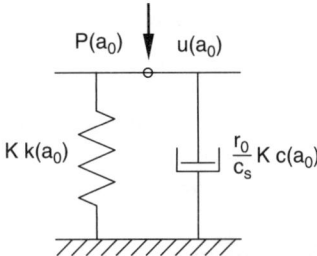

Figure 1.6 Physical interpretation of dynamic-stiffness coefficient for harmonic excitation as spring and dashpot in parallel with frequency-dependent coefficients

The dynamic-stiffness coefficient is complex, with real and imaginary parts. When only one quantity is to be addressed, for instance when checking the accuracy, the magnitude

$$|S(a_0)| = K\sqrt{k^2(a_0) + [a_0 c(a_0)]^2} \tag{1.6}$$

is often chosen, representing the largest value of the force $P(a_0)$ for a unit $u(a_0)$. (This is analogous to the reasoning used when a response spectrum analysis is applied.)

The *effective foundation input motion* is equal to the motion of the unbounded soil containing an excavation with a rigid interface (soil system ground) caused by the seismic waves (Fig. B.2), and is described by the amplitudes $\{u_0^g(\omega)\}$. Only a qualitative description without any equations is feasible at this stage. As already mentioned, only waves propagating vertically in the free field are considered. In general, the motion will amplify when propagating towards the free surface. The free-field analysis procedure for harmonic excitation is described in Appendix B.2. The horizontal earthquake with horizontal particle motion is shown schematically on the left-hand side of Fig. 1.2a. The rigid interface along the excavation will lead to a horizontal motion 'averaged' over the embedment depth, and, as the free-field motion at the free surface will differ from that at the level of the base, also to a rotation (rocking). These two components of the effective foundation input motion are shown on the right-hand side of Fig. 1.2a. The vertical earthquake with vertical particle motion in the free field is presented on the left-hand side of Fig. 1.2b. The rigid interface will lead to a vertical component 'averaged' over the embedment depth for the effective foundation input motion, as indicated on the right-hand side of Fig. 1.2b. For a surface foundation the effective foundation input motion equals the corresponding free-field displacement at the free surface. No rotational component occurs.

The effective foundation input motion with amplitudes $\{u_0^g(\omega)\}$ appears on the right-hand side of the basic equation of motion for the soil-structure-interaction analysis derived in Appendix B.1 (Eq. B.4).

The dynamic-stiffness matrix and effective foundation input motion are calculated based on a one-dimensional strength-of-materials approach. The latter is explained here in general terms addressing the dynamic-stiffness coefficient in the vertical direction of a surface foundation. It also applies to the other degrees of freedom, and can be extended to an embedded foundation, where, in addition to the dynamic stiffness, the effective foundation input motion is formulated with the same assumptions. Only the basic notion without any details or derivations can be provided at this stage.

To construct a strength-of-materials approach, physical approximations such as specifying the deformation behaviour ('plane sections remain plane') are introduced, which at the same time simplify the mathematical formulation. The one-dimensional strength-of-materials approach addresses *conical bars and beams*, which are termed *cones* in the following discussion.

The vertical degree of freedom of a circular foundation (also called a disk) on the surface of an elastic homogeneous soil half-space is addressed (Fig. 1.7a). To determine the interaction force-displacement relationship of the disk on the half-space and its dynamic stiffness, the disk's displacement is prescribed and the corresponding interaction force (the load acting on the disk) is calculated. The half-space below the disk is modelled as a *truncated semi-infinite* bar with its area varying as in a *cone* with the same material properties as the half-space. A load applied to the disk on the free surface of a half-space leads to stresses acting on an area that increases with depth due to geometric spreading, which is also the case for the cone. A *wave propagates in this so-called initial cone away from the loaded disk* outwards (which is in this case downwards) *with the cross-sectional area increasing in the direction of wave propagation* towards infinity, yielding axial strains (Fig. 1.7a). This is taken as the first building block in constructing the computational procedure.

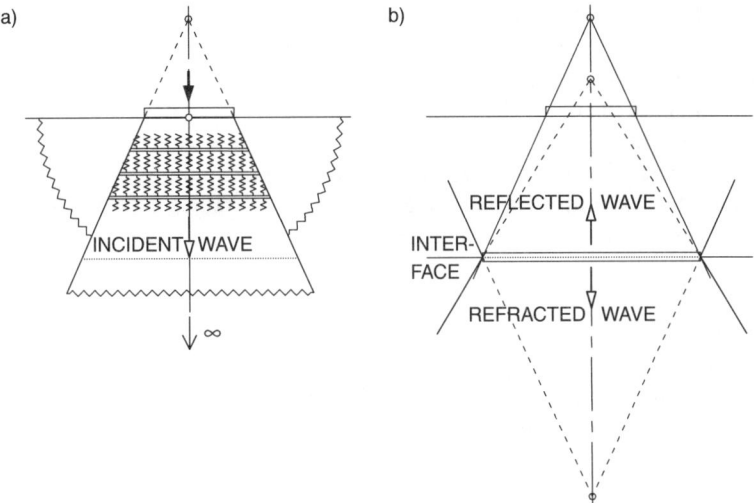

Figure 1.7 Wave propagation in cones. a) Initial cone with outward wave propagation. b) Reflected and refracted waves at a material discontinuity propagating in their own cones

When the half-space is layered, the propagating wave in the *initial cone* will encounter a material discontinuity, which corresponds to a layer interface, as an *incident wave* (Fig. 1.7a). This creates two new waves (Fig. 1.7b), a *reflected wave* propagating upwards and a *refracted wave* propagating downwards, each propagating in their own (initial) cone, again with the cross-sectional area increasing in the direction of wave propagation. This is the second building block. These generated waves will also encounter material discontinuities as incident waves at a later stage, yielding additional reflections and refractions. By *tracking the reflection and refraction* of each incident wave sequentially, the *superimposed wave pattern* can be established for a layered site up to a certain stage.

Within the constraints of the strength-of-materials assumptions, the computational procedure for a surface foundation is analytical. No numerical approximation needs to be introduced.

The use of cone models does indeed lead to some loss of precision compared to applying rigorous methods based on three-dimensional elastodynamics. However, this is more than compensated for by the many advantages. The strength-of-materials approach using cones leads to *physical insight with conceptual clarity*, is *simple to use* and solve, as the mathematical solution is simplified, and provides *sufficient generality* (layered site, embedment, all frequencies) and acceptable *engineering accuracy*. The accuracy of any analysis is limited anyway, because of the many uncertainties, some of which can never be eliminated (for instance the definition of the dynamic loads and the values of the dynamic soil properties in the analysis of a machine foundation). Parametric studies are thus essential. The analysis with cones fits the size and economics of practical engineering projects. Thus, cone models should be applied for foundation vibration and dynamic soil-structure-interaction analysis in a design office whenever possible.

1.2 Organisation of the text

To achieve the goal of the book, presenting a strength-of-materials approach with cones to analyse embedded foundation vibrations and fully describing and integrating a computer program for practical cases with as few prerequisites as possible, the following procedure is adopted.

Chapter 1 states the objective of foundation vibration analysis, defines the problem and indicates how the results, the dynamic stiffness and the effective foundation input motion, can be processed further. A brief glimpse at cone models is also provided.

Chapter 2 introduces the concepts of cone models. After describing the characteristics of rigorous methods not addressed in this book, wave propagation in cones is examined in the time domain, emphasising radiation damping, which occurs in semi-infinite cones. Only those aspects that are essential to enable the reader to feel at ease are introduced. Two conceptual building blocks, which form the basis of the computational procedure, are outlined. The features of the cone models are also discussed.

Chapters 3, 4 and 5 form the core of the book, providing a detailed derivation of all relations. The specialist in foundation vibration analysis can start with Chapter 3, skipping Chapter 2.

Chapter 3 addresses the initial cone with outward wave propagation, already illustrated in Fig. 1.7a, the first of the two building blocks of the computational procedure. This permits calculation of the dynamic stiffness of a foundation on the surface of a homogeneous half-space.

Chapter 4 examines wave reflection and refraction at a material discontinuity representing a layer interface, which is the second building block (Fig. 1.7b). The dynamic stiffness of a foundation on the surface of a layer overlying a flexible half-space is determined. A computer program implementing the two building blocks is also introduced.

Chapter 5 describes the spatial discretisation for a foundation embedded in a layered half-space with a stack of embedded disks. After determining the dynamic flexibility of the free field with respect to these disks, which is then inverted, the dynamic stiffness and the effective foundation input motion of the embedded foundation are calculated by enforcing the rigid body motion of the wall and subtracting the trapped mass of the excavated soil. The computer implementation is also discussed, with simple illustrative examples.

Chapter 6 evaluates the accuracy. A parametric study is performed for a foundation both on the surface of and embedded in a multiple-layered half-space, which may also be fixed at its base. A soil half-space with a very large number of layers is addressed, as well as the behaviour below the cutoff frequency of an undamped site fixed at its base. A hemi-ellipsoid and a fully-embedded sphere are also investigated.

Chapter 7 presents engineering applications, addressing the foundation vibration of a reciprocating machine, a seismic soil-structure-interaction analysis and the dynamic analysis of a wind turbine tower with a suction caisson foundation.

Chapter 8 contains concluding remarks. The strength-of-materials approach with cones achieves sufficient engineering accuracy of the dynamic stiffness and effective foundation input motion for surface and embedded foundations for a vast variation of parameters. Cone models based on one-dimensional wave propagation with reflections and refractions at material discontinuities are well suited for everyday practical foundation vibration analyses, as sufficient generality exists, good accuracy results, and the procedure provides physical insight and is easy to use.

Appendix A develops the key elements of the frequency domain analysis, which is based on harmonic excitation.

Appendix B derives the basic equations of motion of dynamic soil-structure interaction with the contribution of the soil consisting of the dynamic stiffness and the effective foundation input motion for seismic waves.

Appendix C discusses wave propagation in a semi-infinite prismatic bar with material discontinuities analysed in the time domain.

Appendix D reviews the pioneering research which leads to the cone models, with selected references. It is not appropriate in an introductory book to acknowledge sources of knowledge throughout the text, as this can lead to a description of the history of the development of the method, but not of the method itself.

Appendix E addresses the executable computer program CONAN (CONe ANalysis) which can be used for the dynamic analysis of actual foundations. A complete User's Guide, including details of input and output, is provided, along with examples.

Appendix F provides complete MATLAB implementations of all procedures addressed in Chapters 3, 4 and 5, with detailed descriptions of how each works. The MATLAB procedures are well commented, and a web address is provided which allows the reader to obtain updated versions of the code.

Appendix G describes the analysis with cones directly in the time domain as an extension.

Finally a dictionary translates the key technical expressions into various languages.

2

Concepts of the cone model

This chapter discusses the fundamentals of the one-dimensional strength-of-materials approach with conical bars and beams, referred to here as cones. Key concepts of wave propagation in cones are explained. An overview of the computational procedure to determine the dynamic response of a foundation embedded in a layered half-space is provided. The treatment of the various items in this chapter is, to a large extent, only descriptive without specifying equations. Selected aspects are illustrated in some detail, providing formulas developed mostly in the time domain. This permits the reader to gain physical insight, and, at the same time, builds up confidence in the procedure. The rigorous and systematic derivation of all required equations is delayed until Chapters 3, 4 and 5.

Section 2.1 characterises the rigorous analysis and compares it with the strength-of-materials approach using cones. Section 2.2 develops wave propagation in a truncated semi-infinite homogeneous cone. By constructing the cone appropriately, a surface foundation on a homogeneous half-space can be analysed. Section 2.3 addresses wave propagation with reflection and refraction occurring at a material discontinuity in a cone. This permits a surface foundation on a layer fixed at its base to be formulated. Section 2.4 examines wave propagation in cone segments for a disk on the surface of a layered half-space. Section 2.5 defines a double cone that permits the wave propagation in cone segments for a disk embedded in a layered half-space to be described. Section 2.6 summarises the computational procedure to calculate the dynamic stiffness of an embedded foundation by combining the elements of wave propagation developed in Sections 2.2 to 2.5. In particular, the importance of the two building blocks of an analysis with cones, the initial cone with outward wave propagation (Section 2.2) and the wave reflection and refraction at a material discontinuity in a cone (Section 2.3) are stressed. Section 2.7 discusses the features and requirements of cone models.

To provide physical insight in a straightforward manner, all demonstrations in this chapter are performed in the time domain, the natural domain of dynamics. The latter follows the development of the solution through time.

2.1 Rigorous methods

Rigorous procedures to calculate the dynamic stiffness and the effective foundation input motion for seismic excitation exist. These include the boundary element method, sophisticated finite element methods such as the thin-layer method (the consistent boundary method), the scaled boundary finite-element method and the Dirichlet-to-Neuman method.

These rigorous methods require a formidable theoretical background. A considerable amount of expertise in idealising the dynamic system is necessary. Significant data preparation and interpretation of the results must be performed. The computational expense for just one run is large, making it difficult from an economic point of view to perform the necessary parametric studies and to investigate alternative design schemes. The rigorous methods can thus provide a false sense of security to the user. The mathematical complexity obscures the physical insight. The rigorous methods belong more to the discipline of applied computational mechanics than to civil engineering. Engineers tend to be intimidated by these procedures. These methods should be used only for large projects or critical facilities such as nuclear power plants, bunkered military constructions or dams, with the corresponding budget and available time. In addition, they have to be used in those cases that are not covered by the strength-of-materials approach.

The vast majority of foundation vibration analyses will thus not be performed using rigorous methods, but with strength-of-materials approaches. Various formulations exist, which are described in Ref. [37]. Three different types of models are available: lumped-parameter models, consisting of a few degrees of freedom connected by springs, dashpots and masses with constant frequency-independent coefficients; representations based on prescribed wave patterns in the horizontal plane of one-dimensional body and surface waves and cylindrical waves; and cone models. Only the latter, which can be regarded as a one-dimensional strength-of-materials approach, are described in this book.

In a strength-of-materials approach the theory of three-dimensional elastodynamics is regarded as a too detailed description. It is sufficient to postulate certain deformation behaviour. For simple shapes, such as bars and beams, a one-dimensional description enforcing 'plane sections remain plane' is applied, resulting in an approximate method. All variables are referred to the axis. In structural engineering, most stress analysis is performed based on the strength-of-materials approach using bars and beams that have cross-sectional properties which can vary along the axis. The approach is very well developed, permitting complicated structural systems such as, for instance, a curved skewed prestressed bridge with moving loads to be modelled with one-dimensional bars and beams. In structural engineering the load bearing elements to be analysed tend to have a dominant direction determining the axis and cross-sectional properties. In geotechnical engineering, however, the soil and rock are three-dimensional media. The choice of the axis, and especially the cross-sectional properties (tributary section), which must be able to represent with the prescribed deformation behaviour all essential features, is thus more difficult in geotechnical engineering. This challenge can be advantageous, as the freedom of choice can be used to enforce a desired property. This is demonstrated in the next section where the opening angle of the cone model leading to the cross-sectional properties is determined by equating the static stiffness of the one-dimensional cone to that of the exact three-dimensional solution of the disk on a half-space. Thus, for statics, cone models reproduce the exact response of the disk.

2.2 Wave propagation in a truncated semi-infinite homogeneous cone

A circular foundation on the surface of a homogeneous half-space is investigated in this section. The massless rigid foundation is also called a disk. All degrees of freedom are examined. This introduces the first building block of the computational procedure.

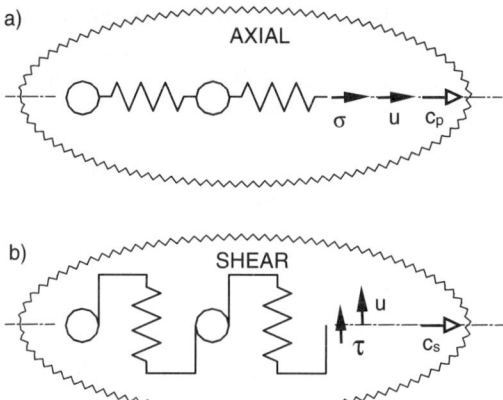

Figure 2.1 Body waves in full-space. a) Dilatational waves with particle motion parallel to direction of wave propagation. b) Shear waves with particle motion perpendicular to direction of wave propagation

As already mentioned when discussing the objective of foundation vibration analysis in Section 1.1, the unboundedness of the half-space leads to wave propagation towards infinity. A simple semi-infinite system, the prismatic bar, is investigated in Appendix C. For a prescribed axial displacement (Fig. C.1a) a wave propagates from this source of disturbance towards infinity, causing axial strains, with the wave velocity in the bar c specified in Eq. C.4.

In a three-dimensional homogeneous isotropic full-space, two types of body waves exist, dilatational waves (P-waves) and shear waves (S-waves). It is helpful to consider the medium as consisting of mass particles connected by springs. When a disturbance acts on a mass particle, it is transmitted to the adjacent mass particle via the connecting spring (Fig. 2.1). When the particle motion is parallel to the direction of wave propagation (Fig. 2.1a), a situation similar to that of axial waves in the bar exists. Axial displacements u, axial strains and normal stresses σ are present. The waves propagate with the dilatational-wave velocity

$$c_p = \sqrt{\frac{E_c}{\rho}} \tag{2.1}$$

with the constrained modulus E_c and the mass density ρ. The index p indicates that the *dilatational waves* are also called P-waves (primary waves). E_c can be expressed as

$$E_c = \lambda + 2G = \frac{1-\nu}{(1+\nu)(1-2\nu)} E = \frac{1-\nu}{1-2\nu} 2G \tag{2.2a}$$

where the Lamé constants λ and G (shear modulus) are

$$\lambda = \frac{\nu}{(1+\nu)(1-2\nu)} E \tag{2.2b}$$

$$G = \frac{1}{2(1+\nu)} E \tag{2.2c}$$

with the modulus of elasticity E and the Poisson's ratio ν. Note that for $\nu \to 1/2$ (incompressible material), $E_c \to \infty$ and $c_p \to \infty$. In this chapter only compressible material will be considered.

When the particle motion is perpendicular to the direction of wave propagation (Fig. 2.1b), the prismatic bar of Fig. C.1a can still be used for illustration. The prescribed displacement is then applied perpendicular to the axis, leading to shear strains and shear stresses. In the full-space, lateral displacements u, shear strains and shear stresses τ are present. These *shear waves* (*S*-waves or secondary waves) propagate with the shear-wave velocity

$$c_s = \sqrt{\frac{G}{\rho}} \tag{2.3}$$

These two types of body waves will be used when constructing cone models.

As an example, the vertical degree of freedom of a disk of radius r_0 on the surface of a homogeneous half-space is addressed (Fig. 2.2). This dynamic system is modelled in the one-dimensional strength-of-materials approach as a *truncated semi-infinite* bar with a vertical axis. The area increases with depth as in a *cone*. This choice of a conical bar is based on the fact that when a load is applied to the disk on the free surface of a half-space, stresses act on an area that increases with depth due to geometric spreading. This is also the case for the cone (although the details of the spatial variation will be different). Dilatational waves with displacements $u(z, t)$ propagate along the axis with the dilatational-wave velocity c_p. Constant axial strains inside the cross-section, corresponding to the deformation behaviour 'plane sections remain plane' are present. The exact complicated three-dimensional wave pattern of the half-space with body and surface waves and three different velocities is replaced by the approximate simple one-dimensional wave propagation governed by the one constant dilatational-wave velocity of the cone. For the cone modelling the vertical degree of freedom, it is appropriate to select as material properties c_p, ν and ρ with the same values as for the half-space. To be able to construct the cone, the opening angle expressed by the aspect ratio z_0/r_0 (Fig. 2.2a) must be determined. Instead of choosing it more or less arbitrarily, it is possible to enforce a desirable property in the cone model. The cone's opening angle is calculated by equating the static-stiffness coefficient of the truncated semi-infinite cone to that of the disk on a half-space determined exactly using three-dimensional elasticity theory. It turns out that the opening angle depends only on Poisson's ratio ν (and the degree of freedom). As a consequence, for a static loading case, the cone model will result in the exact displacement of the disk on the surface (surface foundation). Applying a dynamic load $P_0(t)$ to the disk leads to

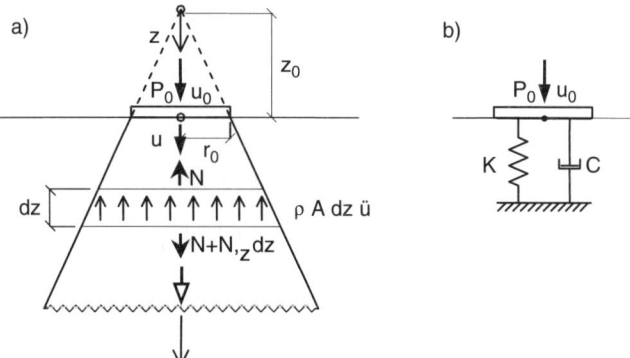

Figure 2.2 Disk on surface of homogeneous half-space. a) Truncated semi-infinite cone for vertical motion with outward wave propagation and equilibrium of infinitesimal element. b) Lumped-parameter model consisting of ordinary spring and ordinary dashpot in parallel

wave propagation downwards towards infinity, yielding *radiation damping*. This is analogous to the semi-infinite prismatic bar discussed in Appendix C. A loaded disk representing a disturbance leads to a wave originating at the disk and *propagating outwards away from the disturbance with the cross-sectional area increasing in the direction of wave propagation*. In the context of the computational procedure, this cone adjacent to a disk with a specified disturbance is called the *initial cone*.

In detail, the analysis of the semi-infinite cone with apex height z_0 and radius r_0 at the truncated section, which is used to model the vertical degree of freedom of a disk of radius r_0 on the surface of a half-space, proceeds as follows. The area at depth z is $A(z) = (z/z_0)^2 A_0$ with $A_0 = \pi r_0^2$, where z is measured from the apex. The constrained modulus E_c follows from c_p and ρ as specified in Eq. 2.1. $u(z, t)$ represents the axial displacement and $N(z, t)$ the normal force. The dynamic equilibrium equation of an infinitesimal element considering the inertial loads (Fig. 2.2a) is formulated as

$$-N(z, t) + N(z, t) + N(z, t),_z dz - \rho A(z) dz \ddot{u}(z, t) = 0 \qquad (2.4)$$

$N(z, t),_z$ denotes the partial derivative of $N(z, t)$ with respect to $z (= \partial N(z, t)/\partial z)$. Substituting the force-displacement relationship (using Eq. 2.1)

$$N(z, t) = \rho c_p^2 A(z) u(z, t),_z \qquad (2.5)$$

leads, for A a given function of z, to the equation of motion in the time domain for the cone

$$u(z, t),_{zz} + \frac{2}{z} u(z, t),_z - \frac{1}{c_p^2} \ddot{u}(z, t) = 0 \qquad (2.6)$$

Equation 2.6 may be rewritten as the one-dimensional wave equation in terms of the function $zu(z, t)$ as

$$(zu(z, t)),_{zz} - \frac{1}{c_p^2} (z \ddot{u}(z, t)) = 0 \qquad (2.7)$$

The static cone is addressed to determine the aspect ratio z_0/r_0. For a prescribed displacement u_0 of the disk, the interaction force P_0 of the disk is calculated, which is equal to the negative value of the normal force $N(z = z_0)$. With $\ddot{u}(z) = 0$, Eq. 2.7 leads to

$$(z u(z)),_{zz} = 0 \qquad (2.8)$$

the solution of which is

$$z u(z) = c_1 + c_2 z \qquad (2.9)$$

with integration constants c_1 and c_2. Enforcing the boundary conditions

$$u(z = z_0) = u_0 \qquad (2.10a)$$
$$u(z \to \infty) = 0 \qquad (2.10b)$$

results in

$$u(z) = \frac{z_0}{z} u_0 \qquad (2.11)$$

With

$$P_0 = -N(z = z_0) = -E_c \pi r_0^2 \, u_{,z}(z = z_0) \tag{2.12}$$

and substituting the derivative calculated from Eq. 2.11 yields

$$P_0 = \frac{E_c \pi r_0^2}{z_0} u_0 \tag{2.13}$$

This defines the static-stiffness coefficient of the cone as

$$K = \frac{E_c \pi r_0^2}{z_0} \tag{2.14}$$

The well-known closed-form solution for the static-stiffness coefficient of a disk on a half-space derived using the exact theory of elasticity is

$$K_{exact} = \frac{4G r_0}{1 - \nu} \tag{2.15}$$

Matching the static-stiffness coefficients by equating Eqs 2.14 and 2.15 yields, using Eqs 2.2a and 2.2c,

$$\frac{z_0}{r_0} = \frac{\pi}{2} \frac{(1-\nu)^2}{1-2\nu} \tag{2.16}$$

The opening angle of the cone thus depends only on Poisson's ratio ν.

Returning to the wave equation of the cone (Eq. 2.7), it is observed that this partial differential equation is in the same form as that of the prismatic bar (Eq. C.3), but for the function $z\,u(z,t)$ and not $u(z,t)$. The familiar solution of the wave equation (Eq. C.8) will thus lead to

$$z\,u(z,t) = z_0 f\!\left(t - \frac{z - z_0}{c_p}\right) + z_0\, g\!\left(t + \frac{z - z_0}{c_p}\right) \tag{2.17a}$$

or

$$u(z,t) = \frac{z_0}{z} f\!\left(t - \frac{z - z_0}{c_p}\right) + \frac{z_0}{z} g\!\left(t + \frac{z - z_0}{c_p}\right) \tag{2.17b}$$

The constant z_0/c_p is introduced into the argument to ensure that at $t=0$ and $z=z_0$ the argument vanishes. As discussed in connection with Eq. C.8, f and g represent waves propagating in the positive and negative z-directions respectively, with constant velocity c_p. In contrast to wave propagation in a prismatic bar, in a cone the wave shape does not remain constant, but is inversely proportional to z, the distance from the apex.

For the loaded disk shown in Fig. 2.2, only waves propagating downwards (positive z-direction) are present ($g = 0$), resulting in (Eq. 2.17b)

$$u(z, t) = \frac{z_0}{z} f\left(t - \frac{z - z_0}{c_p}\right) \qquad (2.18)$$

The interaction force-displacement relationship of the disk is now addressed. Differentiating Eq. 2.18 with respect to z, the spatial derivative of the displacement is obtained as

$$u(z, t),_z = -\frac{z_0}{z^2} f\left(t - \frac{z - z_0}{c_p}\right) - \frac{z_0}{c_p z} f'\left(t - \frac{z - z_0}{c_p}\right) \qquad (2.19)$$

where f' denotes differentiation of f with respect to the argument $t - (z - z_0)/c_p$. Equation 2.5 is now evaluated at $z = z_0$, allowing the interaction force $P_0(t)$ to be determined as

$$P_0(t) = -N(z = z_0, t) = -E_c \pi r_0^2 u(z = z_0, t),_z = \rho c_p^2 \pi r_0^2 \left(\frac{1}{z_0} f(t) + \frac{1}{c_p} f'(t)\right) \qquad (2.20)$$

For a prescribed displacement at the disk of $u_0(t)$, the boundary condition

$$u(z = z_0, t) = u_0(t) \qquad (2.21)$$

yields from Eq. 2.18

$$f(t) = u_0(t) \qquad (2.22)$$

Substituting Eq. 2.22, Eq. 2.20 is rewritten (since $u_0'(t) = \dot{u}_0(t)$) as

$$P_0(t) = \frac{\rho c_p^2 \pi r_0^2}{z_0} u_0(t) + \rho c_p \pi r_0^2 \dot{u}_0(t) \qquad (2.23)$$

or

$$P_0(t) = K u_0(t) + C \dot{u}_0(t) \qquad (2.24)$$

In this interaction force-displacement relationship K and C are constant coefficients of the spring and dashpot

$$K = \frac{\rho c_p^2 \pi r_0^2}{z_0} \qquad (2.25a)$$

$$C = \rho c_p \pi r_0^2 \qquad (2.25b)$$

Thus, an ordinary spring and an ordinary dashpot with the coefficients specified in Eq. 2.25 in parallel forming a lumped-parameter model represent rigorously the cone (Fig. 2.2b). The coefficient of the dashpot C (Eq. 2.25b) is the same as that of the semi-infinite prismatic bar (Eq. C.18 with appropriate indices added). Again, as discussed in Appendix C, in this dashpot the energy of the wave radiating towards infinity is dissipated (*radiation damping*).

In the high frequency limit ($\omega \to \infty$), the spring force $K u_0(t) = K u_0(\omega) e^{i\omega t}$ can be neglected as it is much smaller than the corresponding dashpot force $C \dot{u}_0(t) = i \omega C u_0(\omega) e^{i\omega t}$. This means that wave propagation as in a prismatic bar, i.e. perpendicular to the disk, exists, which is the exact

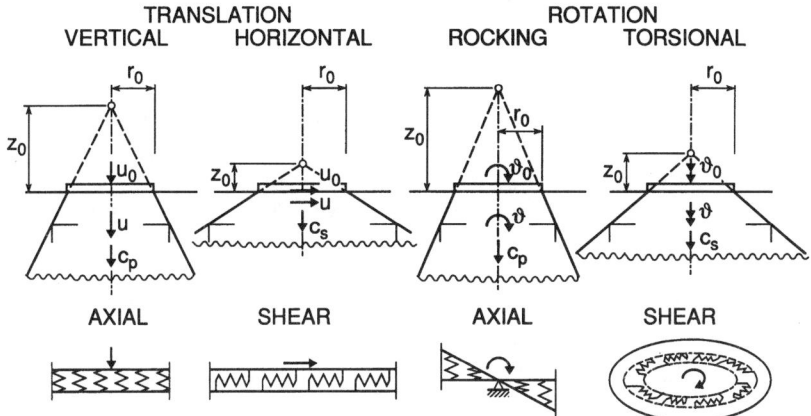

Figure 2.3 Cones for various degrees of freedom with corresponding apex ratio (opening angle), wave-propagation velocity and distortion

wave pattern of a disk on a half-space in the high frequency limit. Thus, the cone model also yields exact results for $\omega \to \infty$. As the opening angle of the cone is calculated by matching the static-stiffness coefficients, a *doubly-asymptotic approximation* results for the cone, correct both for zero frequency (the static case) and for the high frequency limit dominated by the radiation dashpot C.

The other degrees of freedom are modelled analogously. Translational and rotational cones exist. The disk on the half-space is idealised for each degree of freedom as a truncated semi-infinite cone with its own aspect ratio z_0/r_0 (Fig. 2.3). Matching the static-stiffness coefficient of the cone to the corresponding exact value of the disk on a half-space determines z_0/r_0 and thus the opening angle. Applying a load or moment to the disk supported on the free surface leads to stresses in the half-space acting on an area that increases with depth, which is represented approximately by the cone. The cone is regarded as a bar or beam with the displacement pattern over the cross-section determined by the corresponding value on the axis of the cone. The theory of strength of materials is thus used ('plane sections remain plane'). The medium of the half-space outside the cone is disregarded. Depending on the nature of the deformation, it is necessary to distinguish between the translational cone for vertical and horizontal motion, and the rotational cone for rocking and torsion. As indicated in Fig. 2.3, the appropriate wave propagation velocities are c_s for the horizontal and torsional cases, which deform in shear, and c_p for the vertical and rocking cases, which deform axially. Poisson's ratio ν and the mass density ρ are the same as for the half-space.

The equations governing wave propagation in these truncated semi-infinite homogeneous cases are derived for harmonic excitation for all degrees of freedom in Chapter 3.

2.3 Wave reflection and refraction at a material discontinuity in a cone

The wave mechanism generated at a material discontinuity in a cone corresponding to an interface between two layers is discussed qualitatively in this section. This introduces the second building block of the computational procedure. As an example, the vertical degree of freedom of a disk on the surface of a layer of depth d with Poisson's ratio ν, dilatational-wave velocity c_p, and mass density ρ overlying a half-space with the respective material properties ν', c'_p and ρ' is addressed (Fig. 2.4a).

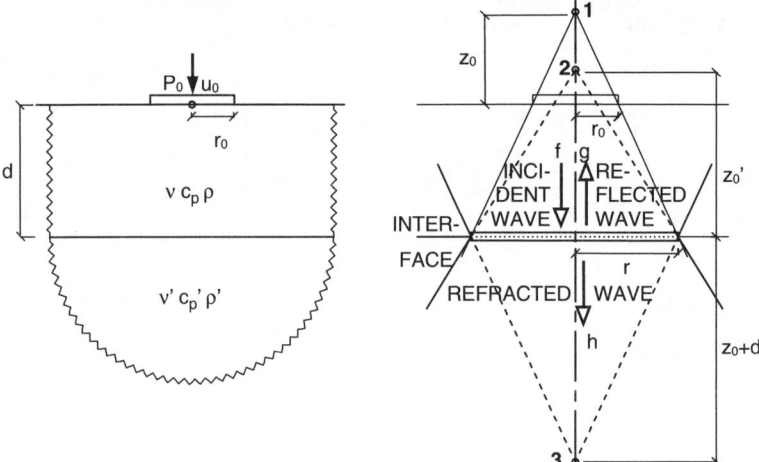

Figure 2.4 Disk on layer overlying half-space. a) Geometry with nomenclature. b) Resulting outward propagating waves in their own cones at material discontinuity

If the half-space is not homogeneous but is layered, the wave propagation adjacent to the loaded disk (which is the source of the disturbance) is initially unaffected. When the wave starts propagating it has no 'knowledge' that after a distance d a discontinuity in material properties will be encountered. Thus, the wave propagates downwards in a cone segment called the initial cone, away from the disk. The properties of this cone, the apex of which will be designated as 1 (Fig. 2.4b), are the same as for a homogeneous half-space with the material properties of the layer. The wave velocity is equal to c_p and the aspect ratio follows from Eq. 2.16 with Poisson's ratio ν of the layer. When a discontinuity of the material properties corresponding to a layer interface is encountered, a source of disturbance is created which can be conceptualised as a fictitious disk. The incident wave f in the initial cone will result in a *refracted wave h* propagating downwards and a *reflected wave g* propagating upwards, both propagating away from the source of the disturbance in their own cones. Formulating displacement compatibility and equilibrium at the interface permits determination of the reflected and refracted waves for the given incident wave, which in turn can be regarded as two new incident waves propagating in their own initial cones. The detailed derivation, which is performed in the frequency domain, is presented in Chapter 4. (It turns out that the so-called reflection coefficient is frequency dependent.) It is analogous to the procedure discussed for the prismatic bar at the end of Appendix C (Fig. C.1d, Eq. C.24 with Eq. C.25). The radius of the fictitious disk at the interface is $r = ((z_0 + d)/z_0) r_0$ (Fig. 2.4b). The refracted wave h propagates in its own cone, the apex of which is designated by 2 in Fig. 2.4b, with the wave velocity c'_p and the aspect ratio z'_0/r specified by Eq. 2.16 with Poisson's ratio ν'. The reflected wave f propagates in its own cone, the apex of which is designated by 3 in Fig. 2.4b, with wave velocity c_p. The aspect ratio $(z_0 + d)/r$ is the same as that of the first cone with apex 1, z_0/r_0, as both model wave propagation in the same material.

The arrows in Fig. 2.4b indicate the propagation directions of the incident wave f, the reflected wave g and the refracted wave h. The *area of the cone increases in the direction of wave propagation* modelling the *spreading* of the disturbance in a medium. Thus, the *cones* (cone segments) *are 'radiating'*.

A free surface and a fixed boundary in a half-space are special cases of the material discontinuity, as discussed below.

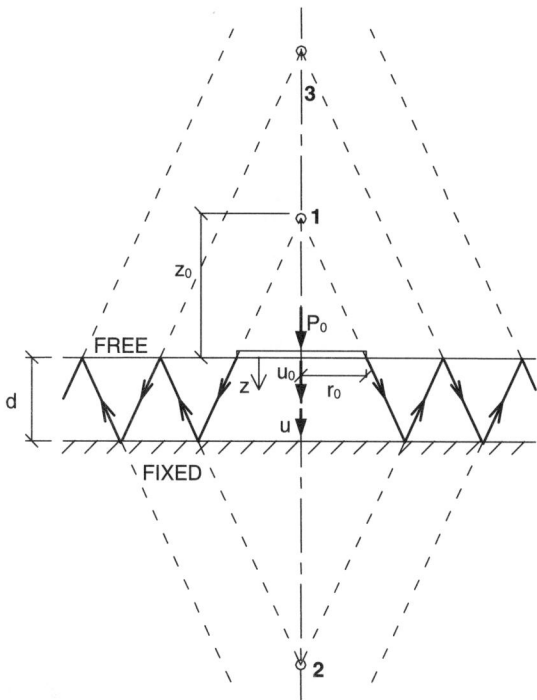

Figure 2.5 Disk loaded vertically on surface of layer fixed at base with wave pattern generated by reflections

In a multiple layered half-space, a large number of reflections and refractions are generated. In principle, each reflected and refracted wave (propagating in its own cone away from the interface outwards) acts as an additional incident wave, which will generate its own reflected and refracted waves when encountering another discontinuity in material properties. Tracking the reflection and refraction of each incident wave sequentially and superposing leads to the resulting wave pattern.

For illustration, the wave pattern as a function of time in a layer of depth d fixed at its base is determined, resulting from applying a time varying load $P_0(t)$ to a surface disk of radius r_0 (Fig. 2.5). The force $P_0(t)$ causes dilatational waves to emanate from beneath the disk which propagate vertically with the velocity $c(=c_p$, with the index p omitted for conciseness), reflecting back and forth between the fixed boundary and the free surface, spreading and decreasing in amplitude.

The following nomenclature is introduced for the displacements. The displacement of the cone modelling a disk with load $P_0(t)$ on a homogeneous half-space with the material properties of the layer (Fig. 2.6) is denoted with an overbar as $\bar{u}(z, t)$ with the value under the disk $\bar{u}_0(t)$. For the disk loaded by the same force $P_0(t)$ on the layer, the displacements calculated by superposing the contributions of all the cones involved are $u(z, t)$ and $u_0(t)$. It will become apparent that the surface motion of the cone representing the homogeneous half-space $\bar{u}_0(t)$ can be used to generate the motion of the layer $u(z, t)$ with its surface value $u_0(t)$. Thus, $\bar{u}_0(t)$ can also be called the *generating function*.

The wave pattern in the layer consisting of the superposition of the contributions of the various cones (Fig. 2.5) is now discussed in detail (Fig. 2.7). The force $P_0(t)$ produces dilatational waves propagating downwards from the disk along a cone with apex 1 and height z_0. This initial part

20 Foundation Vibration Analysis: A Strength-of-Materials Approach

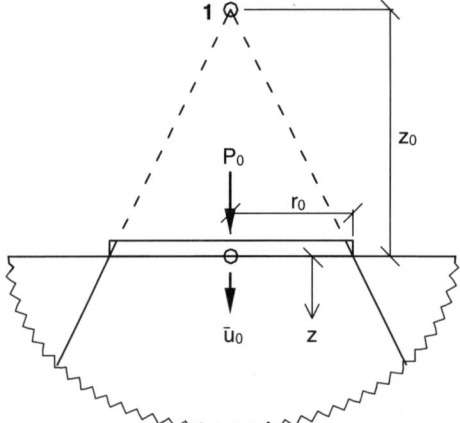

Figure 2.6 Disk loaded vertically on surface of homogeneous half-space with material properties of layer generating incident wave in initial cone

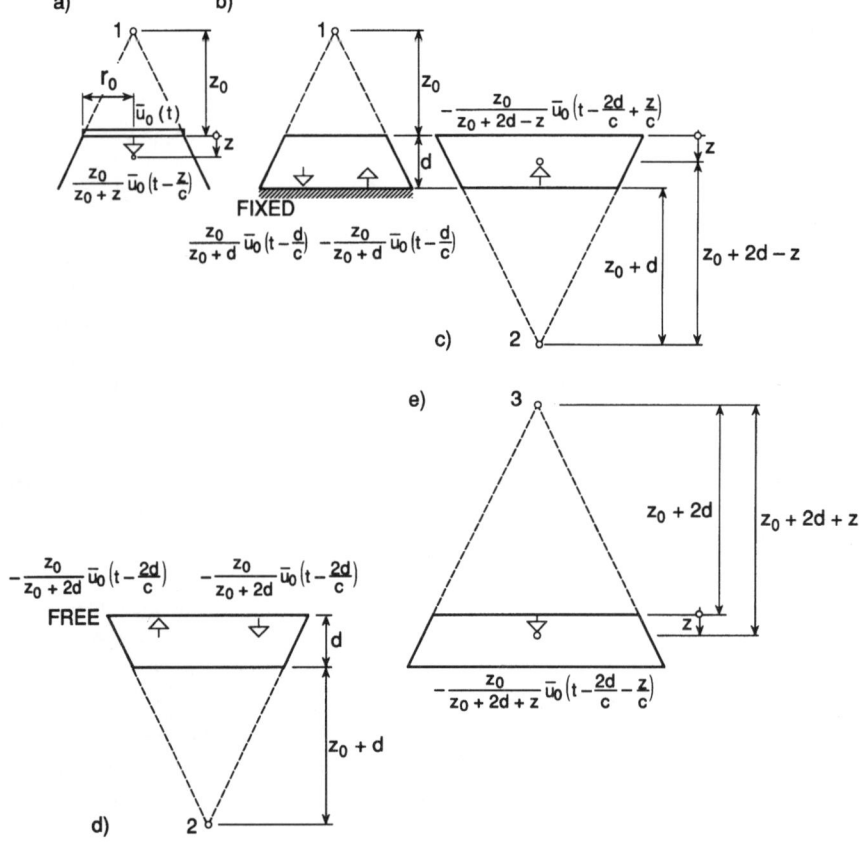

Figure 2.7 Wave pattern and corresponding cones. a) Generating wave in half-space cone. b) Incident and reflected waves at fixed base. c) Upwave in layer from reflection at fixed base. d) Incident and reflected waves at free surface. e) Downwave in layer from reflection at free surface

of the wave pattern, called the incident wave, and the cone along which it propagates, will be the same if the soil is a layer or a half-space, as the wave generated beneath the disk does not 'know' if at a specific depth a fixed interface will be encountered or not, as already mentioned. The same z_0 thus also represents the apex height of the cone of the homogeneous half-space (Fig. 2.6), specified in Eq. 2.16. The interaction force-displacement relationship of the disk on a homogeneous half-space specified in Eq. 2.24 also applies, with $\bar{u}_0(t)$ replacing $u_0(t)$, leading to

$$C\dot{\bar{u}}_0(t) + K\bar{u}_0(t) = P_0(t) \tag{2.26}$$

with K and C specified in Eq. 2.25. For a general variation of $P_0(t)$, a numerical integration procedure can be used, or the analytical solution is applied for this first-order ordinary differential equation with constant coefficients to determine $\bar{u}_0(t)$ as

$$\bar{u}_0(t) = \frac{1}{C}\int_0^t e^{-\frac{K}{C}(t-\tau)} P_0(\tau)\,d\tau \tag{2.27}$$

The function $\bar{u}_0(t)$ vanishes for a negative argument, that is, for $t < 0$.

It is shown in Eq. 2.18 that the displacement, which is a function of $t - z/c$ for propagation in the positive z-direction, is inversely proportional to the distance from the apex $z_0 + z$ in the cone. Notice that the origin of the coordinate z representing the depth is taken to be on the surface of the layer (Fig. 2.5) and not, as in Eq. 2.18 and Fig. 2.2, at the apex. This *incident wave* pattern propagating along the cone with apex 1 shown in Fig. 2.7a is thus formulated as

$$\bar{u}(z, t) = \frac{z_0}{z_0 + z}\bar{u}_0\left(t - \frac{z}{c}\right) \tag{2.28}$$

At the fixed boundary the displacement of the incident wave derived by substituting $z = d$ in Eq. 2.28 (Fig. 2.7b) is

$$\bar{u}(d, t) = \frac{z_0}{z_0 + d}\bar{u}_0\left(t - \frac{d}{c}\right) \tag{2.29}$$

As the total displacement vanishes at the fixed boundary, the reflected wave must exhibit the same displacement as in Eq. 2.29, but with the other sign. This reflected wave will propagate upwards along its own cone with apex 2 (Fig. 2.7c). As the cone's opening angle depends only on Poisson's ratio of the soil (Eq. 2.16), the opening angle of the cone with apex 2 (and of all cones to be introduced subsequently) will be the same as that of the original cone with apex 1. The apex height of the cone with apex 2 is thus $z_0 + d$. The distance from apex 2 to a point at depth z is $z_0 + 2d - z$; the displacement will again be inversely proportional to this distance. For this *upwave* propagating in the negative z-direction, the displacement will be a function of $t + z/c$ as in Eq. 2.17b. A constant will also arise in the displacement's argument, which is determined by equating the argument at the fixed boundary to $t - d/c$ (Eq. 2.29). The displacement of the upwave is

$$-\frac{z_0}{z_0 + 2d - z}\bar{u}_0\left(t - \frac{2d}{c} + \frac{z}{c}\right) \tag{2.30}$$

At the free surface of the layer the displacement of the upwave derived by substituting $z = 0$ in Eq. 2.30 (Fig. 2.7d) is

$$-\frac{z_0}{z_0 + 2d}\bar{u}_0\left(t - \frac{2d}{c}\right) \tag{2.31}$$

As is the case at a free boundary of a prismatic bar (see the end of Appendix C), the reflected wave in the cone will exhibit the same displacement as in Eq. 2.31. A verification follows (Eq. 2.33).

This reflected wave will propagate downwards along its own cone with apex 3 (Fig. 2.7e). With the geometric relationship shown in this figure, and enforcing compatibility of the amplitude and of the argument of the reflected wave's displacement at the free surface, the displacement of the *downwave* is formulated as

$$-\frac{z_0}{z_0 + 2d + z}\bar{u}_0\left(t - \frac{2d}{c} - \frac{z}{c}\right) \qquad (2.32)$$

The displacement corresponding to the sum of Eqs 2.30 and 2.32 must satisfy the free-surface boundary condition of vanishing force, and thus of zero axial strain for $z = 0$. The latter is equal to the partial derivative of the displacement with respect to z. The derivative of Eq. 2.30 at $z = 0$ equals

$$-\frac{z_0}{(z_0 + 2d)^2}\bar{u}_0\left(t - \frac{2d}{c}\right) - \frac{z_0}{z_0 + 2d}\frac{1}{c}\bar{u}_0'\left(t - \frac{2d}{c}\right) \qquad (2.33)$$

where $\bar{u}_0'$ denotes differentiation with respect to the total argument of the generating function: $\bar{u}_0' = \partial \bar{u}_0/\partial()$ with $() = t - 2d/c + z/c$. The derivative of Eq. 2.32 results in the same value, but with a sign change. The sum will thus be zero.

This process of generating waves will continue. The downwave described by Eq. 2.32 will be reflected at the fixed boundary (Fig. 2.5). The newly created upwave propagating along its own cone will be reflected at the free surface, giving rise to a new downwave. At each fixed and free boundary the apex height of the cone along which the reflected wave propagates increases. *The waves in the layer thus decrease in amplitude and spread resulting in radiation of energy in the horizontal direction.*

The resulting displacement in the layer $u(z, t)$ is equal to the *superposition* of the contributions of all cones; the displacements of the *incident wave* (Eq. 2.28, Fig. 2.7a), of the upwave (Eq. 2.30, Fig. 2.7c), and of the downwave (Eq. 2.32, Fig. 2.7e), and of *all* subsequent *upwaves* and *downwaves* are summed. In general the displacement $u(z, t)$ of the layer at depth z and time t may be expressed as the wave pattern

$$u(z, t) = \overset{\text{incident wave}}{\frac{z_0}{z_0 + z}\bar{u}_0\left(t - \frac{z}{c}\right)} + \sum_{j=1}^{\infty}(-1)^j \left[\overset{\substack{\text{upwaves from} \\ \text{fixed boundary}}}{\frac{z_0\bar{u}_0\left(t - \frac{2jd}{c} + \frac{z}{c}\right)}{z_0 + 2jd - z}} + \overset{\substack{\text{downwaves from} \\ \text{free surface}}}{\frac{z_0\bar{u}_0\left(t - \frac{2jd}{c} - \frac{z}{c}\right)}{z_0 + 2jd + z}}\right] \qquad (2.34)$$

As $\bar{u}_0(t)$ vanishes for a negative argument, the sum shown in Eq. 2.34 extending to infinity is limited.

Note that, for all waves, propagation in the vertical direction occurs in *radiating cones* (cone segments), that is, the area of the cones always increases in the direction of wave propagation. The layer cannot be idealised by an ordinary non-radiating cone segment (a finite element of a tapered bar). In an ordinary cone segment, waves reflected at the fixed boundary are focused in the narrowing neck. Figure 2.8 should be compared to Fig. 2.7c. The reflected upwave cannot spread and radiate energy horizontally. This means that no radiation damping occurs when using an ordinary cone segment, as a bounded medium is analysed.

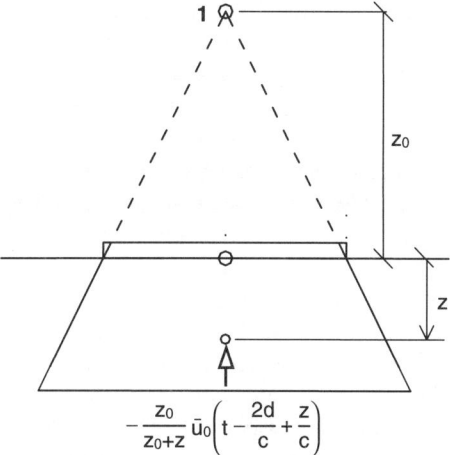

Figure 2.8 Upwave in layer from reflection at fixed base propagating in ordinary cone and not in radiating cone

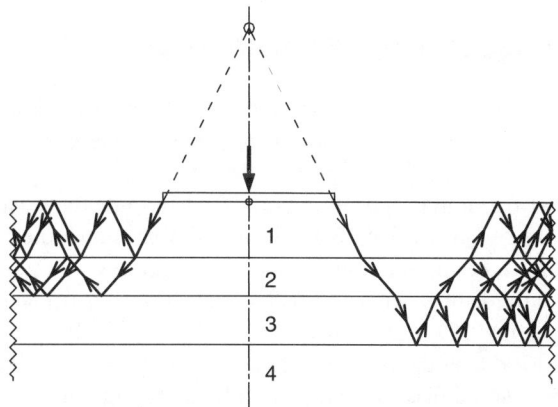

Figure 2.9 Downwards and upwards wave propagation in cone segments for surface disk on layered half-space

2.4 Disk on the surface of a layered half-space

Based on the building blocks described in Sections 2.2 and 2.3, the wave propagation caused by a loaded disk on the surface of a layered half-space can be described. (The vertical motion is illustrated in Fig. 2.9.) The particle motion and propagation occur vertically downwards and upwards on the axes of cone segments with cross-sectional areas increasing in the direction of wave propagation.

The initial wave propagates downwards in an initial cone with an aspect ratio determined by Poisson's ratio of layer 1. When the incident wave encounters the interface between layers 1 and 2, a reflected wave propagating upwards in layer 1 and a refracted wave propagating downwards in layer 2 occur, each propagating in its own cone segment. These waves act as incident waves in new initial cones, each causing additional reflected and refracted waves when impinging on an interface. This mechanism repeats itself. For instance, the refracted wave propagating in layer 2 yields a reflected wave in layer 2 (left-hand side of Fig. 2.9) and a refracted wave in layer 3 (right-hand side). The waves propagate in both directions with the dilatational-wave velocity of the layer. The opening angles of all cone segments corresponding to a layer for wave propagation in both directions are equal. On the left-hand side of Fig. 2.9 the wave pattern in the first two layers is sketched without the influence of upwaves from layer 3 and layers further down. On the right-hand side of the same figure the wave pattern caused by the 'first' wave in layer 3 in the first three layers is illustrated not addressing the upwaves from layer 4 and layers further down. Tracking the reflections and refractions of each incident wave sequentially and superposing leads to the resulting wave pattern up to a certain stage.

The procedure to calculate quantitatively the resulting displacements in the layered half-space is, in principle, the same as explained in addressing the example of Fig. 2.7. The generating function $\bar{u}_0(t)$ follows from the interaction force-displacement relationship of the first initial cone (Eqs 2.26 and 2.27). Outward wave propagation as in Eq. 2.28 results in the incident wave f at the interface (Eq. 2.29). Formulating displacement compatibility and equilibrium yields the reflected wave g and refracted wave h (Fig. 2.4b), which are new incident waves that propagate outwards. The detailed derivation, which is performed for harmonic excitation, is presented in Chapters 3 and 4.

2.5 Disk embedded in a layered half-space

In the computational procedure to analyse an embedded foundation described in Section 2.6, besides the wave pattern of displacements caused by a loaded surface disk, that occurring from a loaded disk embedded in the layered half-space is also required.

The vertical degree of freedom is addressed here. To determine the wave pattern in a layered half-space corresponding to a surface disk (Fig. 2.9), the starting point is the truncated semi-infinite cone modelling a surface footing on a homogeneous half-space with the material properties of the (first) layer adjacent to the disk (Fig. 2.10a). The aspect ratio z_0/r_0 determining the opening angle of the cone is specified in Eq. 2.16, with the displacement following from Eq. 2.18 using Eq. 2.24. Analogously, to address the wave pattern in a layered half-space corresponding to a disk embedded in a specific layer, a disk embedded in a homogeneous full-space with the material properties of this layer is examined. This situation may be approximated by a *double-cone model*

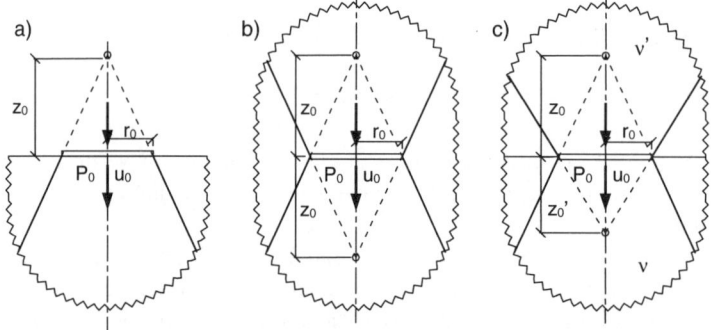

Figure 2.10 Disk with corresponding cone model. a) Disk on surface of half-space with one-sided cone model. b) Disk embedded in homogeneous full-space with double-cone model. c) Disk embedded at interface of two homogeneous half-spaces forming a full-space with double-cone model

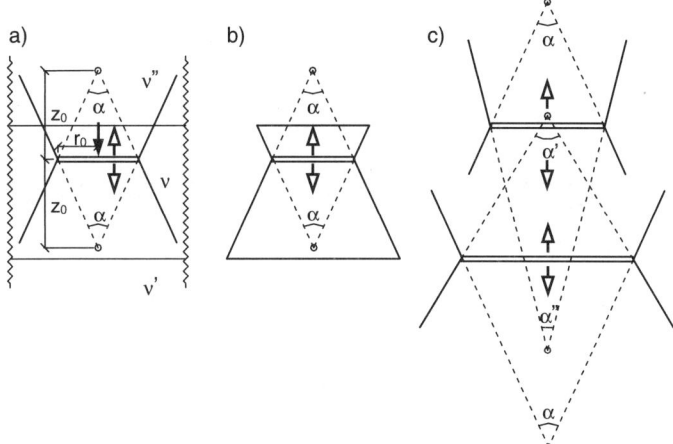

Figure 2.11 Wave propagation in double cone embedded in layered full-space. a) Initial truncated semi-infinite cones. b) Incident waves in initial cone segments. c) Reflected and refracted waves at material discontinuities propagating outwards in their own cone segments

(shown in Fig. 2.10b) with the same proportions as for the one-sided cone model of Fig. 2.10a. The applied load is resisted half in tension (upper cone) and half in compression (lower cone), with the waves propagating away from the disk outwards. Stated differently, the stiffness of the double cone is doubled compared to that of the one-sided cone. The interaction force-displacement relationship of the double cone is thus (see Eq. 2.24)

$$P_0(t) = 2(K\,u_0(t) + C\dot{u}_0(t)) \tag{2.35}$$

with K and C as specified in Eq. 2.25.

For an embedded disk, the wave pattern in a layered half-space in constructed as for a surface disk, the difference being that two initial cones with incident waves exist. As an example a disk embedded in a layer with Poisson's ratio v overlying a layer with Poisson's ratio v' and underlying another layer with v'' is examined (Fig. 2.11). The directions of wave propagation are indicated by arrows in the cones with their opening angles α, α' and α'' determined by v, v' and v'' respectively (Eq. 2.16).

When the embedded disk coincides with the interface of two layers (Fig. 2.10c) the concept of the double cone applies with each single cone determined by the properties of the corresponding half-space. The interaction force-displacement relationship is

$$P_0(t) = (K + K')\,u_0(t) + (C + C')\dot{u}_0(t) \tag{2.36}$$

with K' and C' specified in Eq. 2.25, but using the properties of the upper half-space denoted by a prime.

For a multiple-layered site, the wave pattern for an embedded disk is presented in Fig. 2.12. On the left-hand side the upwards and downwards wave propagation in the first five cone segments created by the first initial cone in layer 2 is indicated. On the right-hand side the downwards and upwards wave propagation in the first five cones segments arising from the first initial cone in layer 3 are shown.

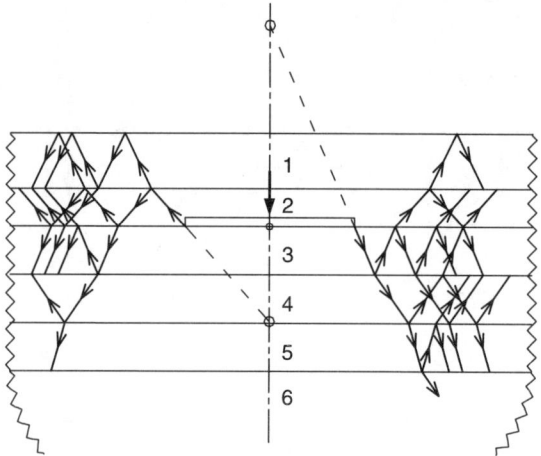

Figure 2.12 Downwards and upwards wave propagation in cone segments for embedded disk in layered half-space.

Instead of determining the aspect ratio by using the static-stiffness coefficient of a disk on a half-space, as expressed in Eq. 2.16, the exact static-stiffness coefficient of a disk embedded in a homogeneous full-space available in closed form can be used. For the vertical degree of freedom the latter is

$$K_{exact} = \frac{32(1-\nu)}{3-4\nu} G r_0 \qquad (2.37)$$

Matching Eq. 2.37 to the static-stiffness coefficient of the double cone (Eq. 2.14)

$$K = 2\frac{E_c \pi r_0^2}{z_0} \qquad (2.38)$$

yields, using Eqs 2.2a and 2.2c,

$$\frac{z_0}{r_0} = \frac{\pi}{8}\frac{3-4\nu}{1-2\nu} \qquad (2.39)$$

This calibration based on a disk in a full-space leads to aspect ratios which are somewhat smaller than those based on two disks on a half-space ($\nu = 1/4$: 1.571 compared to 1.767; $\nu = 1/3$: 1.963 compared to 2.094).

For foundations partially-embedded in a half-space where a free surface is present, the calibration with respect to the half-space values is preferable. This will not hold for fully-embedded foundations, especially at large depth.

2.6 Foundation embedded in a layered half-space

The concept of the computational procedure to calculate the dynamic stiffness and the effective foundation input motion of a rigid cylindrical foundation is described in this section.

The vertical degree of freedom is used for illustration. Figure 2.13 shows a cylindrical soil region with radius r_0 extending to a depth e into the layered half-space. This soil region, which

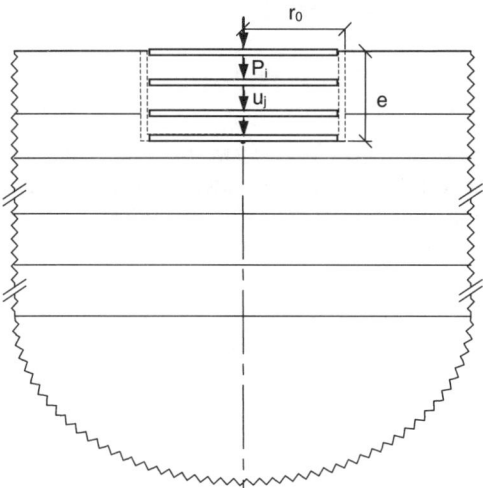

Figure 2.13 Stack of disks with redundant forces to model embedded foundation

will later be excavated, is viewed as a sandwich consisting of rigid disks separated by soil layers. The disks will coincide with the interfaces of the half-space. In addition, further disks are selected to adequately represent the dynamic behaviour. At least ten nodes (disks) per wave length are recommended. On the disks on the surface of and embedded in the layered half-space, unknown vertical forces P_i are applied, which can be regarded as redundants. The primary system, which is indeterminate, addresses the vertical displacements u_j of the disks on the surface of and embedded in the layered half-space without excavation, that is, the free field. As in the force method of structural analysis, the relationship between the displacements of the disks and the forces acting on the disks is established. This is based on wave propagation in cone segments considering layering, as described for a surface disk and an embedded disk in Sections 2.4 and 2.5. This dynamic-flexibility relationship is inverted, yielding a dynamic-stiffness matrix for the disks on the surface of and embedded in the layered half-space (free field), expressing the forces (redundants) as a function of the displacements. The disks and the soil trapped between them are constrained to execute rigid-body motion, as the walls of the foundation are rigid. For the vertical degree of freedom this is enforced by equating the vertical displacements of all disks. This also results in the force acting on the rigid foundation being equal to the sum of all the vertical forces acting on the disks. The trapped soil is analytically 'excavated' by subtracting the mass of the soil times acceleration of the rigid interior cylinder of soil from the force. This leads to the dynamic stiffness of the rigid foundation embedded in a layered half-space.

To calculate the effective foundation input motion the free-field motion is determined (Fig. 1.2b, Appendix B.2). Using the dynamic stiffness of the disks in the free field and the rigid-body constraint, both mentioned above, the effective foundation input motion follows from the condition that the resultant force acting on the rigid foundation vanishes (Appendix B.1).

All analysis is performed for harmonic excitation, as derived in detail in Chapter 5. The key relationship consists of the dynamic flexibility of the free field, that is the layered half-space before excavating, with respect to the disks. It is established using wave propagation in cone segments with the two building blocks, the initial cone with outward wave propagation (Chapter 3), and the wave reflection and refraction at a material discontinuity in a cone (Chapter 4).

2.7 Features of the cone model

Cone models, the one-dimensional strength-of-materials theory with conical bars and beams, meet the following requirements.

1. *Conceptual clarity with physical insight.* As an example, consider the foundation on the surface of a layer fixed at its base (Fig. 2.5). The wave pattern is clearly postulated. The one-dimensional waves propagate with the dilatational-wave velocity, reflecting back and forth, spreading and decreasing in amplitude and thus radiating energy towards infinity horizontally.
2. *Simplicity in physics and exact mathematical solution as well as in practical application.* Due to the deformation behaviour enforced in cones, one-dimensional models arise which can be solved exactly in closed form, and which can in many cases be analysed with a hand calculator.
3. *Sufficient generality concerning layering and embedment for all degrees of freedom and all frequencies.*
4. *Sufficient engineering accuracy.* A deviation of $\pm 20\%$ of the results of cone models from those of the rigorous methods for one set of input parameters is, in general, sufficient, since many of the uncertainties can never be eliminated.
5. *Satisfaction of physical features.* Certain properties are enforced exactly when constructing the cone model, such as the doubly-asymptotic behaviour of the truncated semi-infinite cone (Section 2.2). Others are very closely modelled as a consequence of the assumed wave pattern. For instance, no radiation damping occurs in a layer fixed at its base in the frequency range below the so-called cutoff frequency (the fundamental frequency of the layer), as discussed in Section 4.3.

All the features mentioned above allow the cone models to be applied for everyday *practical foundation vibration and dynamic soil-structure-interaction analyses* in a design office. This strength-of-materials approach, which is so successful in stress analysis of structural engineering, allows the mathematical complexity (far beyond that typical for civil engineering practice) of rigorous solutions in applied mechanics to be avoided. The simplicity and low cost of a single analysis permit parametric studies, varying the parameters with large uncertainty (such as soil properties), to be performed, and the key parameters of the dynamic system to be identified, as well as alternative designs to be investigated. Finally, the cone models may be used to check the results of more rigorous procedures determined with sophisticated computer programs.

As already mentioned at the end of Section 1.1, cone models do lead to some loss of precision. However, it cannot be the aim of the engineer to calculate the complex reality as closely as possible. A well-balanced design, one that is both safe and economical, does not call for rigorous results (of which the accuracy is limited anyway because of the uncertainties that cannot be eliminated) in a standard project. Cone models capture the salient features, and are based on experience gained from rigorous analyses. They are not the first attempt to capture the physics of the problem! Cone models are quite dependable, incorporating implicitly much more than meets the eye. Cone models should be applied wherever possible, as '*Simplicity that is based on rationality is the ultimate sophistication*' (A.S. Veletsos).

3

Initial cone with outward wave propagation

The first of the two building blocks used to calculate the vibrations of an embedded foundation consists of a disk embedded in a layered half-space, modelled as a double cone, with each cone exhibiting outward wave propagation. Such a model is shown in Fig. 2.10b for the vertical degree of freedom. As a special case the disk is placed on the surface of the half-space, which leads to a one-sided cone model, as presented in Fig. 2.10a. The displacement of the disk u_0 is calculated for a specified force P_0. The wave propagation $u(z)$ away from the disk representing the source of the disturbance is analysed. This outward wave propagation occurs (initially) in a truncated semi-infinite homogeneous cone with the material properties of the layer adjacent to the disk, called the initial cone. A foundation that can be represented as an equivalent disk on the surface of a homogeneous half-space can be modelled directly with one initial cone (Fig. 2.2). The same concepts apply to the horizontal and rotational degrees of freedom.

Section 3.1 discusses the translational one-sided cone. The hypotheses and construction of the one-dimensional strength-of-materials approach using a tapered bar with the cross-section increasing in the direction of wave propagation (as in a cone) is addressed. The interaction force P_0-displacement u_0 relationship of the loaded disk is derived, and the displacement $u(z)$ along the cone as a function of u_0 is established. The latter is required to calculate the incident wave f at a material discontinuity (Fig. 2.4b). Section 3.2 derives the analogous relationships for the rotational one-sided cone, which is further discussed in Section 3.3. Section 3.4 addresses the modifications necessary for the vertical and rocking degrees of freedom when Poisson's ratio approaches 1/2, which yields an infinite dilatational-wave velocity. Section 3.5 presents the dynamic-stiffness coefficients of a (circular) foundation on the surface of a homogeneous half-space, which are compared with the results of a rigorous analysis. Section 3.6 examines a disk embedded in a homogeneous full-space, yielding an alternative procedure to construct the double-cone model. Again, a comparison with analytical results is performed.

Chapter 3 (and Chapters 4 and 5) are limited to those relationships which are required to calculate the dynamic-stiffness coefficients and the effective foundation input motion. As the analysis is performed in the frequency domain, the derivation is also performed for harmonic excitation in the elastic medium. To introduce hysteretic material damping, the real shear modulus and constrained modulus are multiplied in the final results by $1 + 2i\zeta$ (where ζ is the damping ratio), making them complex (in accordance with the correspondence principle). To make the chapter self-contained, certain equations addressed in Chapter 2 are repeated. Cross-referencing to the basics discussed

3.1 Translational cones

In this section the vertical degree of freedom is addressed first (Fig. 3.1). A disk of radius r_0 on the surface of a homogeneous half-space with shear modulus G, Poisson's ratio ν, mass density ρ and hysteretic damping ratio ζ is modelled as a (one-sided) truncated semi-infinite translational cone with the same material properties as the half-space. The constrained modulus $E_c = \rho c_p^2$ (with c_p representing the dilatational-wave velocity), and ν and ρ are selected as the representative material properties. When $\nu \to 1/2$ (the incompressible case), c_p (and E_c) $\to \infty$, and so the wave velocity must be limited to permit a cone model to be used (Section 3.4). Cases with Poisson's ratio between $1/3$ and $1/2$ will be referred to as nearly-incompressible. To allow the same equations to be applied for both the compressible and the nearly-incompressible cases, the derivation is performed for an axial-wave velocity c with the corresponding modulus of elasticity expressed as ρc^2. The cone's opening angle is specified by the aspect ratio z_0/r_0 (apex height z_0), which must be determined in the solution process. A vertical load with amplitude $P_0(\omega)$ is applied to the disk, resulting in a vertical displacement of the disk with amplitude $u_0(\omega)$. The downward wave propagation with amplitude $u(z, \omega)$ is to be determined. Within the cone representing a bar with its area increasing with depth z (measured from the apex) as

$$A(z) = \left(\frac{z}{z_0}\right)^2 A_0 \tag{3.1}$$

(with $A_0 = \pi r_0^2$), axial strains are present ('plane sections remain plane'). For a discussion of the concept, the general description of Fig. 2.2a can be consulted.

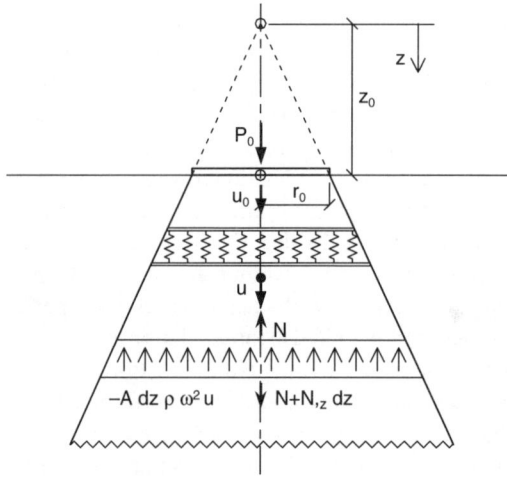

Figure 3.1 Translational truncated semi-infinite cone with vertical motion, axial distortion and equilibrium of infinitesimal element

For harmonic excitation the complex amplitude form (Appendix A.1) is

$$P_0(t) = P_0(\omega)e^{i\omega t} \tag{3.2a}$$

$$u(t) = u(\omega)e^{i\omega t} \tag{3.2b}$$

To establish the equation of motion, the equilibrium of an infinitesimal element is formulated (Fig. 3.1). This yields, with the normal force with amplitude $N(z, \omega)$ and the inertial load (calculated as the mass of the element times acceleration acting in the negative z-direction) with amplitude $-A(z)\,dz\,\rho\omega^2 u(z, \omega)$ (since $\ddot{u}(z, \omega) = -\omega^2 u(z, \omega)$),

$$-N(z, \omega) + N(z, \omega) + N(z, \omega),_z \, dz + \omega^2 A(z)\rho \, dz\, u(z, \omega) = 0 \tag{3.3a}$$

or

$$N(z, \omega),_z + \omega^2 A(z)\rho u(z, \omega) = 0 \tag{3.3b}$$

Substituting the normal force-displacement relationship with the axial distortion with amplitude $u(z, \omega),_z$

$$N(z, \omega) = \rho c^2 A(z) u(z, \omega),_z \tag{3.4}$$

in Eq. 3.3b using Eq. 3.1 results in the equation of motion of the translational cone

$$u(z, \omega),_{zz} + \frac{2}{z} u(z, \omega),_z + \frac{\omega^2}{c^2} u(z, \omega) = 0 \tag{3.5}$$

which may be written as a one-dimensional wave equation in the function $zu(z, \omega)$ as

$$(zu(z, \omega)),_{zz} + \frac{\omega^2}{c^2}(zu(z, \omega)) = 0 \tag{3.6}$$

The solution of this ordinary differential equation is postulated as

$$zu(z, \omega) = e^{i\gamma z} \tag{3.7}$$

Substituting Eq. 3.7 into Eq. 3.6 yields

$$-\gamma^2 + \frac{\omega^2}{c^2} = 0 \tag{3.8a}$$

resulting in

$$\gamma = \pm \frac{\omega}{c} \tag{3.8b}$$

The solution is thus

$$zu(z, \omega) = c_1 e^{i\frac{\omega}{c}z} + c_2 e^{-i\frac{\omega}{c}z} \tag{3.9}$$

with the integration constants c_1 and c_2.

As the solution in the time domain is equal to the amplitude specified in Eq. 3.9 multiplied by $e^{i\omega t}$ (Eq. 3.2b), for instance, the second term $e^{i\omega(t-z/c)}$ corresponds to a wave propagating in the positive z-direction with the velocity c. This is verified as follows. If t is increased by any value $\bar{t}$

and simultaneously z is increased by $c\bar{t}$, the value $e^{i\omega(t+\bar{t}-(z+c\bar{t})/c)} = e^{i\omega(t-z/c)}$ is not altered. Similarly, the first term corresponds to a wave propagating in the negative z-direction.

Applying a load to the disk, the wave will propagate outwards away from the source of the disturbance, that is, downwards in the positive z-direction. Thus, only the second term with the integration constant c_2 is present, yielding

$$zu(z, \omega) = c_2 e^{-i\frac{\omega}{c}z} \tag{3.10}$$

Enforcing the boundary condition

$$u(z = z_0, \omega) = u_0(\omega) \tag{3.11}$$

leads to

$$u(z, \omega) = \frac{z_0}{z} e^{-i\frac{\omega}{c}(z-z_0)} u_0(\omega) \tag{3.12}$$

The amplitude of the displacement is thus inversely proportional to the distance from the apex of the cone. Equation 3.12 specifies the displacement amplitude at the distance $z - z_0$ from the disk with the displacement amplitude $u_0(\omega)$. It serves to calculate the displacement amplitude of the incident wave impinging at a material discontinuity.

The interaction force-displacement relationship of the disk is now addressed. Equilibrium of the disk yields

$$P_0(\omega) = -N(z = z_0, \omega) \tag{3.13}$$

Substituting the derivative $u(z, \omega)_{,z}$ calculated from Eq. 3.12 into Eq. 3.4, which is evaluated at $z = z_0$, leads in Eq. 3.13 to

$$P_0(\omega) = \left(\frac{\rho c^2 A_0}{z_0} + i\omega \rho A_0\right) u_0(\omega) \tag{3.14}$$

or

$$P_0(\omega) = S(\omega) u_0(\omega) \tag{3.15}$$

with the *dynamic-stiffness coefficient*

$$S(\omega) = \frac{\rho c^2 A_0}{z_0} + i\omega \rho c A_0 \tag{3.16}$$

With the spring coefficient K (which is the static-stiffness coefficient) defined as

$$K = \frac{\rho c^2 A_0}{z_0} \tag{3.17}$$

the dashpot coefficient C defined as

$$C = \rho c A_0 \tag{3.18}$$

and with

$$\dot{u}_0(\omega) = i\omega u_0(\omega) \tag{3.19}$$

Eq. 3.14 is reformulated as

$$P_0(\omega) = K u_0(\omega) + C \dot{u}_0(\omega) \tag{3.20}$$

This corresponds to the interaction force-displacement relationship formulated in Eq. 2.24 in the time domain (with $c = c_p$) for the spring-dashpot model presented in Fig. 2.2b.

Equation 3.14 permits the displacement amplitude of the disk $u_0(\omega)$ to be calculated for a specified load amplitude $P_0(\omega)$ acting on the disk in the presence of a one-sided cone.

Using the dimensionless frequency defined with respect to the properties of the cone

$$b_0 = \frac{\omega z_0}{c} \quad (3.21)$$

the dynamic-stiffness coefficient of Eq. 3.16, decomposed in analogy to Eq. 1.5, is

$$S(b_0) = K[k(b_0) + ib_0 \, c(b_0)] \quad (3.22)$$

with the dimensionless spring and damping coefficients (which turn out to be independent of b_0 for the translational cone)

$$k(b_0) = 1 \quad (3.23a)$$
$$c(b_0) = 1 \quad (3.23b)$$

Introducing the standard dimensionless frequency of the disk

$$a_0 = \frac{\omega r_0}{c_s} \quad (3.24)$$

with c_s denoting the shear-wave velocity, Eq. 3.16 is rewritten as

$$S(a_0) = K[k(a_0) + ia_0 \, c(a_0)] \quad (3.25)$$

with

$$k(a_0) = 1 \quad (3.26a)$$
$$c(a_0) = \frac{z_0}{r_0} \frac{c_s}{c} \quad (3.26b)$$

To determine the aspect ratio z_0/r_0 and hence the cone's opening angle, the *static-stiffness coefficient of the truncated semi-infinite cone* (Eq. 3.17) *is matched to the corresponding exact solution of three-dimensional elasticity theory for the disk on a half-space*, which is

$$K_{exact} = \frac{4\rho c_s^2 r_0}{1-\nu} \quad (3.27)$$

(Eq. 2.15 with the shear modulus $G = \rho c_s^2$). Equating Eqs 3.17 and 3.27 yields

$$\frac{z_0}{r_0} = \frac{\pi}{4}(1-\nu)\left(\frac{c}{c_s}\right)^2 \quad (3.28)$$

which is a function of Poisson's ratio ν. For $c = c_p$, Eq. 2.16 results.

The horizontal degree of freedom, modelled with its own cone, is now addressed (Fig. 3.2a). The elastic material properties of the truncated semi-infinite translational cone are selected as $G = \rho c_s^2$, ν and ρ, which are set equal to those of the half-space. A horizontal load with amplitude $P_0(\omega)$ is applied, resulting in a horizontal displacement of the disk with amplitude $u_0(\omega)$.

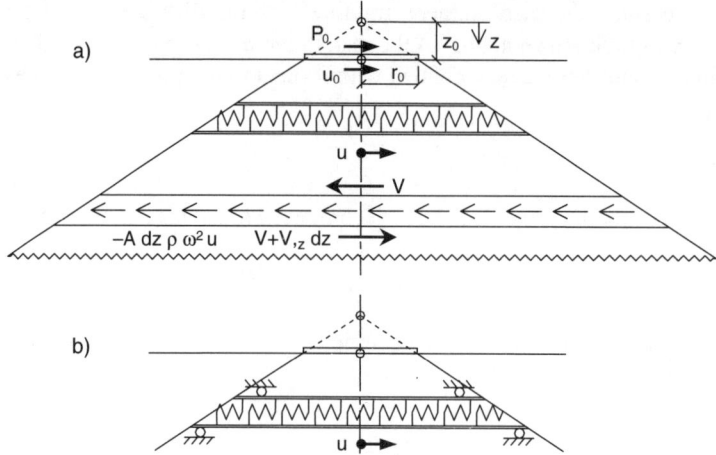

Figure 3.2 Translational truncated semi-infinite cone with horizontal motion, shear distortion and equilibrium of infinitesimal element, where rocking motion is prevented with: a) infinite flexural rigidity; and b) rollers for horizontal motion

Equation 3.1 still applies with $A(z)$ denoting the shear area, as shear strains are present. The amplitude of the shear force (analogous to Eq. 3.4) is

$$V(z,\omega) = \rho c_s^2 A(z) u(z,\omega)_{,z} \quad (3.29)$$

with the amplitude of the lateral displacement $u(z,\omega)$. $u(z,\omega)_{,z}$ represents the amplitude of the shear distortion. The equilibrium equation follows from Eq. 3.3b, replacing $N(z,\omega)$ by $V(z,\omega)$. The derivation is analogous, but with $c = c_s$. The displacement amplitude of the incident wave follows from Eq. 3.12, and the dynamic-stiffness coefficient from Eq. 3.16 (and Eqs 3.22 and 3.25) after setting $c = c_s$. Equation 3.26b leads to $c(a_0) = z_0/r_0$. *Matching the exact horizontal static-stiffness coefficient of a disk on a half-space*

$$K_{exact} = \frac{8\rho c_s^2 r_0}{2-\nu} \quad (3.30)$$

to the corresponding solution of the truncated semi-infinite cone specified in Eq. 3.17 with $c = c_s$ yields

$$\frac{z_0}{r_0} = \frac{\pi}{8}(2-\nu) \quad (3.31)$$

The opening angle thus depends on the degree of freedom. Figures 3.1 and 3.2 correspond to $\nu = 1/3$.

While the cone model for the vertical degree of freedom is based on the strength-of-materials approach of a tapered bar, this does not really apply for the horizontal degree of freedom. Coupling with the rocking degree of freedom arises. In the exact solution for a disk on a half-space, determined from three-dimensional elasto-dynamics, coupling between the horizontal and rocking degrees of freedom does occur, but is small and is thus often neglected. In the cone model for the horizontal degree of freedom, the flexural rigidity is assumed to be infinite, preventing any rocking of the cross-section (Fig. 3.2a). As the flexural rigidity for a beam with a circular cross-section equals $\rho c_p^2 \pi r_0^4/4$ (or $\rho c^2 \pi r_0^4/4$), this value is not infinite. Infinite flexural rigidity can be postulated, which, as already mentioned, results in zero rocking, but from considering equilibrium a bending moment is still present (Fig. 3.2a). Alternatively, rollers acting in the horizontal direction can be added (Fig. 3.2b). In this case, not only is rocking prevented, but the reaction moment generated in the rollers cancels the moment from the applied load, yielding a zero resultant moment.

To consider material damping, the correspondence principle is applied. The latter states that the damped solution is obtained from the elastic one by replacing the elastic moduli with the corresponding complex ones

$$E_c \to E_c^* = \rho c_p^2 (1 + 2i\zeta) \tag{3.32a}$$

$$G \to G^* = \rho c_s^2 (1 + 2i\zeta) \tag{3.32b}$$

where, for the sake of simplicity, the same hysteretic damping ratio ζ is assumed for axial and shear distortions. Note that this modification also affects the wave velocities

$$c_s \to c_s^* = c_s \sqrt{1 + 2i\zeta} \tag{3.33a}$$

$$c_p \to c_p^* = c_p \sqrt{1 + 2i\zeta} \tag{3.33b}$$

and the dimensionless frequency

$$a_0 \to a_0^* = \frac{a_0}{\sqrt{1 + 2i\zeta}} \tag{3.34}$$

in all equations, but that Poisson's ratio ν is not affected. The aspect ratios z_0/r_0, being based on statics, are not changed. However, when the dynamic-stiffness coefficient is plotted against a_0^*, only the real part of a_0^* is used, which is taken as a_0.

3.2 Rotational cones

The torsional degree of freedom is addressed first in this section (Fig. 3.3). A disk of radius r_0 on the surface of a homogeneous half-space with shear modulus $G = \rho c_s^2$, mass density ρ and hysteretic damping ratio ζ is modelled as a one-sided truncated semi-infinite rotational cone with the same material properties as the half-space. A torsional moment (torque) with an amplitude $M_0(\omega)$ is applied to the disk, resulting in a torsional rotation (twisting) of the disk with amplitude $\vartheta_0(\omega)$. The downward wave propagation with amplitude $\vartheta(z, \omega)$ is to be determined. The cone's opening angle, determined by the aspect ratio z_0/r_0, must also be calculated. Within the cone representing a bar with its polar moment of inertia increasing with depth z (measured from the apex) as

$$I(z) = \left(\frac{z}{z_0}\right)^4 I_0 \tag{3.35}$$

(with $I_0 = \pi r_0^4/2$), shear strains are present.

The complex amplitude form for harmonic excitation (Appendix A.1) is

$$M_0(t) = M_0(\omega) e^{i\omega t} \tag{3.36a}$$

$$\vartheta(t) = \vartheta(\omega) e^{i\omega t} \tag{3.36b}$$

To establish the equation of motion, the equilibrium of an infinitesimal element is formulated (Fig. 3.3). This yields, with the torsional moment with amplitude $T(z, \omega)$ and the twisting inertial

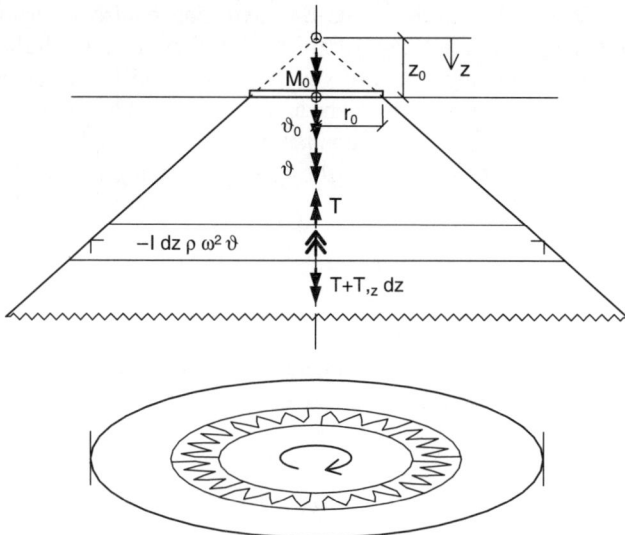

Figure 3.3 Rotational truncated semi-infinite cone with torsional motion, equilibrium of infinitesimal element and torsional (twisting) distortion

moment load (calculated as the polar mass moment of inertia of the element times acceleration of the torsional rotation acting in the negative z-direction) with amplitude $-I(z)\,dz\,\rho\omega^2\vartheta(z,\omega)$ (since $\ddot{\vartheta}(z,\omega) = -\omega^2\vartheta(z,\omega)$),

$$-T(z,\omega) + T(z,\omega) + T(z,\omega),_z\,dz + \omega^2 I(z)\rho\,dz\,\vartheta(z,\omega) = 0 \quad (3.37a)$$

or

$$T(z,\omega),_z + \omega^2 I(z)\rho\vartheta(z,\omega) = 0 \quad (3.37b)$$

Substituting the torsional moment-rotation relationship

$$T(z,\omega) = \rho c_s^2 I(z)\vartheta(z,\omega),_z \quad (3.38)$$

in Eq. 3.37b using Eq. 3.35 results in the equation of motion of the rotational cone

$$\vartheta(z,\omega),_{zz} + \frac{4}{z}\vartheta(z,\omega),_z + \frac{\omega^2}{c_s^2}\vartheta(z,\omega) = 0 \quad (3.39)$$

The solution is

$$\vartheta(z,\omega) = c_2 \left(\frac{1}{z^3} + i\frac{\omega}{c_s}\frac{1}{z^2}\right) e^{-i\frac{\omega}{c_s}z} \quad (3.40)$$

with the integration constant c_2. This term with $e^{-i\omega z/c_s}$ corresponds to a wave propagating in the positive z-direction with the velocity c_s. The verification is the same as described in the paragraph following Eq. 3.9. The other term with $e^{i\omega z/c_s}$ (not included in Eq. 3.40) corresponds to a wave propagating in the negative z-direction, i.e. towards the disk, which is physically impossible. Enforcing the boundary condition

$$\vartheta(z = z_0, \omega) = \vartheta_0(\omega) \tag{3.41}$$

leads to

$$\vartheta(z, \omega) = \frac{(z_0^3/z^3) + \mathrm{i}(\omega/c_s)(z_0^3/z^2)}{1 + \mathrm{i}(\omega/c_s)z_0} e^{-\mathrm{i}\frac{\omega}{c}(z-z_0)} \vartheta_0(\omega) \tag{3.42}$$

Equation 3.42 specifies the rotation amplitude at the distance $z - z_0$ from the disk with the rotation amplitude $\vartheta_0(\omega)$. It serves to calculate the rotation amplitude of the incident wave impinging at a material discontinuity.

The interaction moment-rotation relationship of the disk is now addressed. Equilibrium of the disk yields

$$M_0(\omega) = -T(z = z_0, \omega) \tag{3.43}$$

Substituting the derivative $\vartheta(z, \omega)_{,z}$ calculated from Eq. 3.42 into Eq. 3.38, which is evaluated at $z = z_0$, leads in Eq. 3.43 to

$$M_0(\omega) = S_\vartheta(b_0)\vartheta_0(\omega) \tag{3.44}$$

where the *dynamic-stiffness coefficient* is

$$S_\vartheta(b_0) = K_\vartheta[k_\vartheta(b_0) + \mathrm{i}b_0 c_\vartheta(b_0)] \tag{3.45}$$

with the static-stiffness coefficient K_ϑ (noting that $I_0 = \pi r_0^4/2$) defined as

$$K_\vartheta = \frac{3\rho c_s^2 I_0}{z_0} \tag{3.46}$$

the dimensionless spring and damping coefficients

$$k_\vartheta(b_0) = 1 - \frac{1}{3}\frac{b_0^2}{1+b_0^2} \tag{3.47a}$$

$$c_\vartheta(b_0) = \frac{1}{3}\frac{b_0^2}{1+b_0^2} \tag{3.47b}$$

and the dimensionless frequency defined with respect to the properties of the cone

$$b_0 = \frac{\omega z_0}{c_s} \tag{3.48}$$

Equation 3.44 permits the disk's rotation with amplitude $\vartheta_0(\omega)$ to be determined for a specified moment with amplitude $M_0(\omega)$ acting on the disk in the presence of a one-sided cone.

Introducing the standard dimensionless frequency of the disk

$$a_0 = \frac{\omega r_0}{c_s} \tag{3.49}$$

Equation 3.45 is rewritten as

$$S_\vartheta(a_0) = K_\vartheta [k_\vartheta(a_0) + i a_0 \, c_\vartheta(a_0)] \tag{3.50}$$

with

$$k_\vartheta(a_0) = 1 - \frac{1}{3} \frac{a_0^2}{(r_0/z_0)^2 + a_0^2} \tag{3.51a}$$

$$c_\vartheta(a_0) = \frac{1}{3} \frac{z_0}{r_0} \frac{a_0^2}{(r_0/z_0)^2 + a_0^2} \tag{3.51b}$$

Note that the high frequency limit $\omega \to \infty$ of $S_\vartheta(a_0)$ is

$$\lim_{\omega \to \infty} S_\vartheta(a_0) = i \omega C_\vartheta \tag{3.52}$$

where the dashpot with the high frequency limit of the radiation damping is

$$C_\vartheta = \rho c_s I_0 \tag{3.53}$$

To determine the aspect ratio z_0/r_0 and hence the cone's opening angle, *the static-stiffness coefficient K_ϑ of the truncated semi-infinite cone* (Eq. 3.46) *is matched to the corresponding exact solution of three-dimensional elasticity theory for the disk on a half-space*, which is

$$K_{\vartheta\,exact} = \frac{16}{3} \rho c_s^2 r_0^3 \tag{3.54}$$

Equating Eqs 3.46 and 3.54 yields

$$\frac{z_0}{r_0} = \frac{9\pi}{32} \tag{3.55}$$

The opening angle is, as expected for the torsional degree of freedom, independent of Poisson's ratio ν.

The rocking degree of freedom, modelled with its own cone, is now addressed (Fig. 3.4). As for the vertical degree of freedom discussed at the beginning of Section 3.1, to allow equations to be derived which also cover the nearly-incompressible case, the derivation is performed for an axial-wave velocity c, with the corresponding modulus of elasticity expressed as ρc^2. For the compressible case c is equal to the dilatational-wave velocity c_p and E is equal to the constrained modulus E_c. Besides c, Poisson's ratio ν and the mass density ρ are selected as the elastic material properties. A bending moment with amplitude $M_0(\omega)$ is applied to the disk, resulting in a rocking

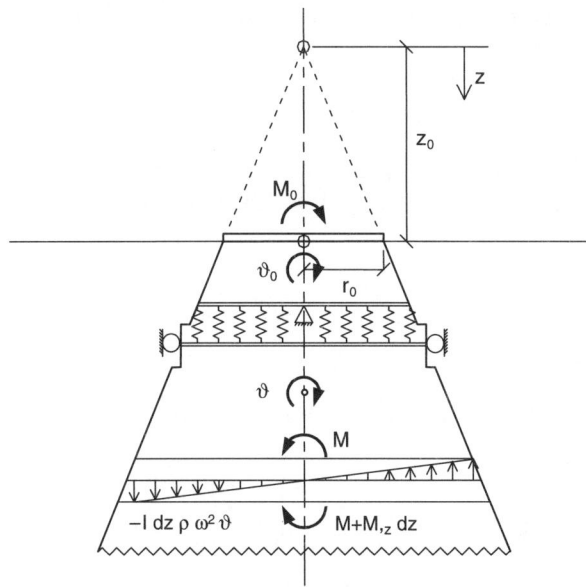

Figure 3.4 Rotational truncated semi-infinite cone with rocking motion, rotational axial distortion and equilibrium of infinitesimal element, where horizontal motion is prevented with support on axis or with rollers for vertical motion

rotation of the disk with amplitude $\vartheta_0(\omega)$. Within the cone representing a beam with its moment of inertia increasing with depth z (measured from the apex) as

$$I(z) = \left(\frac{z}{z_0}\right)^4 I_0 \qquad (3.56)$$

(with $I_0 = \pi r_0^4/4$), axial strains are present, antimetric with respect to the axis. The amplitude of the bending moment with the curvature with amplitude $\vartheta(z,\omega)_{,z}$ (analogous to Eq. 3.38) is

$$M(z, \omega) = \rho c^2 I(z) \vartheta(z, \omega)_{,z} \qquad (3.57)$$

with the amplitude of the rocking rotation $\vartheta(z, \omega)$. The equilibrium equation follows from Eq. 3.37b, replacing $T(z, \omega)$ by $M(z, \omega)$. The derivation is analogous, replacing c_s by c. The rotation amplitude of the incident wave follows from Eq. 3.42, and the dynamic-stiffness coefficient from Eq. 3.45 with b_0 in Eq. 3.48, replacing c_s by c. The static-stiffness coefficient (with $I_0 = \pi r_0^4/4$) is

$$K_\vartheta = \frac{3\rho c^2 I_0}{z_0} \qquad (3.58)$$

The dashpot with the high frequency limit of the radiation damping is

$$C_\vartheta = \rho c I_0 \qquad (3.59)$$

As a_0 is still defined with respect to c_s (Eq. 3.49), the dimensionless spring and damping coefficients are (Eq. 3.51)

$$k_\vartheta(a_0) = 1 - \frac{1}{3}\frac{a_0^2}{(r_0 c)^2/(z_0 c_s)^2 + a_0^2} \tag{3.60a}$$

$$c_\vartheta(a_0) = \frac{1}{3}\frac{z_0}{r_0}\frac{c_s}{c}\frac{a_0^2}{(r_0 c)^2/(z_0 c_s)^2 + a_0^2} \tag{3.60b}$$

with Eq. 3.50 applying. *Matching the exact rocking static-stiffness coefficient of a disk on a half-space*

$$K_{\vartheta\,exact} = \frac{8\rho c_s^2 r_0^3}{3(1-\nu)} \tag{3.61}$$

to the corresponding solution of the truncated semi-infinite cone specified in Eq. 3.58 yields

$$\frac{z_0}{r_0} = \frac{9\pi}{32}(1-\nu)\left(\frac{c}{c_s}\right)^2 \tag{3.62}$$

which is a function of ν. The opening angle shown in Fig. 3.4 corresponds to $\nu = 1/3$.

The rocking model described above possesses certain unique strength-of-materials assumptions. Although an understanding of these is not required to formulate the model, they are addressed in the next section for completeness.

3.3 Interpretation of the rocking cone

The cone model for the torsional degree of freedom (Fig. 3.3) is based on the strength-of-materials approach of a tapered bar. The question arises as to whether the model for the rocking degree of freedom (Fig. 3.4) can be interpreted as a beam model, or not. Horizontal motion of the cross-section is prevented, either by a support on the axis enforcing antimetric vertical motion, or by placing rollers for vertical motion, as indicated in the figure. In this valid model, it follows from formulating equilibrium that the horizontal shear force on the cross-section vanishes. Rotations of the rigid cross-sections with mass do occur, but this does not lead to lateral horizontal displacements. Such displacements would lead to horizontal inertial loads, yielding shear forces and thus coupling between the rocking and horizontal degrees of freedom.

To investigate the question further, the effect of shear deformation and rotary inertia on the lateral displacement of a beam is addressed (Fig. 3.5). Figure 3.5a illustrates the forces, moments and loads acting on an infinitesimal element. It is a combination of the infinitesimal elements for lateral displacement of the axis with amplitude $u(z, \omega)$ (Fig. 3.2a) and for rotation of the

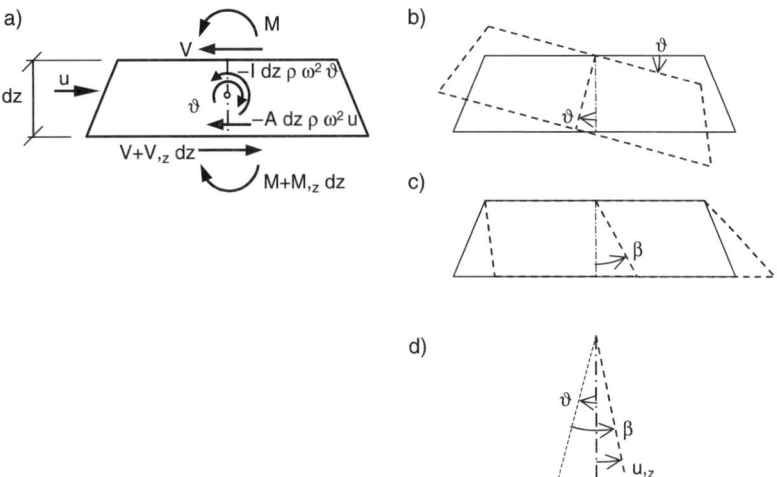

Figure 3.5 Infinitesimal beam element with shear deformation and rotary inertia. a) Forces, moments and loads. b) Deformation from rotation of cross-section. c) Deformation from shear distortion. d) Final deformed axis yielding slope

cross-section with amplitude $\vartheta(z, \omega)$ (Fig. 3.4). Formulating the moment equilibrium

$$-M(z, \omega) + M(z, \omega) + M(z, \omega),_z \, dz - V(z, \omega) \, dz + \omega^2 I(z)\rho \, dz \, \vartheta(z, \omega) = 0 \quad (3.63)$$

leads to

$$M(z, \omega),_z - V(z, \omega) + \omega^2 I(z)\rho \vartheta(z, \omega) = 0 \quad (3.64)$$

and the force equilibrium

$$-V(z, \omega) + V(z, \omega) + V(z, \omega),_z \, dz + \omega^2 A(z)\rho \, dz \, u(z, \omega) = 0 \quad (3.65)$$

yields

$$V(z, \omega),_z + \omega^2 A(z)\rho u(z, \omega) = 0 \quad (3.66)$$

The bending moment-cross-sectional rotation relationship (Eq. 3.57) is

$$M(z, \omega) = \rho c^2 I(z)\vartheta(z, \omega),_z \quad (3.67)$$

and the shear force-shear distortion relationship (see Eq. 3.29) is formulated as

$$V(z, \omega) = \rho c_s^2 A(z)\beta(z, \omega) \quad (3.68)$$

with the shear distortion with amplitude $\beta(z, \omega)$.

The deformed infinitesimal elements caused by the rotation of the cross-section and by the shear distortion are presented in Figs 3.5b and 3.5c, respectively. Figure 3.5d illustrates the kinematic relationship between the various angles and the amplitude of the slope of the axis, which is the derivative of the lateral displacement with amplitude $u(z, \omega)_{,z}$, where

$$u(z, \omega)_{,z} = -\vartheta(z, \omega) + \beta(z, \omega) \qquad (3.69)$$

Solving Eq. 3.69 for $\beta(z, \omega)$ and substituting into Eq. 3.68 results in

$$V(z, \omega) = \rho c_s^2 A(z)(\vartheta(z, \omega) + u(z, \omega)_{,z}) \qquad (3.70)$$

Substituting Eqs 3.67 and 3.70 in Eqs 3.64 and 3.66 leads to the equations of motion expressed in $\vartheta(z, \omega)$ and $u(z, \omega)$

$$(\rho c^2 I(z)\vartheta(z, \omega)_{,z})_{,z} - \rho c_s^2 A(z)(\vartheta(z, \omega) + u(z, \omega)_{,z}) + \omega^2 I(z)\rho\vartheta(z, \omega) = 0 \qquad (3.71)$$

$$(\rho c_s^2 A(z)(\vartheta(z, \omega) + u(z, \omega)_{,z}))_{,z} + \omega^2 A(z)\rho u(z, \omega) = 0 \qquad (3.72)$$

To be able to construct the equations governing the model shown in Fig. 3.4, the shear modulus $G = \rho c_s^2$ is set equal to zero. From Eq. 3.70

$$V(z, \omega) = 0 \qquad (3.73)$$

indicating the shear force vanishes, and from Eq. 3.72

$$u(z, \omega) = 0 \qquad (3.74)$$

indicating the lateral displacement is also zero. Equation 3.69 leads to

$$\beta(z, \omega) = \vartheta(z, \omega) \qquad (3.75)$$

meaning that the shear distortion is equal to the cross-sectional rotation, and, considering the different sign conventions for these two angles, the axis does not displace horizontally (Fig. 3.5d). Equation 3.71 results in

$$(\rho c^2 I(z)\vartheta(z, \omega)_{,z})_{,z} + \omega^2 I(z)\rho\vartheta(z, \omega) = 0 \qquad (3.76)$$

which corresponds to Eq. 3.37b with $T(z, \omega)$ replaced by $M(z, \omega)$ (Eq. 3.57).

Thus, the model shown in Fig. 3.4 corresponds to a tapered beam with shear deformation and rotary inertia in which the shear modulus is zero. For a rotation of the disk caused by an applied moment, no shear force and no lateral displacement occur. The effect of the cross-sectional rotation on the lateral displacement is cancelled by the effect of the shear distortion, which does not vanish.

Equations 3.71 and 3.72 can also be used to verify the translational cone for the horizontal degree of freedom shown in Fig. 3.2b. In the moment equilibrium equation of Eq. 3.71 an additional term representing the reaction moment from the rollers is present, and this equation is used to determine this reaction moment. The rollers enforce $\vartheta(z, \omega) = 0$. Substitution of this relationship into Eq. 3.72 yields

$$(\rho c_s^2 A(z)u(z, \omega)_{,z})_{,z} + \omega^2 A(z)\rho u(z, \omega) = 0 \qquad (3.77)$$

which corresponds to Eq. 3.3b with $N(z, \omega)$ replaced by $V(z, \omega)$ (Eq. 3.29).

Formally, the formulation of the translational and rotational cones can be unified as follows. The cone with apex height z_0, depth z and surface area $A_0 = \pi r_0^2$ is shown for the vertical degree of freedom in Fig. 3.6. In the horizontal direction the area $A(z)$ is plotted instead of the radius of the cross-section of the cone, which always increases linearly with depth. The cone experiences a vertical or horizontal displacement with amplitude $u(z, \omega)$, the value of the amplitude at the disk at the surface being denoted as $u_0(\omega)$. For rotational motion, the same notation may be preserved if A_0 is interpreted as the surface moment of inertia for rocking, or the polar moment of inertia for twist, and $u(z, \omega)$ is taken to be the amplitude of the angle of rocking or twist. The cross-section increases with depth according to the power law $A(z) = A_0(z/z_0)^n$ with z measured from the apex. The values of the exponent $n = 2$ and $n = 4$ correspond to the translational and rotational cones respectively. If c denotes the appropriate wave velocity (shear-wave velocity c_s for the horizontal and torsional degrees of freedom, axial-wave velocity c for the vertical and rocking degrees of freedom, with the dilatational wave velocity c_p for the compressible case) and ρ stands for the mass density, the quantity ρc^2 is equal to the corresponding elastic modulus (shear modulus G and constrained modulus E_c for the compressible case, respectively). The equilibrium equation for harmonic motion of the infinitesimal element (Fig. 3.6) is

$$-N(z, \omega) + N(z, \omega) + N(z, \omega),_z \, dz + \omega^2 A(z) \rho \, dz \, u(z, \omega) = 0 \qquad (3.78a)$$

or

$$N(z, \omega),_z + \omega^2 A(z) \rho u(z, \omega) = 0 \qquad (3.78b)$$

with the amplitude of the internal force or moment $N(z, \omega)$. Substituting the force-displacement relationship

$$N(z, \omega) = \rho c^2 A(z) u(z, \omega),_z \qquad (3.79)$$

in Eq. 3.78b leads to the governing differential equation of motion

$$u(z, \omega),_{zz} + \frac{n}{z} u(z, \omega),_z + \frac{\omega^2}{c^2} u(z, \omega) = 0 \qquad (3.80)$$

When $n = 2$, Eq. 3.80 coincides with Eq. 3.5 for the translational cone, and when $n = 4$, with Eq. 3.39 for the rotational cone.

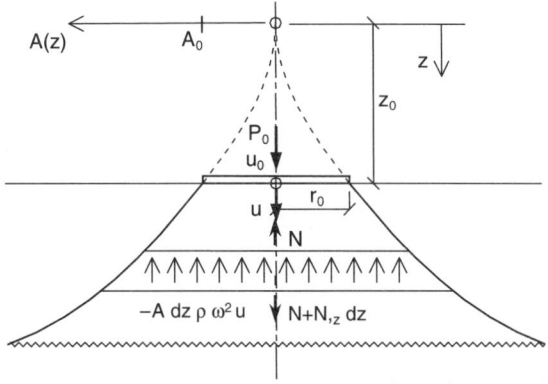

Figure 3.6 Rotational cone interpreted as translational cone in vertical motion with area varying as (polar) moment of inertia

3.4 Incompressible or nearly-incompressible half-space

For the horizontal and torsional degrees of freedom of a disk on the surface of a homogeneous half-space modelled with cones, S-waves occur. The appropriate wave velocity is thus c_s, which remains finite for all values of Poisson's ratio ν. The aspect ratio z_0/r_0 follows from Eq. 3.31 for the horizontal degree of freedom and from Eq. 3.55 for the torsional degree of freedom for all ν.

For the vertical and rocking degrees of freedom, axial distortions occur in the cone models. For low values of ν, P-waves propagating with the dilatational-wave velocity c_p are present. For ν approaching $1/2$, the constrained modulus E_c (Eq. 2.2a) and thus c_p (Eq. 2.1) tend to infinity, and a cone model cannot be constructed in the usual way. Special features are necessary for this case. For the motivation and further details, the reader is referred to Sections 2.1.4 and 2.5 of Ref. [37].

Thus, for the vertical and rocking degrees of freedom, the procedure depends on Poisson's ratio. In the range $0 \le \nu \le 1/3$, the dilatational-wave velocity c_p and the constrained modulus $E_c = \rho c_p^2$ apply. The aspect ratio z_0/r_0 follows from Eq. 3.28 for the vertical degree of freedom and from Eq. 3.62 for the rocking degree of freedom, with $c = c_p$ in both cases. In the range $1/3 < \nu < 1/2$, called the nearly-incompressible case and for $\nu = 1/2$ (the incompressible case), two features are enforced: first, *the axial-wave velocity c is limited to $2c_s$*; and second, a *trapped mass* for the vertical degree of freedom and a *trapped mass moment of inertia* for the rocking degree of freedom are introduced.

3.4.1 Wave velocity

The axial-wave velocity $c = 2c_s$ is enforced. The modulus of elasticity $E = \rho c^2$ is

$$E = 4\rho c_s^2 \qquad 1/3 < \nu \le 1/2 \tag{3.81}$$

The aspect ratios for the vertical and rocking degrees of freedom follow from Eqs 3.28 and 3.62 respectively as

$$\text{vertical} \qquad \frac{z_0}{r_0} = \pi(1-\nu) \qquad 1/3 < \nu \le 1/2 \tag{3.82}$$

$$\text{rocking} \qquad \frac{z_0}{r_0} = \frac{9\pi}{8}(1-\nu) \qquad 1/3 < \nu \le 1/2 \tag{3.83}$$

3.4.2 Trapped mass

A trapped mass of soil beneath the disk is introduced, which moves as a rigid body in phase with the disk. For the vertical degree of freedom the trapped mass is

$$\Delta M = 2.4\left(\nu - \frac{1}{3}\right)\rho A_0 r_0 \tag{3.84}$$

and for the rocking degree of freedom the trapped mass moment of inertia is formulated as

$$\Delta M_\vartheta = 1.2\left(\nu - \frac{1}{3}\right)\rho I_0 r_0 \tag{3.85}$$

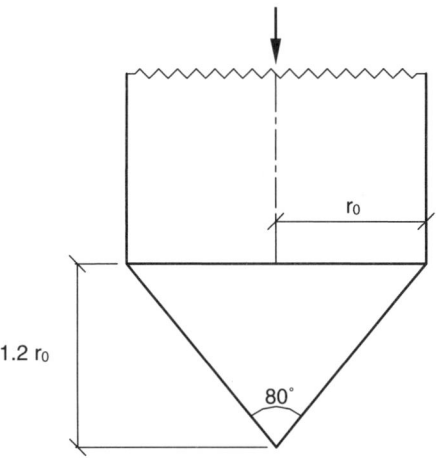

Figure 3.7 Trapped mass of soil forming pointed tip of pile in vertical motion

with $I_0 = \pi r_0^4/4$. According to these formulae, the inclusion of trapped mass begins at $\nu = 1/3$ and increases linearly with ν. For incompressible soil with $\nu = 1/2$, the trapped mass in vertical motion corresponds to an equivalent cylinder of soil with height $0.6r_0$; for rocking, the equivalent cylinder is half as tall (and concentrated in the middle plane of the disk).

Although these quantities of trapped mass are determined by simple curve fitting (see the accuracy evaluation of vertical and rocking dynamic-stiffness coefficients in Section 3.5), they have a physical justification. In engineering practice it has been observed that blunt-ended piles may be driven just as easily as piles with pointed tips. The reason is that the pile creates its own pointed tip (in incompressible soil). As shown in Fig. 3.7, a cone of trapped soil forms, and moves as a rigid body with the pile of radius r_0. Intuitively one would expect the opening angle of the cone to be slightly less than 90°. For 80° the height of the cone is $1.2r_0$. The mass of the cone is $(1/3)\,\pi r_0^2 1.2 r_0 \rho = 0.4 r_0\,\pi r_0^2\,\rho$, in agreement with the trapped mass computed using Eq. 3.84 with $\nu = 1/2$. For the rocking degree of freedom, a physical explanation for the trapped mass can also be constructed based on similar considerations.

To establish the interaction force-displacement relationship for the vertical degree of freedom, the inertial load (trapped mass times acceleration acting in the negative direction) is included in the equilibrium equation of the disk (Eq. 3.13)

$$P_0(\omega) = -N(z = z_0, \omega) - \Delta M \omega^2 u_0(\omega) \tag{3.86}$$

This additional term leads to (Eq. 3.14)

$$P_0(\omega) = \left(\frac{4\rho c_s^2 A_0}{z_0} - \omega^2 \Delta M + i\omega 2\rho c_s A_0\right) u_0(\omega) \tag{3.87}$$

and, with Eqs 3.17 and 3.18, and $\ddot{u}_0(\omega) = -\omega^2 u_0(\omega)$, to

$$P_0(\omega) = K u_0(\omega) + C \dot{u}_0(\omega) + \Delta M \ddot{u}_0(\omega) \tag{3.88}$$

Thus, for $\nu > 1/3$, a mass ΔM is added to the spring-dashpot model presented in Fig. 2.2b.

With the dimensionless frequency a_0 of the disk (Eq. 3.24), the dynamic-stiffness coefficient is decomposed as in Eq. 3.25 with

$$k(a_0) = 1 - 0.6\left(\nu - \frac{1}{3}\right)\frac{z_0}{r_0}a_0^2 \qquad (3.89a)$$

$$c(a_0) = 0.5\frac{z_0}{r_0} \qquad (3.89b)$$

In Eq. 3.89, z_0/r_0 is given by Eq. 3.82.

The interaction bending moment-rotation relationship for the rocking degree of freedom is formulated analogously. Including the inertial moment load (trapped mass moment of inertia times rocking acceleration acting in the negative direction) in the equilibrium equation of the disk (Eq. 3.43 replacing $T(z,\omega)$ with $M(z,\omega)$ of Eq. 3.57) is

$$M_0(\omega) = -M(z = z_0, \omega) - \Delta M_\vartheta \omega^2 \vartheta_0(\omega) \qquad (3.90)$$

This dynamic-stiffness coefficient is decomposed as in Eq. 3.50 with (Eq. 3.60)

$$k_\vartheta(a_0) = 1 - \frac{1}{3}\frac{a_0^2}{4(r_0/z_0)^2 + a_0^2} - 0.1\left(\nu - \frac{1}{3}\right)\frac{z_0}{r_0}a_0^2 \qquad (3.91a)$$

$$c_\vartheta(a_0) = \frac{1}{6}\frac{z_0}{r_0}\frac{a_0^2}{4(r_0/z_0)^2 + a_0^2} \qquad (3.91b)$$

In Eq. 3.91, z_0/r_0 is given by Eq. 3.83.

3.5 Foundation on the surface of a homogeneous half-space

The results of Sections 3.1, 3.2 and 3.4 of this chapter can be used directly to calculate the vibrational behaviour of a circular foundation on the surface of a homogeneous half-space.

In this section the dynamic-stiffness coefficients of a disk of radius r_0 on the surface of a homogeneous undamped half-space with shear modulus G, Poisson's ratio ν and mass density ρ are calculated with cone models for all degrees of freedom, and compared with the rigorous results found in the literature.

The results are presented as a function of the dimensionless frequency a_0 defined in Eq. 3.24. The dynamic-stiffness coefficients are decomposed as specified in Eqs 3.25 and 3.50. For the results of the cone models and the rigorous results, the exact closed-form solutions of the static-stiffness coefficients K and K_ϑ are applied, as specified in Eqs 3.30, 3.27, 3.54 and 3.61 for the horizontal, vertical, torsional and rocking degrees of freedom, respectively.

For the horizontal degree of freedom, the aspect ratio z_0/r_0 follows from Eq. 3.31 and the dimensionless spring and damping coefficients of the cone model from Eq. 3.26 with $c = c_s$, yielding $k = 1$ and $c = z_0/r_0$. These are shown as horizontal lines for $\nu = 1/3$ and $1/2$ in Fig. 3.8. The rigorous values of Ref. [35] are plotted as distinct points. For the two Poisson's ratios the agreement is satisfactory.

For the vertical degree of freedom, cases with $\nu = 1/3$, 0.45 and $1/2$ are processed. z_0/r_0 follows for $\nu = 1/3$ with $c = c_p$ from Eq. 3.28, and for $\nu = 0.45$ and $1/2$ from Eq. 3.82. For $\nu = 1/3$, $k = 1$ and $c = z_0/(2r_0)$ are determined from Eq. 3.26, and for $\nu = 0.45$ and $1/2$, $k(a_0)$

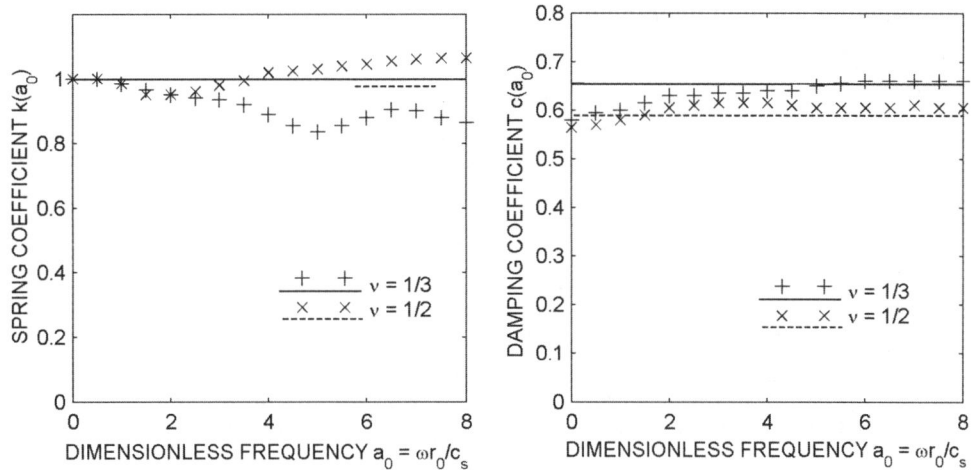

Figure 3.8 Dynamic-stiffness coefficient of disk on homogeneous half-space in horizontal motion for various Poisson's ratios

and $c = z_0/(2r_0)$ follow from Eq. 3.89. The cone results are presented as lines in Fig. 3.9. For the nearly-incompressible case $\nu = 0.45$ and the incompressible case $\nu = 1/2$, $k(a_0)$ is described by a second degree parabola with a downward tendency. The inertial load of the trapped mass ΔM leads to a decrease in $k(a_0)$, resulting in negative values for sufficiently large a_0. The rigorous results (shown as distinct points) are taken from Refs [34] and [15]. The agreement is again satisfactory, and for $k(a_0)$ in the nearly-incompressible and incompressible cases, even excellent.

For the torsional degree of freedom, $z_0/r_0 = 0.884$ is defined by Eq. 3.55, and $k(a_0)$ and $c(a_0)$ are determined from Eq. 3.51. The index ϑ is omitted for conciseness. The results of the cone (drawn as lines) agree quite well with the rigorous results (plotted as distinct points) from Ref. [33] in Fig. 3.10.

Finally, for the rocking degree of freedom, cases with $\nu = 1/3$, 0.45 and 1/2 are calculated. z_0/r_0 follows for $\nu = 1/3$ with $c = c_p(=2c_s)$ from Eq. 3.62, and for $\nu = 0.45$ and 1/2 from Eq. 3.83. For $\nu = 1/3$, $k(a_0)$ and $c(a_0)$ are determined from Eq. 3.60, and for $\nu = 0.45$ and 1/2, $k(a_0)$ and $c(a_0)$ follow from Eq. 3.91. The cone results are presented (as lines) in Fig. 3.11, along with the rigorous solution (shown as distinct points) taken from Refs [35] and [15]. As for the vertical degree of freedom, $k(a_0)$ exhibits a downward-parabolic tendency. The agreement between the cone calculations and the rigorous solution is again satisfactory, and for $k(a_0)$ in the nearly incompressible and incompressible cases, excellent.

For the translational degrees of freedom, the interaction force-displacement relationship of Eq. 3.14 (Eq. 3.20) for harmonic excitation and of Eq. 2.24 in the time domain is modelled exactly by the *spring-dashpot model* presented in Fig. 2.2b. For the torsional degree of freedom, the interaction moment-rotation relationship is specified in Eqs 3.44 and 3.45, with Eqs 3.46 and 3.47 applying. A *spring-dashpot-mass model* with frequency independent coefficients can again be constructed for the rotational cone, but with one additional internal degree of freedom ϑ_1 (Fig. 3.12). The foundation node with the degree of freedom ϑ_0 is connected by a rotational spring with the static-stiffness coefficient K_ϑ (Eq. 3.46) to a rigid support. The additional internal rotational degree of freedom ϑ_1 has its own polar mass moment of inertia

$$M_\vartheta = \rho z_0 I_0 \tag{3.92}$$

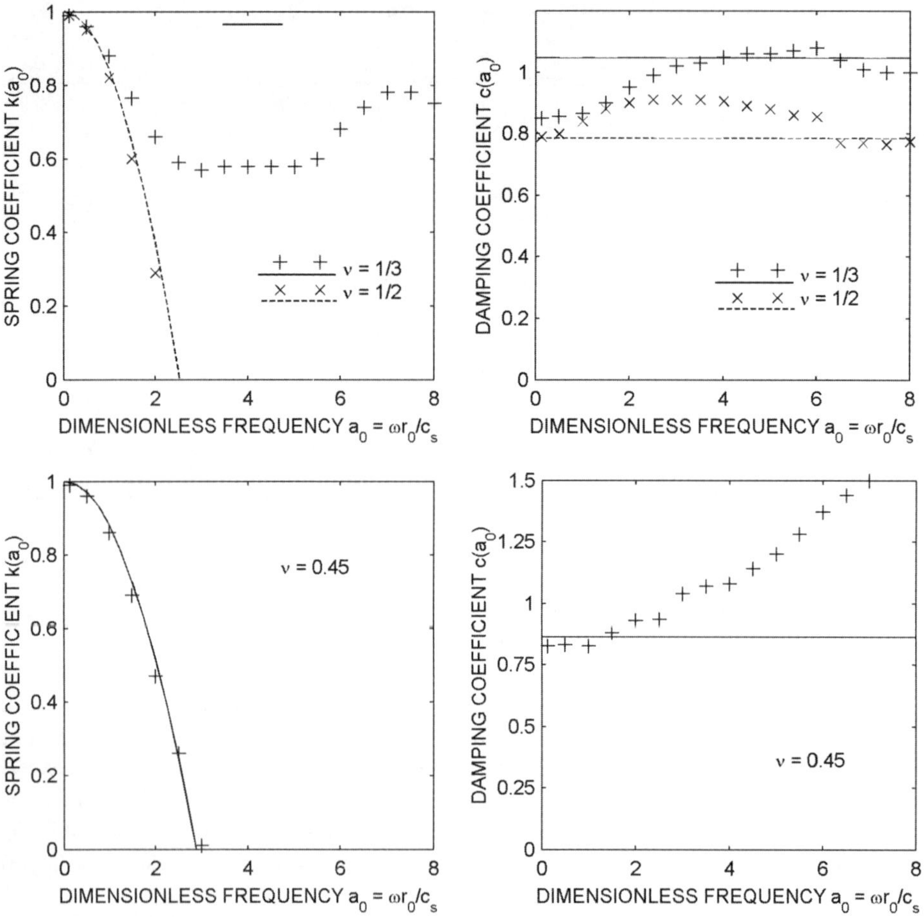

48 Foundation Vibration Analysis: A Strength-of-Materials Approach

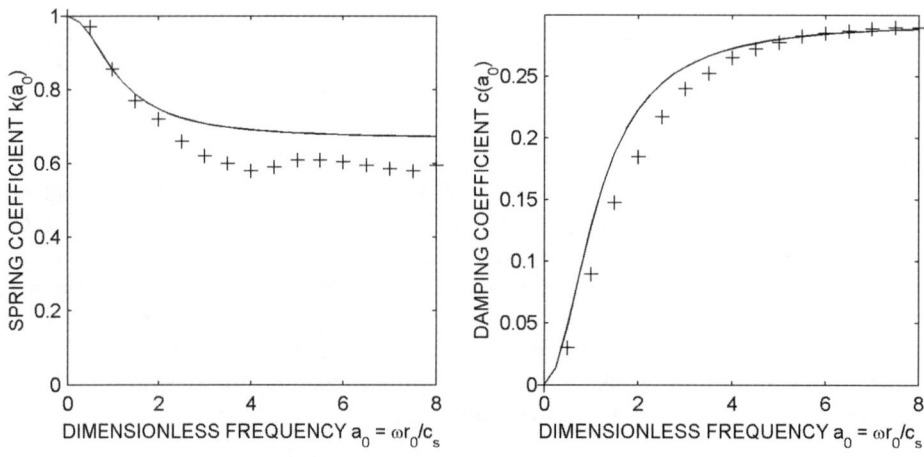

Figure 3.9 Dynamic-stiffness coefficient of disk on homogeneous half-space in vertical motion for various Poisson's ratios

Figure 3.10 Dynamic-stiffness coefficient of disk on homogeneous half-space in torsional motion

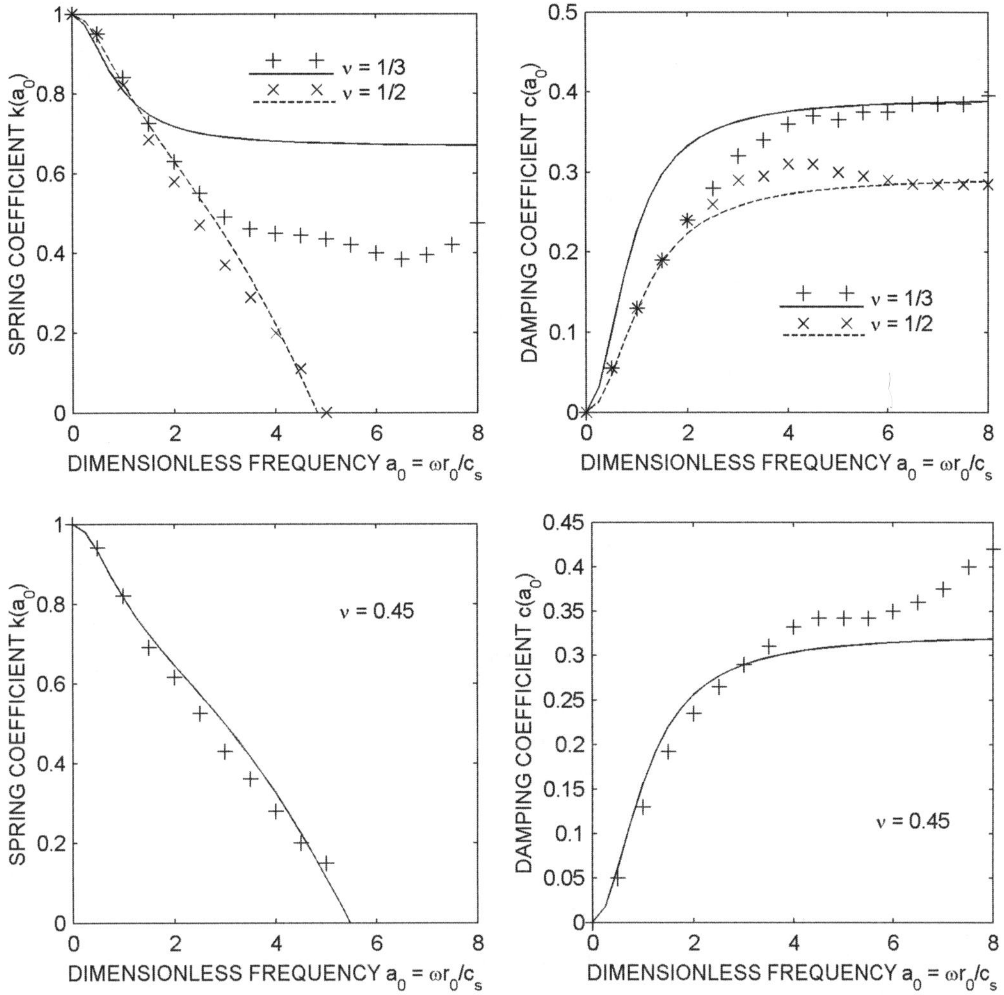

Figure 3.11 Dynamic-stiffness coefficient of disk on homogeneous half-space in rocking motion for various Poisson's ratios

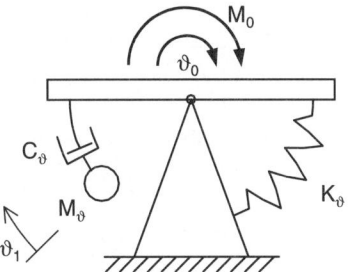

Figure 3.12 Spring-dashpot-mass model with additional internal degree of freedom for rotational degree of freedom

(with $I_0 = \pi r_0^4/2$) which is connected to the foundation node by a rotational dashpot with the high frequency limit of the radiation damping C_ϑ (Eq. 3.53). This model resembles a monkey tail!

The equivalence of the spring-dashpot-mass model of Fig. 3.12 with the interaction moment-rotation relationship of the rotational cone is proved as follows. Formulating equilibrium leads to the two equations of motion in the time domain

$$M_0(t) = C_\vartheta (\dot{\vartheta}_0(t) - \dot{\vartheta}_1(t)) + K_\vartheta \vartheta_0(t) \qquad (3.93a)$$

$$M_\vartheta \ddot{\vartheta}_1(t) + C_\vartheta (\dot{\vartheta}_1(t) - \dot{\vartheta}_0(t)) = 0 \qquad (3.93b)$$

For harmonic excitation, Eq. 3.93 is transformed to

$$M_0(\omega) = i\omega C_\vartheta (\vartheta_0(\omega) - \vartheta_1(\omega)) + K_\vartheta \vartheta_0(\omega) \qquad (3.94a)$$

$$-\omega^2 M_\vartheta \vartheta_1(\omega) + i\omega C_\vartheta (\vartheta_1(\omega) - \vartheta_0(\omega)) = 0 \qquad (3.94b)$$

Eliminating $\vartheta_1(\omega)$ from Eq. 3.94 yields

$$M_0(\omega) = K_\vartheta \left[1 - \frac{\omega^2 (M_\vartheta/K_\vartheta)}{1 + \omega^2 (M_\vartheta^2/C_\vartheta^2)} + i\omega \left(\frac{M_\vartheta}{C_\vartheta} \frac{\omega^2 (M_\vartheta/K_\vartheta)}{1 + \omega^2 (M_\vartheta^2/C_\vartheta^2)} \right) \right] \vartheta_0(\omega) \qquad (3.95)$$

It follows from Eqs 3.46, 3.53, 3.92 and 3.48 that

$$\omega^2 \frac{M_\vartheta}{K_\vartheta} = \frac{1}{3} b_0^2 \qquad (3.96a)$$

$$\omega \frac{M_\vartheta}{C_\vartheta} = b_0 \qquad (3.96b)$$

Substituting Eq. 3.96 into Eq. 3.95 yields the interaction moment-rotation relationship of the rotational cone for the torsional degree of freedom, with its dynamic-stiffness coefficient in Eq. 3.45 with Eq. 3.47.

The equations of motion specified in Eq. 3.93 can be used directly for an analysis of dynamic soil-structure interaction in the time domain. In a practical application the spring-dashpot-mass model with one internal degree of freedom shown in Fig. 3.12 may be attached to the underside of the structure as an exact representation of the rotational cone in a dynamic soil-structure interaction analysis. The complete coupled system may then be analysed directly with a general-purpose computer program.

The model shown in Fig. 3.12 can also be used to represent the rocking cone for the compressible case ($\nu \leq 1/3$). I_0 in Eq. 3.92 then represents the moment of inertia ($=\pi r_0^4/4$). For nearly-incompressible and incompressible soil ($1/3 < \nu \leq 1/2$), a mass moment of inertia with coefficient ΔM_ϑ (Eq. 3.85) is introduced at the foundation node with degree of freedom ϑ_0.

In a dynamic soil-structure-interaction analysis, the simplest way to account for the trapped mass ΔM and the trapped mass moment of inertia ΔM_ϑ is just to assign them to the underside of the structure.

Table 3.1 summarises the modelling information for a foundation on the surface of a homogeneous half-space. For all components of motion a rigid massless foundation with area A_0 and (polar) moment of inertia I_0 on the surface of a homogeneous half-space with Poisson's ratio ν, shear-wave velocity c_s, dilatational-wave velocity c_p and mass density ρ can be modelled with a cone of equivalent radius r_0 (determined by equating A_0 or I_0 of the foundation of general shape

Initial cone with outward wave propagation

Table 3.1 Cone and spring-dashpot-mass model for foundation on surface of homogeneous half-space

Motion	Horizontal	Vertical		Rocking		Torsional
Equivalent radius r_0	$\sqrt{\dfrac{A_0}{\pi}}$	$\sqrt{\dfrac{A_0}{\pi}}$		$\sqrt[4]{\dfrac{4I_0}{\pi}}$		$\sqrt[4]{\dfrac{2I_0}{\pi}}$
Aspect ratio $\dfrac{z_0}{r_0}$	$\dfrac{\pi}{8}(2-\nu)$	$\dfrac{\pi}{4}(1-\nu)\left(\dfrac{c}{c_s}\right)^2$		$\dfrac{9\pi}{32}(1-\nu)\left(\dfrac{c}{c_s}\right)^2$		$\dfrac{9\pi}{32}$
Poisson's ratio ν	all ν	$\nu \leq \tfrac{1}{3}$	$\tfrac{1}{3} < \nu \leq \tfrac{1}{2}$	$\nu \leq \tfrac{1}{3}$	$\tfrac{1}{3} < \nu \leq \tfrac{1}{2}$	all ν
Wave velocity c	c_s	c_p	$2c_s$	c_p	$2c_s$	c_s
Trapped mass ΔM, ΔM_ϑ	0	0	$2.4\left(\nu-\tfrac{1}{3}\right)\rho A_0 r_0$	0	$1.2\left(\nu-\tfrac{1}{3}\right)\rho I_0 r_0$	0
Lumped-parameter model		$K = \rho c^2 A_0/z_0$ $C = \rho c A_0$		$K_\vartheta = 3\rho c^2 I_0/z_0$ $C_\vartheta = \rho c I_0$ $M_\vartheta = \rho I_0 z_0$		

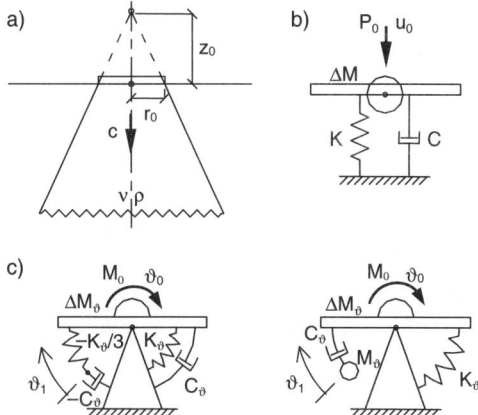

Figure 3.13 Cone model and corresponding lumped-parameter models for foundation on surface of homogeneous half-space. a) Truncated semi-infinite cone. b) Spring-dashpot-mass model for translational degree of freedom. c) Spring-dashpot-mass model for rotational degree of freedom

to that of an equivalent disk), apex height z_0 and wave velocity c (Fig. 3.13a). The translational cone model for the displacement u_0 is dynamically equivalent to the spring K-dashpot C system (Fig. 3.13b). The rotational cone for the rotation ϑ_0 corresponds exactly to the spring K_ϑ-dashpot C_ϑ-mass moment of inertia M_ϑ model with one internal degree of freedom ϑ_1 (illustrated on the right-hand side of Fig. 3.13c). Alternatively, the spring K_ϑ-dashpot C_ϑ model with one internal degree of freedom ϑ_1 (shown on the left-hand side of Fig. 3.13c) can be applied. Note that, in this case, two of the coefficients are negative, which does not present any problems mathematically, although the model cannot be built mechanically. All coefficients are frequency independent.

For the vertical and rocking motions in the case of nearly-incompressible and incompressible soil ($1/3 < \nu \leq 1/2$), c is limited to $2c_s$, and a trapped mass ΔM and a trapped mass moment of inertia ΔM_ϑ assigned to the foundation node arise.

3.6 Double cones

A disk embedded in a full-space is addressed in this section. The rocking degree of freedom is illustrated as an example in Fig. 3.14a. On a disk with radius r_0 embedded in a homogeneous full-space with Poisson's ratio ν, wave velocity c ($c = c_p$ for $\nu \leq 1/3$, $c = 2c_s$ for $1/3 < \nu \leq 1/2$) and mass density ρ, a moment with amplitude $M_0(\omega)$ acts leading to a rotation with amplitude $\vartheta_0(\omega)$. The vertical degree of freedom is also presented in Fig. 2.10b. This situation (which is the first building block used to analyse an embedded foundation) is modelled as a double cone. Thus two initial cones with outward wave propagation exist. Each single (one-sided) cone is constructed as for a disk on a homogeneous half-space. The properties, in particular z_0/r_0, are summarised in Table 3.1. Half of the applied moment is resisted by each of the single cones. The interaction moment-rotation relationship of the embedded disk is (Eq. 3.44)

$$M_0(\omega) = S_\vartheta(a_0)\vartheta_0(\omega) \tag{3.97}$$

where the dynamic-stiffness coefficient for the double-cone model (Eq. 3.50) is

$$S_\vartheta(a_0) = 2K_\vartheta[k_\vartheta(a_0) + ia_0 c_\vartheta(a_0)] \tag{3.98}$$

with the static-stiffness coefficient of the single cone K_ϑ specified in Eq. 3.58, and the dimensionless spring and damping coefficients $k_\vartheta(a_0)$ and $c_\vartheta(a_0)$ specified in Eq. 3.60 for $\nu \leq 1/3$ and in Eq. 3.91 for $1/3 < \nu \leq 1/2$. Thus, the only modification consists of multiplying the static-stiffness coefficient K_ϑ of the single cone by 2 to determine the static-stiffness coefficient of the double cone.

When the full-space consists of two half-spaces (a lower half-space with properties ν, c and ρ, and an upper half-space with corresponding properties ν', c' and ρ') in contact with each other on their 'free' surfaces, and the disk is located at this interface, the concept of the double cone still applies with each single cone determined by the properties of the corresponding half-space.

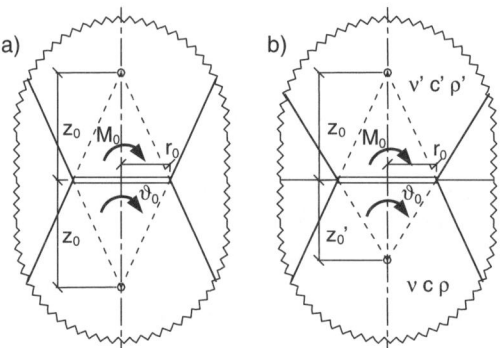

Figure 3.14 Double-cone model. a) Disk embedded in homogeneous full-space. b) Disk embedded at interface of two homogeneous half-spaces forming a full-space

The rocking degree of freedom is illustrated in Fig. 3.14b. The dynamic-stiffness coefficient of the double-cone model is then

$$S_\vartheta(a_0) = K_\vartheta[k_\vartheta(a_0) + ia_0 c_\vartheta(a_0)] + K'_\vartheta[k'_\vartheta(a'_0) + ia'_0 c'_\vartheta(a'_0)] \quad (3.99)$$

with the second term on the right-hand side calculated with the same equations as specified above, but using the material properties of the upper half-space (denoted with a prime). The vertical degree of freedom is also shown in Fig. 2.10c.

The procedure is analogous for the torsional degree of freedom and the translational degrees of freedom. For a disk embedded in a homogeneous full-space, again the only modification consists of multiplying the static-stiffness coefficient K of the single cone by 2. For a disk embedded at the interface between two half-spaces with different properties, Eq. 3.99 still applies for the translational degrees of freedom, omitting the index ϑ.

Instead of determining the aspect ratio of the single cone using the static-stiffness coefficient of a disk on a homogeneous half-space (Table 3.1), the exact static-stiffness coefficient of a disk embedded in a homogeneous full-space available in closed form can be used as an alternative. For the rocking degree of freedom the latter is

$$K_{\vartheta\,exact} = \frac{64(1-\nu)}{3(3-4\nu)} G r_0^3 \quad (3.100)$$

Matching Eq. 3.100 to the static-stiffness coefficient of the double-cone model (Eq. 3.58)

$$K_\vartheta = 2 \frac{3\rho c^2}{z_0} \frac{\pi r_0^4}{4} \quad (3.101)$$

yields

$$\frac{z_0}{r_0} = \frac{9\pi}{128} \frac{3-4\nu}{1-\nu} \frac{c^2}{c_s^2} \quad \text{(rocking)} \quad (3.102)$$

Analogously, for the vertical degree of freedom,

$$K_{exact} = \frac{32(1-\nu)}{3-4\nu} G r_0 \quad (3.103)$$

results in

$$\frac{z_0}{r_0} = \frac{\pi}{16} \frac{3-4\nu}{1-\nu} \frac{c^2}{c_s^2} \quad \text{(vertical)} \quad (3.104)$$

and for the horizontal degree of freedom

$$K_{exact} = \frac{64(1-\nu)}{7-8\nu} G r_0 \quad (3.105)$$

leading to

$$\frac{z_0}{r_0} = \frac{\pi}{32} \frac{7-8\nu}{1-\nu} \quad \text{(horizontal)} \quad (3.106)$$

For the torsional degree of freedom, no change occurs, i.e. the exact static-stiffness coefficient of a disk on a half-space is half that of a disk embedded in a full-space.

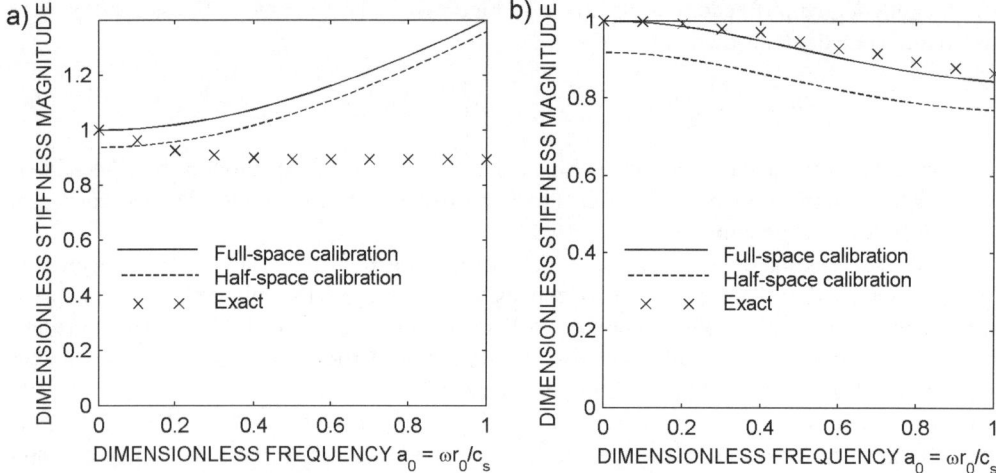

Figure 3.15 Dynamic-stiffness magnitude of disk embedded in homogeneous full-space – effect of calibration method. a) Vertical degree of freedom, $\nu = 1/3$. b) Rocking degree of freedom, $\nu = 0.3$.

All other information on constructing the single (one-sided) cone model presented in Table 3.1 still applies. For the incompressible case ($\nu = 1/2$) both calibration methods yield the same aspect ratio.

The effect of the calibration method on the magnitude of the dynamic-stiffness coefficient is illustrated in Fig. 3.15. A disk embedded in a homogeneous full-space is considered. In each case the magnitude of the dynamic-stiffness coefficient (Eq. 1.6) is normalised by division by the exact static-stiffness coefficient of the embedded disk for the degree of freedom under consideration. For the vertical degree of freedom Poisson's ratio is selected as $1/3$, permitting direct comparison with the exact results of Ref. [29]. For the two calibrations the dynamic stiffnesses are plotted as continuous lines in Fig. 3.15a, while the 'exact' results are indicated as discrete points. In the case of the vertical degree of freedom use of the full-space calibration technique does not yield any significant improvement, with the exception of the very small frequency range.

For the rocking degree of freedom Poisson's ratio is selected as 0.3, allowing comparison with Ref. [30]. The results are plotted in Fig. 3.15b. In this case use of the full-space calibration technique significantly improves the agreement between the computed dynamic stiffness and the exact values.

4

Wave reflection and refraction at a material discontinuity

The second of the two building blocks required to calculate the vibrations of an embedded foundation addresses the wave mechanism generated at a material discontinuity corresponding to an interface between two layers. When the incident wave propagating in the initial cone discussed in Chapter 3 encounters the discontinuity, a reflected wave and a refracted wave, each propagating in its own cone, are created. Enforcement of compatibility of displacement and equilibrium of the interface permits the reflected and refracted waves to be expressed as functions of the incident wave. The reflection coefficient, defined as the ratio of the reflected wave to the incident wave, depends on the properties of the two materials present at the interface and the frequency of the wave.

Section 4.1 calculates the reflection coefficient for the translational cone, and Section 4.2 for the rotational cone. The derivations are performed for harmonic excitation. Section 4.3 determines the dynamic-stiffness coefficients of a circular foundation on the surface of a layer overlying a flexible half-space for illustration. A closed-form solution formulated on one line results. Section 4.4 examines a disk embedded in a homogeneous half-space. Section 4.5 discusses representation of the problem for analysis by a computer program, and describes an efficient recursive implementation of such a program, including MATLAB listings of the key functions. The issue of a termination criterion (deciding the stage at which further reflections and refractions can be ignored) is addressed in Section 4.6. Both Sections 4.5 and 4.6 present worked examples.

4.1 Reflection coefficient for a translational cone

The wave propagation in the various cones is illustrated in Fig. 4.1a. Two materials are present, with the properties of the first denoted as Poisson's ratio v, wave velocity c and mass density ρ, and the corresponding properties for the second indicated with a prime (v', c' and ρ'). A disk of radius r_0 in contact with the first material is located at a distance d from the interface between the two materials. The amplitude of the disk's displacement is denoted as $u_0(\omega)$. In Fig. 4.1a a displacement in the vertical direction is indicated, although the derivation also applies for the horizontal direction. The initial cone with the apex indicated as 1 in Fig. 4.1a and apex height z_0 is specified. The construction of the initial cone and the outward wave propagation are described in Chapter 3, with the essentials for the horizontal and vertical degrees of freedom summarised

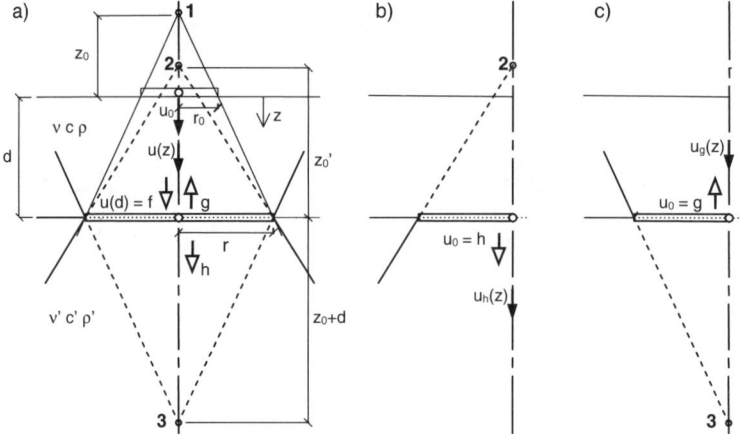

Figure 4.1 Wave propagation across cone segment and at material discontinuity interface. a) Incident wave impinging on interface generating reflected and refracted waves. b) Refracted wave as new incident wave. c) Reflected wave as new incident wave

in Table 3.1. $u_0(\omega)$ is determined in any one of three ways. If the disk is on the free surface and is loaded by a force with amplitude $P_0(\omega)$, $u_0(\omega)$ follows from Eq. 3.15 with Eq. 3.25 and Eq. 3.26 or Eq. 3.89. If the disk is embedded and is loaded by a force with amplitude $P_0(\omega)$, modelling with a double cone applies, as described in Section 3.6 (only one initial cone is shown in Fig. 4.1a). Finally, $u_0(\omega)$ can be equal to the amplitude of the reflected or refracted wave, as discussed below.

The displacement amplitude $u_0(z, \omega)$ of the outward propagating wave (away from the disk downwards in the positive z-direction) is specified as a function of $u_0(\omega)$ in Eq. 3.12. It is appropriate to move the origin of the z-axis from the apex (Fig. 3.1) to the disk (Fig. 4.1a) yielding

$$u(z,\omega) = \frac{z_0}{z_0 + z} e^{-i\frac{\omega}{c}z} u_0(\omega) \tag{4.1}$$

Note that in the argument of the exponential function the distance from the disk is present, while in the denominator the distance from the apex is used. The amplitude of the incident wave at the interface $f(\omega)$ follows from Eq. 4.1 with $z = d$ as

$$f(\omega) = u(d,\omega) = \frac{z_0}{z_0 + d} e^{-i\frac{\omega}{c}d} u_0(\omega) \tag{4.2}$$

The radius of the cone at the interface, r, is determined from geometric considerations as

$$r = \frac{z_0 + d}{z_0} r_0 \tag{4.3}$$

or

$$r = r_0 + d\frac{r_0}{z_0} \tag{4.4}$$

where r_0/z_0 is the reciprocal of the aspect ratio of the cone. This allows Eq. 4.2 to be written in the form

$$f(\omega) = \frac{r_0}{r} e^{-i\frac{\omega}{c}d} u_0(\omega) \tag{4.5}$$

which will be used in the computer procedure described later in this chapter.

Eliminating $u_0(\omega)$ from Eqs 4.1 and 4.2 results in

$$u_f(z, \omega) = \frac{z_0 + d}{z_0 + z} e^{-i\frac{\omega}{c}(z-d)} f(\omega) \qquad (4.6)$$

where the index f is added to denote the incident wave.

The interface with a discontinuity in material properties represents the source of a disturbance. It can be modelled as a fictitious disk of radius r (Eq. 4.3) which is not loaded. Enforcing compatibility of the displacements and equilibrium of the disk requires that an additional reflected wave with amplitude $g(\omega)$ propagating upwards in the first material, and an additional refracted wave with amplitude $h(\omega)$ propagating downwards in the second material are generated. (The arrows shown in Fig. 4.1a indicate the direction of wave propagation.) Both waves propagate away from the source of disturbance in their own radiating cones. The area of these cones increases in the direction of wave propagation.

The refracted wave propagates in the cone with the apex indicated as 2 in Fig. 4.1a. The opening angle of the cone, determined by the aspect ratio z_0'/r, depends on v' of the second material. The amplitude of this refracted wave is formulated as (Fig. 4.1b)

$$u_h(z, \omega) = \frac{z_0'}{z_0' - d + z} e^{-i\frac{\omega}{c'}(z-d)} h(\omega) \qquad (4.7)$$

The wave propagates with the velocity c' in the positive z-direction, yielding the term $e^{-i\omega z/c'}$. A term $e^{i\omega d/c'}$ is included so that the exponential function equals 1 for $z = d$, and the denominator is equal to the distance from apex 2. At the interface $z = d$, $u_h(d, \omega) = h(\omega)$.

The reflected wave propagates in the first material in the cone with the apex indicated as 3 in Fig. 4.1a, which has the same opening angle as the initial cone with apex 1, since both cones describe wave propagation in the same material. In other words, the aspect ratio of the new cone $(z_0 + d)/r$ is the same as the aspect ratio of the initial cone z_0/r_0, which depends on v. The amplitude of this reflected wave is written as (Fig. 4.1c)

$$u_g(z, \omega) = \frac{z_0 + d}{z_0 + 2d - z} e^{i\frac{\omega}{c}(z-d)} g(\omega) \qquad (4.8)$$

The wave propagates with velocity c in the negative z-direction, yielding the term $e^{i\omega z/c}$. Again, a term $e^{-i\omega d/c}$ is included so that the exponential function equals 1 for $z = d$, and the denominator is equal to the distance from apex 3. At the interface $z = d$, $u_g(d, \omega) = g(\omega)$.

The resultant displacement in the first material equals the sum of the incident and reflected waves with the amplitude $u_f(z, \omega) + u_g(z, \omega)$. At the interface $z = d$ compatibility of the displacements matches this value with the amplitude of the refracted wave in the second material $u_h(z, \omega)$ yielding

$$u_f(d, \omega) + u_g(d, \omega) = u_h(d, \omega) \qquad (4.9)$$

or, after substituting Eqs 4.6, 4.8 and 4.7,

$$f(\omega) + g(\omega) = h(\omega) \qquad (4.10)$$

Equilibrium of the resultant internal force in the first material and of the internal force in the second material at the interface $z = d$ yields

$$N_f(d, \omega) + N_g(d, \omega) = N_h(d, \omega) \qquad (4.11)$$

with the internal forces with amplitudes

$$N_f(z, \omega) = \rho c^2 \pi r^2 u_f(z, \omega)_{,z} \tag{4.12a}$$

$$N_g(z, \omega) = \rho c^2 \pi r^2 u_g(z, \omega)_{,z} \tag{4.12b}$$

$$N_h(z, \omega) = \rho' c'^2 \pi r^2 u_h(z, \omega)_{,z} \tag{4.12c}$$

(ρc^2 and $\rho' c'^2$ represent the elastic moduli of the first and second materials respectively, and the derivatives of the displacements with respect to z are equal to the strains.)

Substituting Eqs 4.6, 4.8 and 4.7 in Eq. 4.12, which is then substituted in Eq. 4.11, results in

$$-\rho c^2 \left(\frac{1}{z_0 + d} + i\frac{\omega}{c}\right) f(\omega) + \rho c^2 \left(\frac{1}{z_0 + d} + i\frac{\omega}{c}\right) g(\omega) = -\rho' c'^2 \left(\frac{1}{z'_0} + i\frac{\omega}{c'}\right) h(\omega) \tag{4.13}$$

Eliminating $h(\omega)$ from Eqs 4.10 and 4.13 leads to

$$\left[\frac{\rho c^2}{z_0 + d} - \frac{\rho' c'^2}{z'_0} + i\omega(\rho c - \rho' c')\right] f(\omega) = \left[\frac{\rho c^2}{z_0 + d} + \frac{\rho' c'^2}{z'_0} + i\omega(\rho c + \rho' c')\right] g(\omega) \tag{4.14}$$

The *reflection coefficient* $-\alpha(\omega)$ is defined as the frequency-dependent complex *ratio of the amplitudes of the reflected and incident waves*, yielding, using Eq. 4.14,

$$-\alpha(\omega) = \frac{g(\omega)}{f(\omega)} = \frac{\dfrac{\rho c^2}{z_0 + d} - \dfrac{\rho' c'^2}{z'_0} + i\omega(\rho c - \rho' c')}{\dfrac{\rho c^2}{z_0 + d} + \dfrac{\rho' c'^2}{z'_0} + i\omega(\rho c + \rho' c')}$$

$$= \frac{\dfrac{\rho c^2 \pi r^2}{z_0 + d}\left(1 + i\dfrac{\omega(z_0 + d)}{c}\right) - \dfrac{\rho' c'^2 \pi r^2}{z'_0}\left(1 + i\dfrac{\omega z'_0}{c'}\right)}{\dfrac{\rho c^2 \pi r^2}{z_0 + d}\left(1 + i\dfrac{\omega(z_0 + d)}{c}\right) + \dfrac{\rho' c'^2 \pi r^2}{z'_0}\left(1 + i\dfrac{\omega z'_0}{c'}\right)} \tag{4.15}$$

Alternatively, with

$$\beta = \rho c^2 \left(\frac{1}{z_0 + d} + i\frac{\omega}{c}\right) \tag{4.16}$$

and

$$\beta' = \rho' c'^2 \left(\frac{1}{z'_0} + i\frac{\omega}{c'}\right) \tag{4.17}$$

this ratio is expressed as

$$-\alpha(\omega) = \frac{\beta - \beta'}{\beta + \beta'} \tag{4.18}$$

The first and second terms in the denominator of the expression on the right-hand side of the third equal sign of Eq. 4.15 represent the dynamic-stiffness coefficients (Eqs 3.17, 3.22, 3.21 and 3.23) of the cones with the reflected and refracted waves at the interface with radius r. The reflection coefficient is thus equal to the ratio of the difference and the sum of these values. From Eq. 4.15

$$g(\omega) = -\alpha(\omega) f(\omega) \tag{4.19}$$

follows, and substituting into Eq. 4.10

$$h(\omega) = (1 - \alpha(\omega)) f(\omega) \tag{4.20}$$

is obtained.

These reflected and refracted waves determine the amplitudes $u_0(\omega)$ of new incident waves propagating in their own initial cones, as shown in Figs 4.1c and 4.1b.

For a fixed boundary $(c' \gg c) - \alpha = -1$. The amplitude of the reflected wave is equal to minus the amplitude of the incident wave, or

$$g(\omega) = -f(\omega) \tag{4.21}$$

For a free boundary $(c' \ll c)$, $-\alpha = +1$. The amplitude of the reflected wave is equal to that of the incident wave, or

$$g(\omega) = f(\omega) \tag{4.22}$$

These results are also derived in the time domain in Section 2.3.

The reflection coefficient can also be determined for two limits. For the static case $(\omega = 0)$, Eq. 4.15 leads to

$$-\alpha(0) = \frac{\dfrac{\rho c^2}{z_0 + d} - \dfrac{\rho' c'^2}{z'_0}}{\dfrac{\rho c^2}{z_0 + d} + \dfrac{\rho' c'^2}{z'_0}} \tag{4.23a}$$

which for $\nu = \nu'$ simplifies to

$$-\alpha(0) = \frac{\rho c^2 - \rho' c'^2}{\rho c^2 + \rho' c'^2} \tag{4.23b}$$

ρc^2 expresses the corresponding elastic modulus. The reflection coefficient for the static limit is thus equal to the ratio of the difference and the sum of the static-stiffness coefficients of the cones with the reflected and refracted waves.

For the high frequency limit $(\omega \to \infty)$

$$-\alpha(\infty) = \frac{\rho c - \rho' c'}{\rho c + \rho' c'} \tag{4.24}$$

is obtained, with the impedance ρc. Equation 4.24 corresponds to a material discontinuity in a prismatic bar, which is discussed in Appendix C (Eq. C.25). The reflection coefficient for the high frequency limit is thus equal to the ratio of the difference and the sum of the impedances of the cones with the reflected and refracted waves.

4.2 Reflection coefficient for a rotational cone

The derivation of the reflection coefficient in a rotational cone is analogous to that in a translational cone, discussed in Section 4.1. Figure 4.1 still applies, replacing the displacements with amplitudes $u_0(\omega)$ and $u(z, \omega)$ by the rotations with amplitudes $\vartheta_0(\omega)$ and $\vartheta(z, \omega)$. Of course, the properties of the cones, such as the opening angle and the appropriate wave velocity, which depend on the degree of freedom (Table 3.1), will be different. The rotation in a cone decays with depth (Eq. 3.42) as a more complicated function of frequency than the translation (Eq. 3.12) does. This is the only significant difference.

Rocking is addressed in the following, but the derivation applies for the torsional degree of freedom also. Again $\vartheta_0(\omega)$ of the disk of radius r_0 in contact with the first material with ν, c and ρ is determined in one of three ways. If the disk is on the free surface and is loaded by a moment with amplitude $M_0(\omega)$, $\vartheta_0(\omega)$ follows from Eq. 3.44 with Eq. 3.50 and Eq. 3.60 or Eq. 3.91. If the disk is embedded and is loaded by a moment with amplitude $M_0(\omega)$, modelling with a double cone applies (Eq. 3.97 with Eq. 3.98 or Eq. 3.99). Finally, $\vartheta_0(\omega)$ can be equal to the amplitude of the reflected or refracted wave.

The rotation amplitude $\vartheta(z, \omega)$ in the initial cone (with apex 1 in Fig. 4.1a) is specified as a function of $\vartheta_0(\omega)$ in Eq. 3.42, replacing c_s by c. Moving the origin of the z-axis from the apex (Fig. 3.4) to the disk (Fig. 4.1a) yields

$$\vartheta(z, \omega) = \frac{\dfrac{z_0^3}{(z_0+z)^3} + i\dfrac{\omega}{c}\dfrac{z_0^3}{(z_0+z)^2}}{1 + i\dfrac{\omega}{c}z_0} e^{-i\frac{\omega}{c}z} \vartheta_0(\omega) \qquad (4.25)$$

The amplitude of the incident wave $f(\omega)$ at the interface with the second material (with properties ν', c' and ρ') located at a distance d follows from Eq. 4.25 with $z = d$ as

$$f(\omega) = \vartheta(d, \omega) = \frac{\dfrac{z_0^3}{(z_0+d)^3} + i\dfrac{\omega}{c}\dfrac{z_0^3}{(z_0+d)^2}}{1 + i\dfrac{\omega}{c}z_0} e^{-i\frac{\omega}{c}d} \vartheta_0(\omega) \qquad (4.26a)$$

or alternatively, in terms of the cone radius r at the interface and the initial radius r_0 (substituting Eq. 4.3 and rearranging), as

$$f(\omega) = \left(\frac{r_0}{r}\right)^2 \left(1 + \frac{\dfrac{r_0}{r} - 1}{1 + i\dfrac{\omega}{c}z_0}\right) e^{-i\frac{\omega}{c}d} \vartheta_0(\omega) \qquad (4.26b)$$

This form will be used in the computer procedure described later in this chapter.

Eliminating $\vartheta_0(\omega)$ from Eqs 4.25 and 4.26a leads to

$$\vartheta_f(z, \omega) = \frac{\dfrac{z_0^3}{(z_0+z)^3} + i\dfrac{\omega}{c}\dfrac{z_0^3}{(z_0+z)^2}}{\dfrac{z_0^3}{(z_0+d)^3} + i\dfrac{\omega}{c}\dfrac{z_0^3}{(z_0+d)^2}} e^{-i\frac{\omega}{c}(z-d)} f(\omega) \qquad (4.27)$$

where the index f is added to indicate the incident wave.

The interface with a discontinuity in material properties represents the source of a disturbance that is modelled as a fictitious unloaded disk of radius r (Fig. 4.1a). A reflected wave with amplitude $g(\omega)$ propagating upwards in the first material and a refracted wave with amplitude $h(\omega)$ propagating downwards in the second material are generated. The refracted wave propagates in the cone with apex denoted as 2 in Fig. 4.1a, with the aspect ratio z'_0/r depending on v' of the second material with the velocity c' in the positive z-direction. Its amplitude is (Fig. 4.1b)

$$\vartheta_h(z,\omega) = \frac{\dfrac{z_0'^3}{(z_0'-d+z)^3} + \mathrm{i}\dfrac{\omega}{c'}\dfrac{z_0'^3}{(z_0'-d+z)^2}}{1+\mathrm{i}\dfrac{\omega}{c'}z_0'} \mathrm{e}^{-\mathrm{i}\frac{\omega}{c'}(z-d)} h(\omega) \qquad (4.28)$$

The reflected wave propagates in the cone with apex denoted as 3 in Fig. 4.1a with the velocity c in the negative z-direction. This cone has the same opening angle as the initial cone (with apex 1). The amplitude of the wave is (Fig. 4.1c)

$$\vartheta_g(z,\omega) = \frac{\dfrac{(z_0+d)^3}{(z_0+2d-z)^3} + \mathrm{i}\dfrac{\omega}{c}\dfrac{(z_0+d)^3}{(z_0+2d-z)^2}}{1+\mathrm{i}\dfrac{\omega}{c}(z_0+d)} \mathrm{e}^{\mathrm{i}\frac{\omega}{c}(z-d)} g(\omega) \qquad (4.29)$$

Compatibility of the displacements at the interface $z = d$ yields

$$\vartheta_f(d,\omega) + \vartheta_g(d,\omega) = \vartheta_h(d,\omega) \qquad (4.30)$$

or, after substituting Eqs 4.27, 4.29 and 4.28,

$$f(\omega) + g(\omega) = h(\omega) \qquad (4.31)$$

The amplitudes of the resultant moments are (modulus of elasticity times moment of inertia multiplied by curvature)

$$M_f(z,\omega) = \rho c^2 \frac{\pi r^4}{4} \vartheta_f(z,\omega)_{,z} \qquad (4.32\mathrm{a})$$

$$M_g(z,\omega) = \rho c^2 \frac{\pi r^4}{4} \vartheta_g(z,\omega)_{,z} \qquad (4.32\mathrm{b})$$

$$M_h(z,\omega) = \rho' c'^2 \frac{\pi r^4}{4} \vartheta_h(z,\omega)_{,z} \qquad (4.32\mathrm{c})$$

At the interface $z = d$ moment equilibrium enforces

$$M_f(d,\omega) + M_g(d,\omega) = M_h(d,\omega) \qquad (4.33)$$

Substituting Eqs 4.27, 4.29 and 4.28 in Eq. 4.32, which is then substituted in Eq. 4.33, results in

$$-\beta f(\omega) + \beta g(\omega) = -\beta' h(\omega) \qquad (4.34)$$

with

$$\beta = \rho c^2 \frac{\dfrac{3}{z_0+d} + 3\mathrm{i}\dfrac{\omega}{c} + \left(\dfrac{\mathrm{i}\omega}{c}\right)^2 (z_0+d)}{1 + \mathrm{i}\dfrac{\omega}{c}(z_0+d)}$$

$$= \frac{3\rho c^2}{z_0+d}\left(1 - \frac{1}{3}\frac{\dfrac{\omega^2}{c^2}(z_0+d)^2}{1+\dfrac{\omega^2}{c^2}(z_0+d)^2} + \mathrm{i}\frac{\omega(z_0+d)}{3c}\frac{\dfrac{\omega^2}{c^2}(z_0+d)^2}{1+\dfrac{\omega^2}{c^2}(z_0+d)^2}\right) \qquad (4.35)$$

and

$$\beta' = \rho' c'^2 \frac{\dfrac{3}{z'_0} + 3\mathrm{i}\dfrac{\omega}{c'} + \left(\dfrac{\mathrm{i}\omega}{c'}\right)^2 z'_0}{1 + \mathrm{i}\dfrac{\omega}{c'}z'_0}$$

$$= \frac{3\rho' c'^2}{z'_0}\left(1 - \frac{1}{3}\frac{\dfrac{\omega^2}{c'^2}z'^2_0}{1+\dfrac{\omega^2}{c'^2}z'^2_0} + \mathrm{i}\frac{\omega z'_0}{3c'}\frac{\dfrac{\omega^2}{c'^2}z'^2_0}{1+\dfrac{\omega^2}{c'^2}z'^2_0}\right) \qquad (4.36)$$

Eliminating $h(\omega)$ from Eqs 4.31 and 4.34 leads to

$$(\beta - \beta') f(\omega) = (\beta + \beta') g(\omega) \qquad (4.37)$$

The reflection coefficient $-\alpha(\omega)$ is again defined as the frequency-dependent complex ratio of the amplitudes of the reflected and incident waves, yielding, using Eq. 4.37,

$$-\alpha(\omega) = \frac{g(\omega)}{f(\omega)} = \frac{\beta - \beta'}{\beta + \beta'} \qquad (4.38)$$

β in Eq. 4.35 and β' in Eq. 4.36 are equal to the dynamic-stiffness coefficients (Eqs 3.58, 3.45, 3.48 and 3.47) of the cones with the reflected and refracted waves at the interface with radius r (divided by $\pi r^4/4$). The reflection coefficient is thus equal to the ratio of the difference and the sum of these values (Eq. 4.38), as for the translational cone. From Eq. 4.38

$$g(\omega) = -\alpha(\omega) f(\omega) \qquad (4.39)$$

and

$$h(\omega) = (1 - \alpha(\omega)) f(\omega) \qquad (4.40)$$

also apply for the rotational cone.

Again, the reflected and refracted waves determine the amplitudes $\vartheta_0(\omega)$ of the new incident waves propagating in their own initial cones, as shown in Figs 4.1c and 4.1b.

For the interface acting as a fixed boundary ($c' \gg c$), $-\alpha = -1$, and for that acting as a free boundary ($c' \ll c$), $-\alpha = +1$. Equations 4.21 and 4.22 are thus also valid for the rotational cone.

The frequency-dependent reflection coefficient reverts to a constant in the two limits. For the low frequency limit, the static case ($\omega = 0$) leads to Eq. 4.23, and the high frequency limit ($\omega \to \infty$) yields Eq. 4.24.

4.3 Dynamic stiffness of a surface foundation on a layer overlying a half-space

In the computational procedure to track the waves in a layered half-space, each cone segment is treated sequentially. In a cone segment the displacement amplitude $u_0(\omega)$ of a disk or fictitious disk is known (Fig. 4.1a). Outward wave propagation in a segment of the initial cone occurs until the wave impinges on an interface, yielding the incident wave $f(\omega)$. At this material discontinuity a new source of disturbance, modelled with a fictitious disk, is generated. Based on the reflection coefficient, the refracted wave $h(\omega)$ (Fig. 4.1b) and reflected wave $g(\omega)$ (Fig. 4.1c) are calculated, which become new amplitudes $u_0(\omega)$ in two additional initial cones. The latter are subsequently treated analogously.

To gain physical insight on the effect of layering in a half-space, a circular foundation on the surface of a layer overlying a (flexible) half-space is addressed (Fig. 4.2). It is possible for this simple example to express the resulting wave pattern in the layer as a function of the incident wave in the initial cone, called the generating function. This approach is also applied in Section 2.3 working in the time domain.

The vertical degree of freedom is addressed (Fig. 4.2). The layer of depth d has the material constants v, ρ and c, while the corresponding properties of the half-space are v', ρ' and c'. A load with amplitude $P_0(\omega)$ acts on the disk of radius r_0 on the surface of the layer. The disk's displacement with amplitude $u_0(\omega)$ calculated in the first initial cone (Fig. 4.3a) follows from Eqs 3.15 and 3.16 for $v \leq 1/3$. It is denoted as $\bar{u}_0(\omega)$ in the following to indicate its role as a generating function, yielding

$$\bar{u}_0(\omega) = \frac{1}{1 + i\omega\dfrac{z_0}{c}} \frac{P_0(\omega)}{K} \tag{4.41}$$

The incident wave propagating downwards in the initial cone with the apex designated as 1 in Fig. 4.2 and apex height z_0 (Eq. 4.1) is

$$u(z, \omega) = \frac{z_0}{z_0 + z} e^{-i\frac{\omega}{c}z} \bar{u}_0(\omega) \tag{4.42}$$

and at the interface ($z = d$) (Eq. 4.2) is

$$f(\omega) = \frac{z_0}{z_0 + d} e^{-i\frac{\omega}{c}d} \bar{u}_0(\omega) \tag{4.43}$$

As the refracted wave $h(\omega)$ (Fig. 4.3b) will propagate downwards towards infinity in the cone with apex designated as 2 in Fig. 4.2 and apex height z'_0, not causing any waves to propagate in the layer, it is not addressed. The reflected wave (Eq. 4.19), after substituting Eq. 4.43, is

$$g(\omega) = -\alpha_1(\omega) f(\omega) = -\alpha_1(\omega) \frac{z_0}{z_0 + d} e^{-i\frac{\omega}{c}d} \bar{u}_0(\omega) \tag{4.44}$$

with the reflection coefficient $-\alpha_1(\omega)$ of the first impingement, with distances of the apexes from the interface $z_0 + d$ and z'_0, specified in Eq. 4.15. $g(\omega)$ is equal to $u_0(\omega)$ propagating upwards in

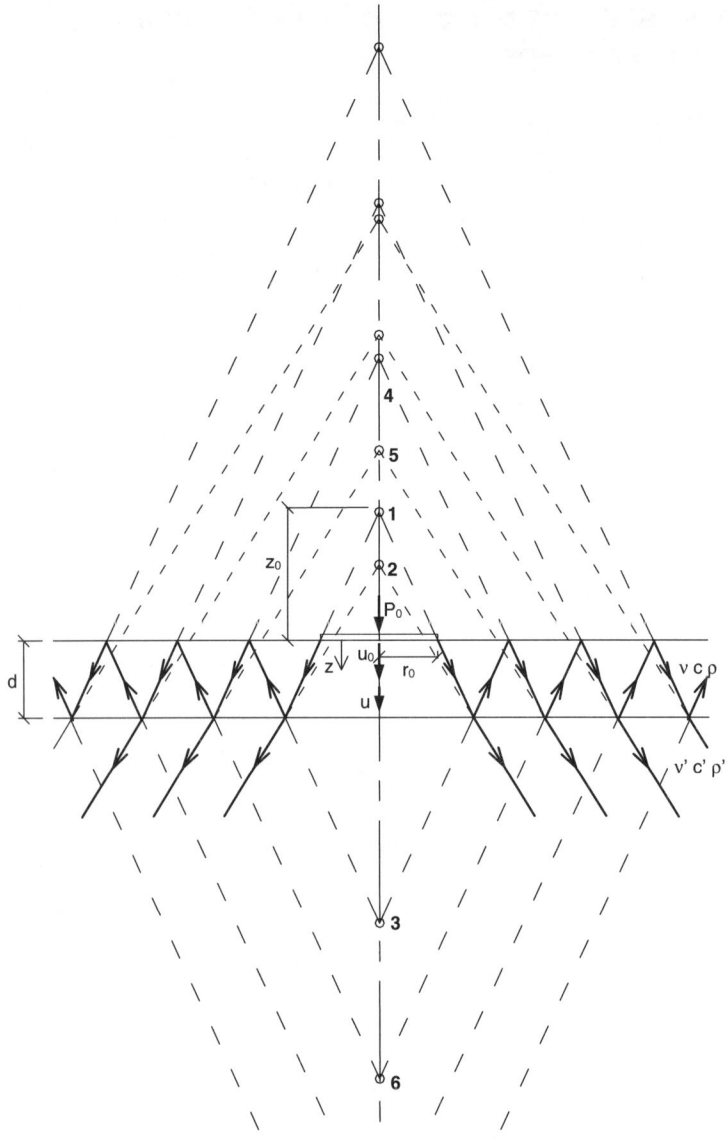

Figure 4.2 Disk loaded vertically on surface of layer overlying half-space with wave pattern generated by reflections and refractions

a new initial cone with the apex designated as 3 in Fig. 4.2 and the apex height $z_0 + d$ (Fig. 4.3b). The incident wave is formulated as (Eq. 4.8)

$$u(z, \omega) = \frac{z_0 + d}{z_0 + 2d - z} e^{i\frac{\omega}{c}(z-d)} g(\omega) \tag{4.45}$$

or, substituting Eq. 4.44,

$$u(z, \omega) = -\alpha_1(\omega) \frac{z_0}{z_0 + 2d - z} e^{i\frac{\omega}{c}(z-2d)} \overline{u}_0(\omega) \tag{4.46}$$

Wave reflection and refraction at a material discontinuity 65

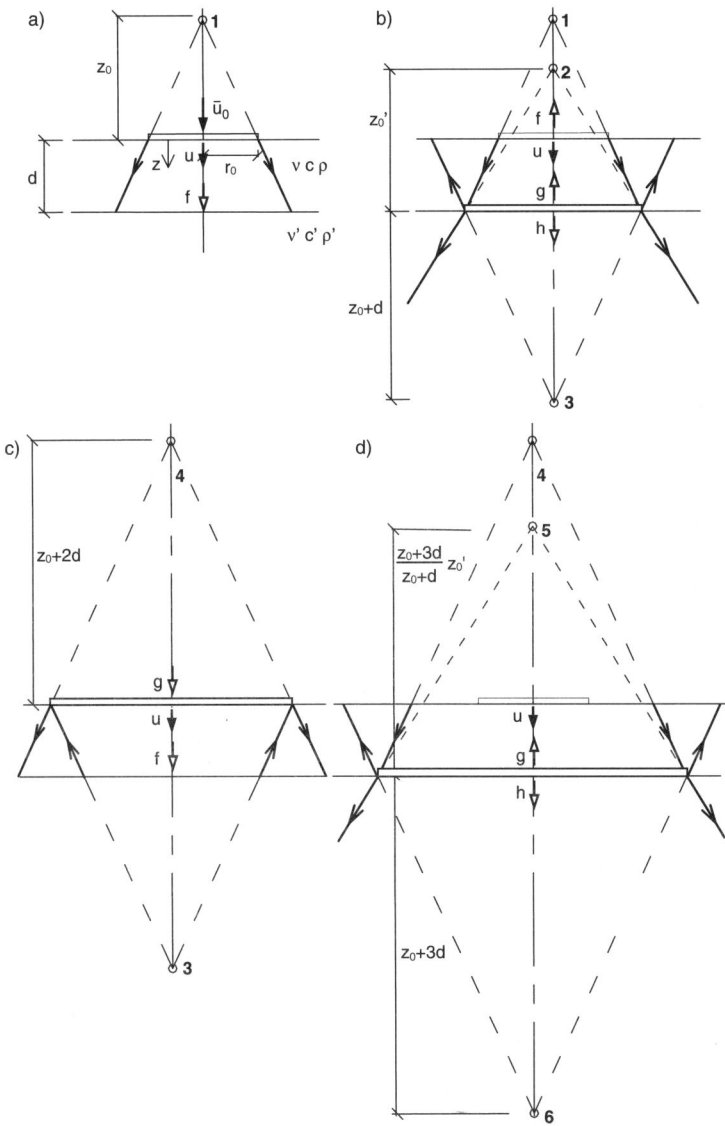

Figure 4.3 Wave pattern and corresponding cones. a) Incident (generating) wave in first initial cone. b) Incident, reflected and refracted waves at first impingement at interface. c) Incident and reflected waves at first impingement at free surface. d) Incident, reflected and refracted waves at second impingement at interface

yielding at the free surface ($z = 0$)

$$f(\omega) = -\alpha_1(\omega)\frac{z_0}{z_0 + 2d}e^{-i\frac{\omega}{c}2d}\overline{u}_0(\omega) \tag{4.47}$$

The reflected wave (Fig. 4.3c) at the free surface (Eq. 4.22) is

$$g(\omega) = f(\omega) = -\alpha_1(\omega)\frac{z_0}{z_0 + 2d}e^{-i\frac{\omega}{c}2d}\overline{u}_0(\omega) \tag{4.48}$$

$g(\omega)$ represents $u_0(\omega)$ propagating downwards in a new initial cone with apex designated as 4 in Fig. 4.2 and apex height $z_0 + 2d$ (Fig. 4.3c). The incident wave is written as

$$u(z,\omega) = \frac{z_0 + 2d}{z_0 + 2d + z} e^{-i\frac{\omega}{c}z} g(\omega) = -\alpha_1(\omega) \frac{z_0}{z_0 + 2d + z} e^{-i\frac{\omega}{c}(z+2d)} \bar{u}_0(\omega) \tag{4.49}$$

leading at the interface ($z = d$) to

$$f(\omega) = -\alpha_1(\omega) \frac{z_0}{z_0 + 3d} e^{-i\frac{\omega}{c}3d} \bar{u}_0(\omega) \tag{4.50}$$

The reflected wave (Fig. 4.3d) is

$$g(\omega) = -\alpha_2(\omega) f(\omega) = (-\alpha_1(\omega))(-\alpha_2(\omega)) \frac{z_0}{z_0 + 3d} e^{-i\frac{\omega}{c}3d} \bar{u}_0(\omega) \tag{4.51}$$

with the reflection coefficient $-\alpha_2(\omega)$ of the second impingement, with distances of the apexes from the interface $z_0 + 3d$ and $z_0'(z_0 + 3d)/(z_0 + d)$ replacing $z_0 + d$ and z_0', respectively, in Eq. 4.15. Again, $g(\omega)$ is equal to $u_0(\omega)$ propagating upwards in a new initial cone with apex designated as 6 in Fig. 4.2, and apex height $z_0 + 3d$ (Fig. 4.3d). The incident wave is formulated as

$$u(z,\omega) = \frac{z_0 + 3d}{z_0 + 4d - z} e^{i\frac{\omega}{c}(z-d)} g(\omega) = (-\alpha_1(\omega))(-\alpha_2(\omega)) \frac{z_0}{z_0 + 4d - z} e^{i\frac{\omega}{c}(z-4d)} \bar{u}_0(\omega)$$
$$\tag{4.52}$$

This process of generating waves will continue.

The resultant displacement in the layer is equal to the superposition of all the waves in the cone segments (sum of Eqs 4.42, 4.46, 4.49, 4.52, etc.). With the reflection coefficient of the jth impingement obtained by generalising Eq. 4.15 as

$$-\alpha_j(\omega) = \frac{\dfrac{\rho c^2}{z_0 + (2j-1)d} - \dfrac{\rho' c'^2}{(z_0 + (2j-1)d)z_0'/(z_0+d)} + i\omega(\rho c - \rho' c')}{\dfrac{\rho c^2}{z_0 + (2j-1)d} + \dfrac{\rho' c'^2}{(z_0 + (2j-1)d)z_0'/(z_0+d)} + i\omega(\rho c + \rho' c')} \tag{4.53}$$

the superposition results in

$$u(z,\omega) = \frac{z_0}{z_0 + z} e^{-i\frac{\omega}{c}z} \bar{u}_0(\omega) + \sum_{j=1}^{\infty} (-\alpha_1(\omega))(-\alpha_2(\omega)) \ldots (-\alpha_j(\omega))$$

$$\times \left[\frac{z_0}{z_0 + 2jd - z} e^{i\frac{\omega}{c}(z-2jd)} + \frac{z_0}{z_0 + 2jd + z} e^{-i\frac{\omega}{c}(z+2jd)} \right] \bar{u}_0(\omega) \tag{4.54}$$

The factor in front of the square bracket represents the product of the first j reflection coefficients. The first term on the right-hand side corresponds to the incident (generating) wave in the first initial cone, the second term with $e^{i\omega z/c}$ to the upwaves from the interface and the third term with $e^{-i\omega z/c}$ to the downwaves from the free surface.

In Eq. 4.54, the multiplication of all previous reflection coefficients demonstrates that the outward propagating incident wave associated with a specific impingement depends on all previous reflections at the interface. This is the procedure followed in the consistent formulation on which the computer program is based. However, numerical experiments demonstrate that the reflection coefficient of the first impingement of the interface $-\alpha_1(\omega)$ can be assumed to apply for all impingements without significant error. This simplification is applied in the following analysis. With $-\alpha(\omega) = -\alpha_1(\omega)$, Eq. 4.54 becomes

$$u(z, \omega) = \left(\frac{z_0}{z_0 + z} e^{-i\frac{\omega}{c}z} + \sum_{j=1}^{\infty} (-\alpha(\omega))^j \left[\frac{z_0}{z_0 + 2jd - z} e^{i\frac{\omega}{c}(z-2jd)} \right. \right.$$
$$\left. \left. + \frac{z_0}{z_0 + 2jd + z} e^{-i\frac{\omega}{c}(z+2jd)} \right] \right) \bar{u}_0(\omega) \quad (4.55)$$

The response of the disk follows from Eq. 4.55 with $z = 0$ as

$$u_0(\omega) = \left(1 + 2 \sum_{j=1}^{\infty} (-\alpha(\omega))^j \frac{z_0}{z_0 + 2jd} e^{-i\frac{\omega}{c}2jd} \right) \bar{u}_0(\omega) \quad (4.56)$$

Introducing the geometric parameter

$$\kappa = \frac{2d}{z_0} \quad (4.57a)$$

and the propagation time parameter

$$T = \frac{2d}{c} \quad (4.57b)$$

Eq. 4.56 becomes

$$u_0(\omega) = \left(1 + 2 \sum_{j=1}^{\infty} (-\alpha(\omega))^j \frac{e^{-i\omega jT}}{1 + j\kappa} \right) \bar{u}_0(\omega) \quad (4.58)$$

with $\bar{u}_0(\omega)$ determined from Eq. 4.41. Thus, the displacement amplitude $u_0(\omega)$ of the loaded disk on the surface of a layer overlying a half-space follows from that of the displacement amplitude $\bar{u}_0(\omega)$ of the disk with the same load on the surface of a homogeneous half-space with the material properties of the layer multiplied by a transfer function formulated as a closed-form expression.

The dynamic-stiffness coefficient of a disk on the surface of a layer overlying a half-space can also be formulated concisely. The interaction force-displacement relationship is formulated as

$$P_0(\omega) = S(\omega) u_0(\omega) \quad (4.59)$$

with the dynamic-stiffness coefficient $S(\omega)$. The corresponding relationship for the disk on a homogeneous half-space is expressed in Eq. 4.41, which is written using Eq. 4.57 as

$$P_0(\omega) = K \left(1 + i\omega \frac{T}{\kappa} \right) \bar{u}_0(\omega) \quad (4.60)$$

with the static-stiffness coefficient K of the disk on a homogeneous half-space with the material properties of the layer. Substituting Eq. 4.58 in Eq. 4.59 and setting the resulting right-hand side

68 Foundation Vibration Analysis: A Strength-of-Materials Approach

equal to that of Eq. 4.60 leads to

$$S(\omega) = K \frac{1 + i\omega \frac{T}{\kappa}}{1 + 2 \sum_{j=1}^{\infty} (-\alpha(\omega))^j \frac{e^{-i\omega jT}}{1 + j\kappa}} \tag{4.61}$$

Obviously, Eq. 4.61 also applies for the horizontal degree of freedom, adopting the corresponding parameters.

For the rotational degrees of freedom, the procedure is analogous. The rotational dynamic-stiffness coefficient of a disk on the surface of a layer overlying a half-space (no trapped mass present) is

$$S_\vartheta(\omega) = K_\vartheta \frac{1 - \frac{1}{3}\frac{(\omega T)^2}{\kappa^2 + (\omega T)^2} + i\frac{\omega T}{3\kappa}\frac{(\omega T)^2}{\kappa^2 + (\omega T)^2}}{1 + \frac{2}{1 + i\frac{\omega T}{\kappa}} \left(\sum_{j=1}^{\infty} (-\alpha(\omega))^j \frac{e^{-i\omega jT}}{(1+j\kappa)^3} + i\frac{\omega T}{\kappa} \sum_{j=1}^{\infty} (-\alpha(\omega))^j \frac{e^{-i\omega jT}}{(1+j\kappa)^2} \right)} \tag{4.62}$$

where K_ϑ denotes the static-stiffness coefficient (Eq. 3.46 or Eq. 3.58) and the numerator the dynamic-stiffness coefficient (Eq. 3.50 with Eq. 3.51 or Eq. 3.60) of a disk on a homogeneous half-space with the material properties of the layer. The denominator describes the transfer function between the amplitude of the rotation of the disk on the half-space $\bar{\vartheta}_0(\omega)$ and that of the disk on the layer overlying a half-space $\vartheta_0(\omega)$

$$\vartheta_0(\omega) = \left(1 + \frac{2}{1 + i\frac{\omega T}{\kappa}} \left(\sum_{j=1}^{\infty} (-\alpha(\omega))^j \frac{e^{-i\omega jT}}{(1+j\kappa)^3} + i\frac{\omega T}{\kappa} \sum_{j=1}^{\infty} (-\alpha(\omega))^j \frac{e^{-i\omega jT}}{(1+j\kappa)^2} \right) \right) \bar{\vartheta}_0(\omega) \tag{4.63}$$

$-\alpha(\omega)$ again represents the reflection coefficient for the first impingement (Eq. 4.38).

As an example the dynamic-stiffness coefficients of a disk of radius r_0 on a layer of depth d fixed at its base are examined (Fig. 2.5). $d = r_0$ and $\nu = 1/3$ are selected. The reflection coefficient becomes $-\alpha(\omega) = -1$. The information necessary to construct the cones is specified in Table 3.1. The dynamic-stiffness coefficient $S(a_0)$ calculated from Eq. 4.61 (and analogously $S_\vartheta(a_0)$) is decomposed as

$$S(a_0) = K[k(a_0) + i a_0 c(a_0)] \tag{4.64}$$

with the static-stiffness coefficient of a disk on the layer K and the dimensionless frequency $a_0 = \omega r_0/c_s$. The spring coefficent $k(a_0)$ and damping coefficient $c(a_0)$ are plotted for the horizontal and vertical degrees of freedom in Figs 4.4a and 4.4b, and for the torsional and rocking degrees of freedom in Figs 4.5a and 4.5b, and are compared with the rigorous solutions of Ref. [11].

In Figs 4.4 and 4.5 another dimensionless frequency is also used, defined as $\omega T = \omega(2d/c)$ with $c = c_s$ for the horizontal and torsional motions and $c = c_p = 2c_s$ for the vertical and rocking motions. T represents the propagation time from the free surface to the fixed boundary and back to the free surface. The values $\omega T = \pi, 3\pi, 5\pi, \ldots$ correspond to the natural frequencies $f = c/(4d), 3c/(4d), 5c/(4d), \ldots$ of the layer fixed at its base. The fundamental frequency has special physical significance. As is visible in Figs 4.4 and 4.5, the radiation damping coefficients become

Wave reflection and refraction at a material discontinuity 69

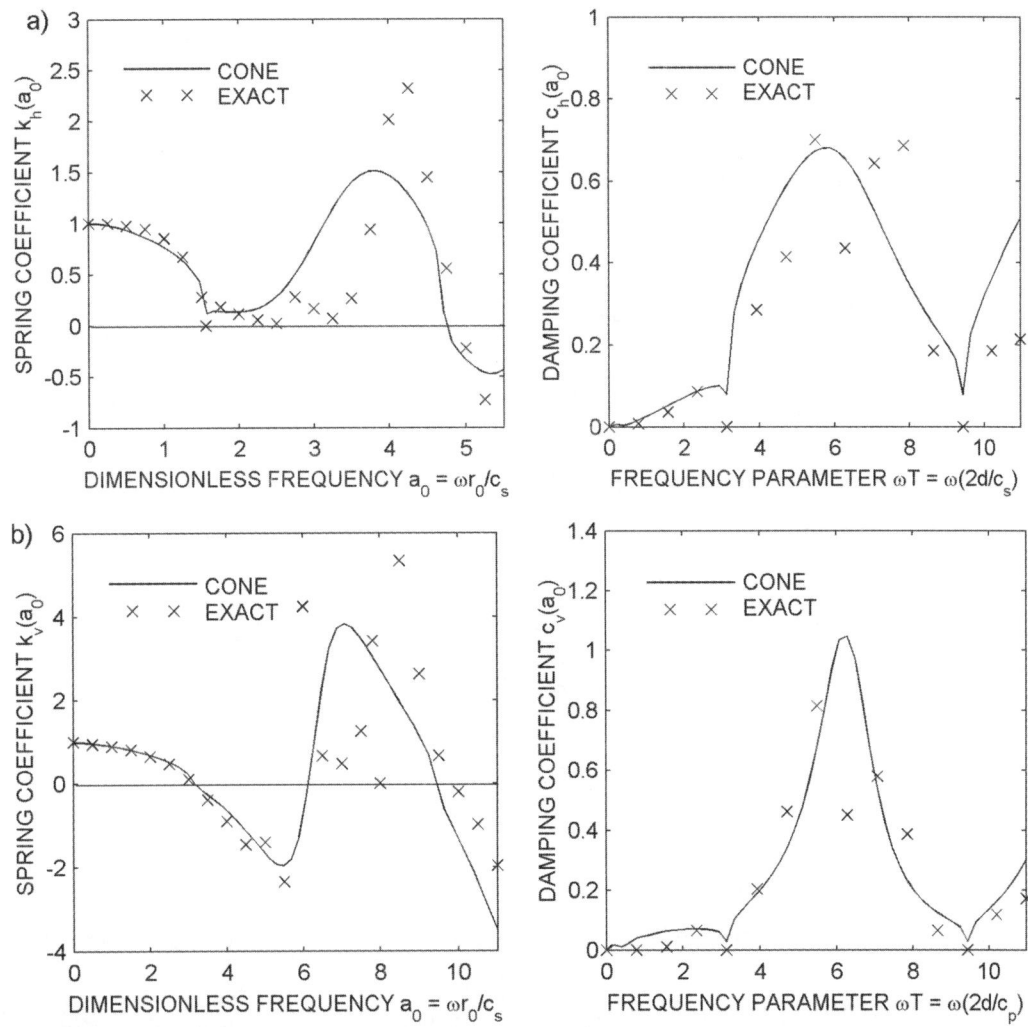

Figure 4.4 Translational dynamic-stiffness coefficients of disk on layer fixed at base. a) Horizontal degree of freedom. b) Vertical degree of freedom

appreciable for values of ωT greater than π, whereas for lower frequencies the damping is comparatively small. In view of its role as a border between damped and undamped behaviour, the (circular) frequency $\omega = \pi/T$ is called the *cutoff frequency*. The cone model captures the phenomenon of the cutoff frequency quite satisfactorily, despite a small amount of precursor damping $c(a_0)$ in the range $\omega < \pi/T$.

Above the cutoff frequency, the cone results yield a good approximation in the sense of a best fit to the rigorous solution, which tends to be irregular. For the translational motions both $k(a_0)$ and $c(a_0)$ should be zero at the (circular) natural frequencies $\omega = \pi/T$ and $3\pi/T$, which correspond to resonance. However, since only a finite number of reflections are processed, the cone results are not quite zero at these frequencies. As an increasing number of reflections are processed, the coefficients approach zero. For the rotational degree of freedom, $k_\vartheta(a_0)$ and $c_\vartheta(a_0)$ become small when $\omega = \pi/T$ and $3\pi/T$.

70 Foundation Vibration Analysis: A Strength-of-Materials Approach

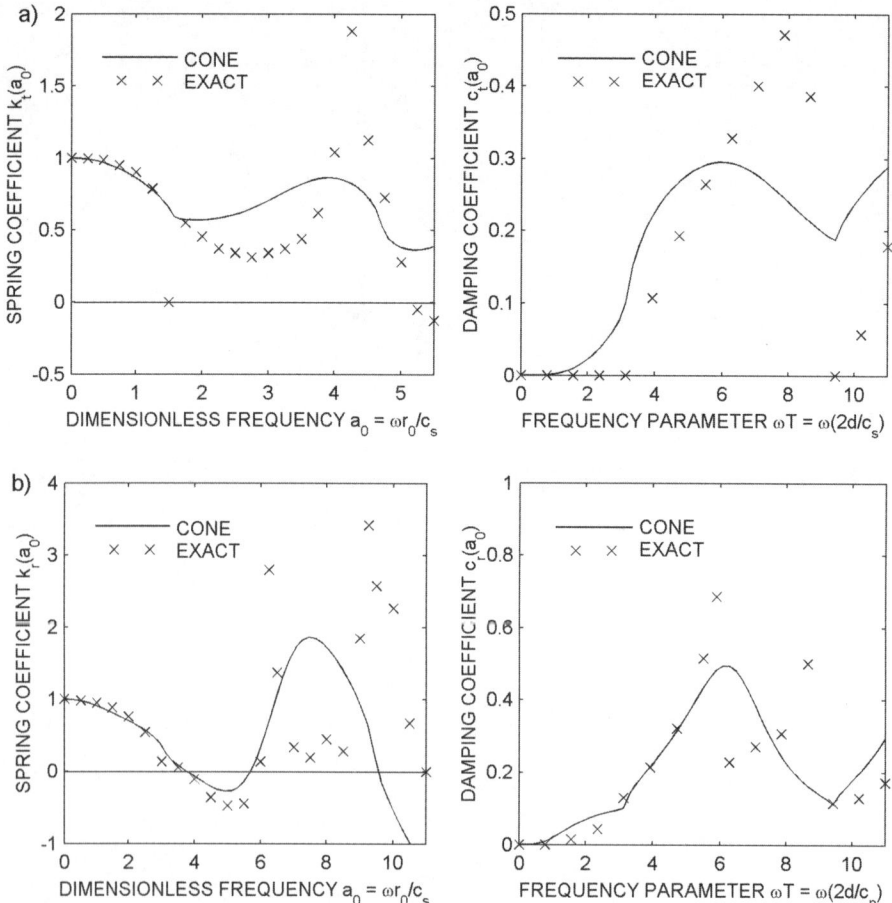

Figure 4.5 Rotational dynamic-stiffness coefficients of disk on layer fixed at base. a) Torsional degree of freedom. b) Rocking degree of freedom

The dynamic behaviour of a layer fixed at its base (Figs 4.4 and 4.5), representing one limiting case of a layered half-space, differs significantly from that of the other limiting case, the homogeneous half-space (Figs 3.8 to 3.11).

4.4 Disk embedded in a homogeneous half-space

The cone model representing a disk of radius r_0 embedded at a depth e in a homogeneous half-space also leads to a closed form solution for the dynamic-stiffness coefficient $S(\omega)$. The model is illustrated in Fig. 4.6 for the vertical degree of freedom, which will be considered first. For this case a double cone is used. The dynamic stiffnesses of the two initial cones are the same. The relationship between the initial disk displacement (the generating function with amplitude $\bar{u}_0(\omega)$) and the force applied to the disk (with amplitude $P_0(\omega)$) becomes (see Eq. 4.41)

$$\bar{u}_0(\omega) = \frac{1}{1 + i\omega \dfrac{z_0}{c}} \frac{P_0(\omega)}{2K} \qquad (4.65)$$

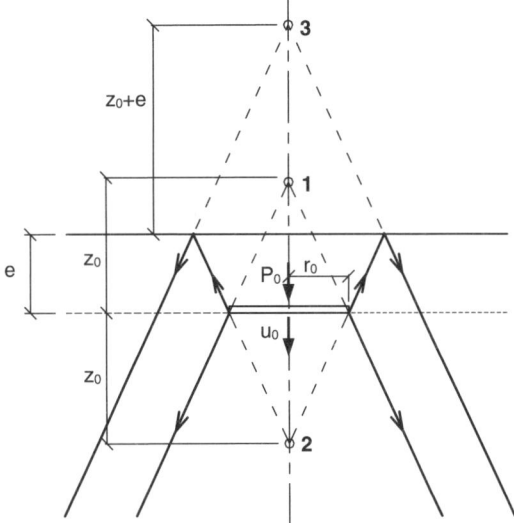

Figure 4.6 Wave propagation from disk embedded in homogeneous half-space

with the static-stiffness coefficient K of the disk on a half-space.

The wave propagating downwards in the initial cone below the disk (with apex denoted 1 in Fig. 4.6) does not lead to any reflections, and need not be processed any further. The wave propagating upwards in the initial cone above the disk (with apex denoted as 2) impinges on the free surface as an incident wave with the amplitude (Eq. 4.2 replacing d by e)

$$f(\omega) = \frac{z_0}{z_0 + e} \, e^{-i\frac{\omega}{c}e} \bar{u}_0(\omega) \tag{4.66}$$

The reflected wave from this surface is

$$g(\omega) = -\alpha(\omega) f(\omega) = \frac{z_0}{z_0 + e} \, e^{-i\frac{\omega}{c}e} \bar{u}_0(\omega) \tag{4.67}$$

since $-\alpha(\omega) = +1$ for a free boundary (Eq. 4.22). This reflected wave propagates in the new initial cone with apex denoted as 3 in Fig. 4.6. When it reaches the depth of the disk its amplitude has decreased to (analogous to Eqs 4.45 and 4.47 with d replaced by e and $-\alpha_1(\omega) = 1$)

$$f(\omega) = \frac{z_0 + e}{z_0 + 2e} \, e^{-i\frac{\omega}{c}e} \left(\frac{z_0}{z_0 + e} \, e^{-i\frac{\omega}{c}e} \bar{u}_0(\omega) \right) = \frac{z_0}{z_0 + 2e} \, e^{-i\frac{\omega}{c}2e} \bar{u}_0(\omega) \tag{4.68}$$

Since the half-space is homogeneous, this wave continues to propagate towards infinity in the truncated semi-infinite cone, and no further reflections occur. The total displacement at the disk is obtained by superposing this wave (Eq. 4.68) and the incident generating wave (Eq. 4.65), yielding the amplitude

$$u_0(\omega) = \left(1 + \frac{z_0}{z_0 + 2e} e^{-i\frac{\omega}{c}2e}\right) \bar{u}_0(\omega) \qquad (4.69)$$

or, using Eq. 4.57 with e replacing d,

$$u_0(\omega) = \left(1 + \frac{e^{-i\omega T}}{1 + \kappa}\right) \bar{u}_0(\omega) \qquad (4.70)$$

The dynamic-stiffness coefficient is obtained by combining Eqs 4.65 and 4.70 with Eq. 4.59, leading to

$$S(\omega) = 2K \frac{1 + i\omega\dfrac{T}{\kappa}}{1 + \dfrac{e^{-i\omega T}}{1 + \kappa}} \qquad (4.71)$$

which should be compared with Eq. 4.61.

As an example, the dynamic-stiffness coefficient for the vertical degree of freedom of a disk of radius r_0 embedded in a homogeneous half-space is investigated for Poisson's ratios $\nu = 1/4$ and $\nu = 1/2$. (For the incompressible case a trapped mass is also employed, as described in Section 3.4.) The spring coefficient $k(a_0)$ and damping coefficient $c(a_0)$ obtained with an embedment ratio $e/r_0 = 2$ are plotted in Fig. 4.7a, and those obtained with $e/r_0 = 5$ are presented in Fig. 4.7b. In both figures the coefficients are normalised (Eq. 4.64) with respect to the static-stiffness coefficient K for the same disk on the surface of the half-space (Eq. 3.27), and exact results from Ref. [26] are plotted as discrete points for comparison. Impressive agreement is evident.

For the rotational degrees of freedom, the procedure is analogous. The rotational dynamic-stiffness coefficient of a disk embedded in a homogeneous half-space (no trapped mass present) is

$$S_\vartheta(\omega) = 2K_\vartheta \frac{1 - \dfrac{1}{3}\dfrac{(\omega T)^2}{\kappa^2 + (\omega T)^2} + i\dfrac{\omega T}{3\kappa}\dfrac{(\omega T)^2}{\kappa^2 + (\omega T)^2}}{1 + \dfrac{e^{-i\omega T}}{1 + i\dfrac{\omega T}{\kappa}}\left(\dfrac{1}{(1+\kappa)^3} + i\dfrac{\omega T}{\kappa}\dfrac{1}{(1+\kappa)^2}\right)} \qquad (4.72)$$

where K_ϑ denotes the static-stiffness coefficient of the disk on a half-space with the same properties (Eq. 3.46 or Eq. 3.58). Equation 4.72 should be compared to Eq. 4.62. The dynamic-stiffness coefficients for the torsional degree of freedom for the disk described above with embedment ratios $e/r_0 = 1$ and $e/r_0 = 2$ are plotted in Fig. 4.8, normalised with respect to the static-stiffness coefficient of the same disk on the surface of the half-space. Exact results from Ref. [27] are plotted as discrete points for comparison.

4.5 Computer implementation

Although closed form solutions can be obtained for certain simple problems, a half-space with two or more layers generates a complex sequence of reflections and refractions (discussed in Section 2.4), which becomes impossible to process manually. However, a simple computer program allows the sequence to be handled quickly and easily. This section addresses computer

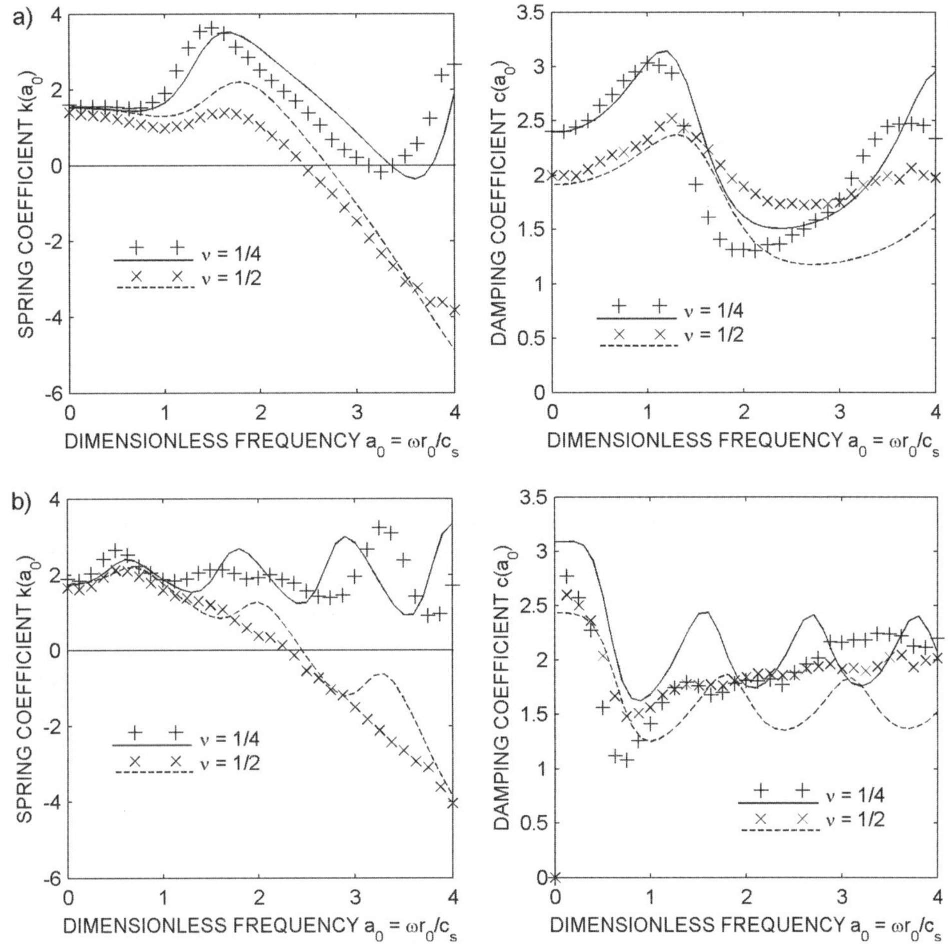

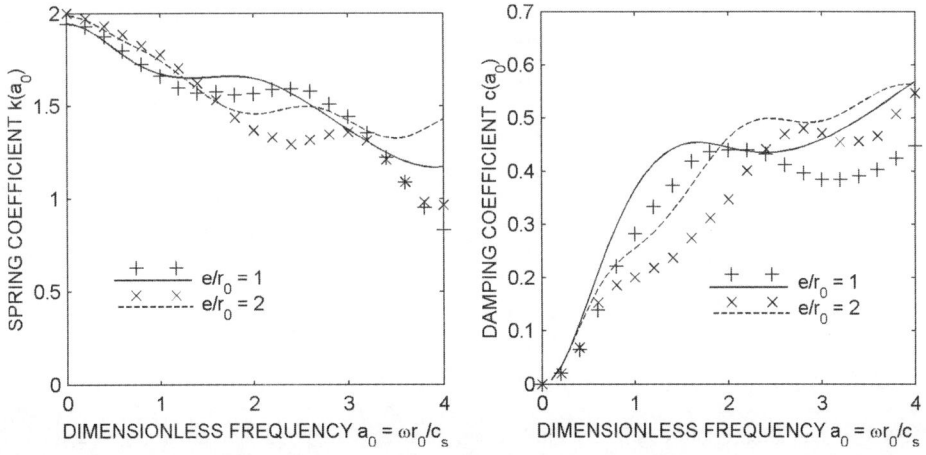

Figure 4.7 Dynamic-stiffness coefficient of disk embedded in homogeneous half-space in vertical motion. a) Embedment ratio = 2. b) Embedment ratio = 5

Figure 4.8 Dynamic-stiffness coefficient of disk embedded in homogeneous half-space in torsional motion

implementation of the procedure computing the dynamic stiffness of a rigid disk embedded in a horizontally layered full-space, using the two building blocks described previously, for a harmonic loading of a single frequency. (If the variation of dynamic stiffness with frequency is required, the procedure is applied repeatedly for a range of frequencies.) Representation of the input data in a form suitable for processing is described first. It is then shown that the reflected and refracted waves generated by a harmonic load on the disk can be represented using a binary tree. Efficient traversal of the tree using a recursive procedure is described, including MATLAB listings of the key functions. An example of a surface foundation on two layers overlying a flexible half-space is analysed to demonstrate the implementation.

The selection of a layered full-space allows a single procedure to be established to deal with all the cases discussed through suitable selection of the material properties of the layers and the half-spaces. The arrangement of the layers and the half-spaces is shown in Fig. 4.9. A homogeneous upper half-space overlies a number of layers of finite thickness and constant material properties. The bottom-most layer overlies a homogeneous lower half-space. The layers of finite thickness are restricted to having non-zero finite shear modulus and mass density, while the upper and lower half-spaces may have zero, non-zero or infinite shear modulus. The rigid disk is located at an interface between two of the layers. As an example, a surface foundation on two layers overlying a flexible half-space (Fig. 4.10a) is represented by assigning a shear modulus of zero to the upper half-space and assigning the properties of the flexible half-space to the lower half-space. The rigid disk is specified to be at the interface between the upper half-space and the upper layer. A second example is presented in Fig. 4.10b, where a disk embedded in a homogeneous layer overlying a rigid base is modelled. In this case the homogeneous layer is split into two adjacent layers with identical properties, and the embedded disk is positioned at the interface between these layers.

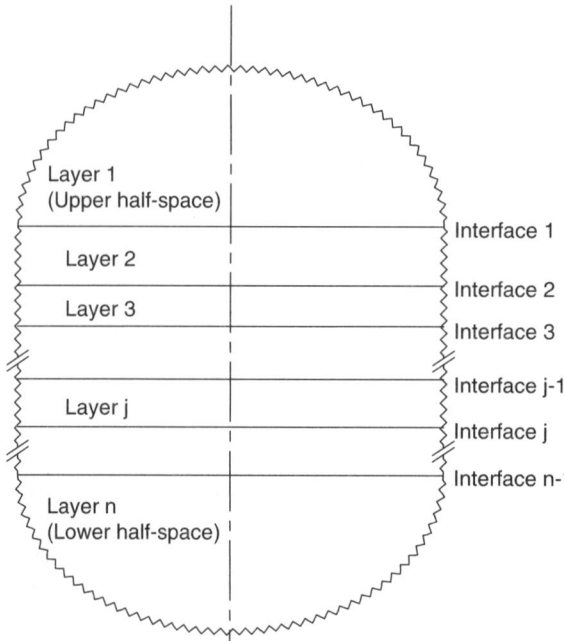

Figure 4.9 General description of layered full-space, showing numbering scheme for layers and interfaces

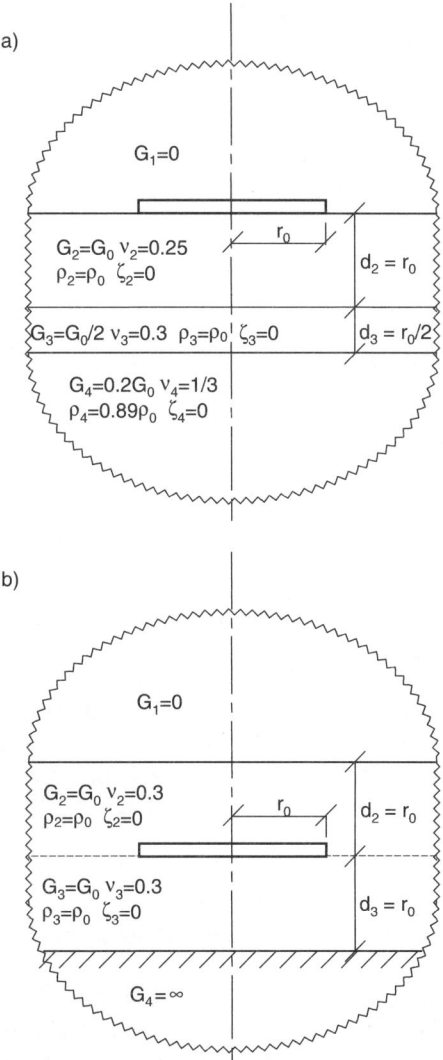

Figure 4.10 Examples illustrating the general description. a) Disk on surface of two layers overlying half-space. b) Disk embedded in homogeneous layer fixed at base

The upper half-space is assigned a shear modulus of zero, and the lower half-space an infinite shear modulus.

For convenience, the upper and lower half-spaces will be referred to as 'layers', despite being of infinite thickness. The model thus consists of n *layers* ($n-2$ of which have finite thickness) with $n-1$ *interfaces*. These are numbered from the top down, as indicated in Fig. 4.9, so that the jth interface underlies the jth layer. This allows all the data describing the model to be stored in a single array. The layers can be categorised as *finite*, *half-space*, or *rigid*, where the designation *rigid* is used to indicate a half-space of infinite shear modulus. For convenience of input, a fourth category of *free* is also introduced to indicate a half-space of zero shear modulus. A foundation radius is specified at each interface to allow extension to embedded foundations (Chapter 5). In

Table 4.1a Data describing example of Fig. 4.10a

layer	type	disk	G	nu	rho	damping	thickness
1	F	r_0	(0)				
2	L	0	G_0	0.25	ρ_0	0	r_0
3	L	0	$0.5\,G_0$	0.3	ρ_0	0	$0.5\,r_0$
4	H	0	$0.2\,G_0$	1/3	$0.89\,\rho_0$	0	

Table 4.1b Data describing example of Fig. 4.10b

layer	type	disk	G	nu	rho	damping	thickness
1	F	0	(0)				
2	L	r_0	G_0	0.3	ρ_0	0	r_0
3	L	0	G_0	0.3	ρ_0	0	r_0
4	R	0	(∞)				

the case of a single embedded disk (dealt with in this chapter), the radius is only non-zero at one interface. For layers categorised as *finite* or *half-space*, the shear modulus G, Poisson's ratio ν, mass density ρ and damping ratio ζ are specified. For finite layers, the layer thickness is also required. Using the character 'F' to represent the free layers, 'L' to represent the finite layers, 'H' to represent homogeneous half-spaces and 'R' to represent rigid half-spaces, the array of data describing the example in Fig. 4.10a is presented in Table 4.1a, while that describing Fig. 4.10b is in Table 4.1b. To allow easy interpretation of the program listings, a data structure is used for each layer. The names of the fields of the data structure are used as column headings in Tables 4.1a and 4.1b.

The task of the computer program is to place a harmonic excitation force on the disk, track all the reflected and refracted waves to a certain stage (beyond which the influence of the additional reflections and refractions is considered to be negligible), and hence to determine the resultant displacement amplitude of the disk. The dynamic stiffness of the disk can then be found by dividing the amplitude of the applied force by the amplitude of the displacement. This process is outlined in Chapter 2.

Every wave impinging on an interface at a material discontinuity generates a reflected wave and a refracted wave. If these waves do not pass into a half-space, each wave propagates across a layer and impinges on another interface, creating two more waves. Thus, each wave generates two 'child' waves when impinging on an interface. If the layer on the far side of the interface (in the sense of the direction of the propagating wave) is a half-space, the refracted child wave need not be processed, as it will cause no further reflections and refractions. All the waves are generated from two initial 'parent' waves radiating from the (embedded) disk. Each wave impinging on an interface in turn becomes a parent. The relationship between the waves indicated in Fig. 4.11a can be represented by the structure illustrated in Fig. 4.11b. In computer science, such a data structure occurs frequently, and is termed a *binary tree*. The points where the tree branches intersect are termed *nodes*. In this application each node represents an incident wave propagating across a layer

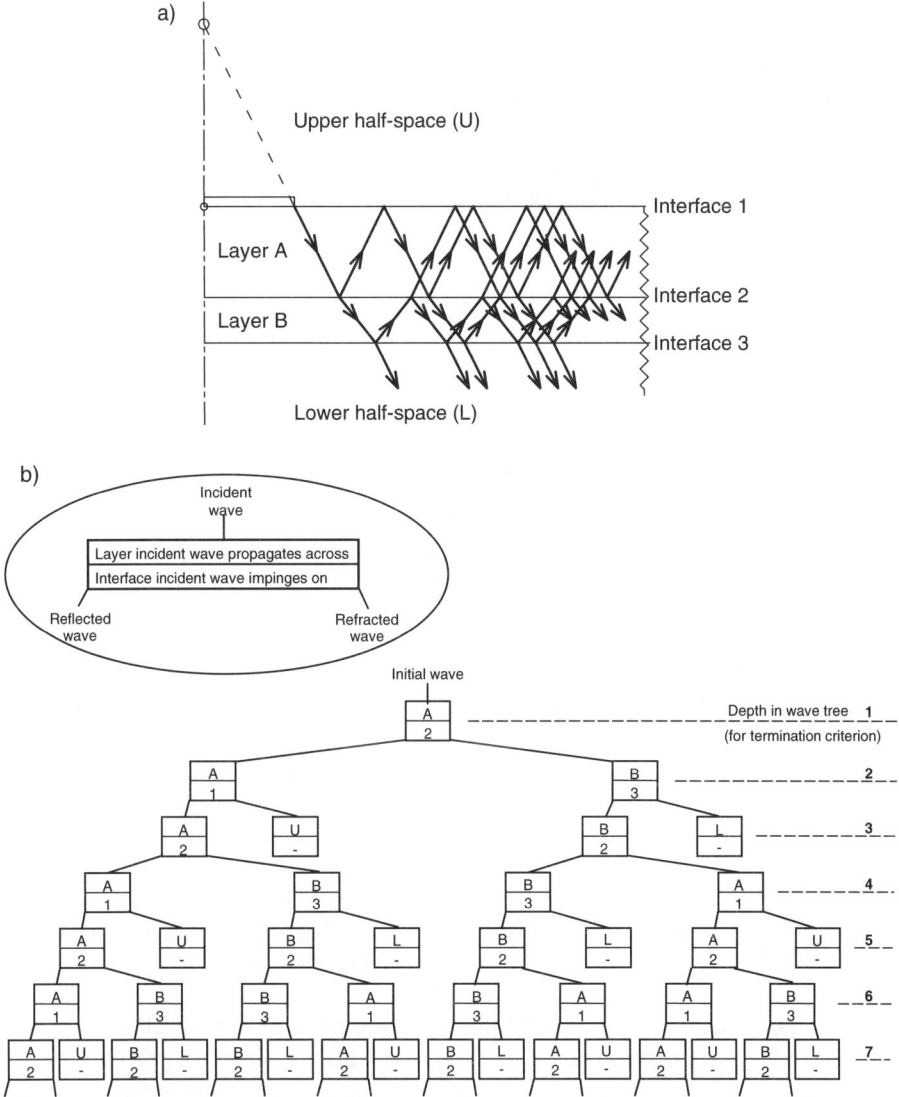

Figure 4.11 Representation of wave pattern as binary tree. a) Wave pattern to depth of 7. b) Corresponding wave tree

and impinging on an interface, generating reflected and refracted waves. This particular binary tree will be referred to as the *wave tree*.

The computer procedure must process every wave in the wave tree, cumulating the contributions of the waves to the displacement amplitude of the disk. (In fact the contributions to the displacement amplitudes at each interface will be cumulated. This allows embedded foundations to be processed. Full details will be provided in Chapter 5.) The process of visiting every node in a tree is termed *traversal*. A binary tree may be traversed efficiently and concisely using *recursion*.

Recursion is a fundamental concept in mathematics and computer science. A simple definition is that a recursive procedure is one that calls itself, and a recursive mathematical function is one that is defined in terms of itself (Ref. [28]). A recursive procedure can be used to traverse the binary tree defining the generated waves propagating in the cone segments (the wave tree) because the reflection and refraction process is identical at each node in the tree, and each process generates two more processes of the same type (unless the wave propagates into a half-space or becomes small enough in amplitude to be neglected).

The recursive procedure (named *Transmit()* in the MATLAB program) used to traverse the wave tree implements the techniques and equations developed in Sections 4.1 and 4.2. The procedure commences with an initial wave of amplitude u_0 propagating from a disk or fictitious disk of radius r_0 in a truncated cone across a layer of finite thickness (termed *layerA* in the program). The properties of this layer are denoted as G, ν, ρ and ζ. The aspect ratio of the cone and the wave velocity are determined from Table 3.1, according to the degree of freedom excited. In the computer program the functions *AspectRatio()* and *WaveVelocity()* compute the corresponding values. (Listings of these simple functions are presented in Appendix F.)

Since *layerA* has finite thickness d, the wave propagates across the layer and impinges on an interface. The radius of the cone r at this interface is computed using Eq. 4.4, and the amplitude of the incident wave f when it impinges on the interface follows from Eq. 4.5 for a translational cone or Eq. 4.26b for a rotational cone. The coefficient relating f to u_0 in these equations is computed in a function called *Attenuation()*, the name deriving from the fact that the coefficient attenuates the wave as it propagates across the layer. (A listing of this procedure is presented in Appendix F.)

The layer on the far side of the interface (in terms of the direction of wave propagation) is termed *layerB* in the program. The properties of this layer are denoted as G', ν', ρ' and ζ'. The incident wave f generates a reflected wave g (Eq. 4.19) and a refracted wave h (Eq. 4.20), computed using the reflection coefficient $-\alpha$ (Eq. 4.18 with Eqs 4.16 and 4.17 or Eqs 4.35 and 4.36, depending on the type of cone). In the program α is computed using the equations indicated in a function called *Alpha()*, listed in Appendix F.

The reflected wave g then propagates in a new initial cone, with $u_0 = g$ and $r_0 = r$ in the opposite direction back across *layerA*. The *Transmit()* procedure calls itself to process this wave. If *layerB* has finite thickness, the refracted wave will lead to further reflections and refractions, and is processed in the same way. The refracted wave propagates in a new initial cone in the same direction as the initial wave, but across *layerB*, with $u_0 = h$ and $r_0 = r$. The *Transmit()* procedure calls itself to process this wave also.

The *Transmit()* procedure (written in MATLAB) is presented in Listing 4.1. It forms the heart of the computer program. On entry to the function, n specifies the interface from which the initial wave propagates. The initial wave, with amplitude $u0$, circular frequency w and degree of freedom *dof*, propagates in the direction indicated by *dirn*, where 1 represents down and -1 represents up. $r0$ specifies the radius of the disk or fictitious disk which is the source of this initial wave.

Once the sequence of wave reflection and refraction has been followed until the amplitude of all waves is sufficiently small (<0.001 of the initial wave magnitude in Listing 4.1), the procedure terminates automatically, leaving the amplitude of the motion of each interface stored in the layer data array.

To permit the dynamic-stiffness coefficient of a single disk to be determined using the recursive *Transmit()* procedure, a driver function named *DiskStiffness()* is implemented. This function is presented in Listing 4.2.

The driver function initiates the sequence of wave reflections and refractions from the disk. On entry the function is called with an array of data structures describing the properties of the layers *layers*, an integer n indicating the interface at which the disk is located, and the circular frequency w and degree of freedom *dof* for which the dynamic-stiffness coefficient is

Wave reflection and refraction at a material discontinuity 79

```
function Transmit(n, dirn, r0, u0, w, dof)
   global laYers % global array of data structures describing soil layers
   if dirn==1 % DOWN
     layerA = laYers(n);
     layerB = laYers(n+1);
   else % dirn==-1 % UP
     layerA = laYers(n+1);
     layerB = laYers(n);
   end
   r = r0 + layerA.thickness/AspectRatio(layerA,dof);        % Eq. 4.4
   f = Attenuation(layerA,r0,r,w,dof)*u0;                    % Eq. 4.5 or 4.26b
   if norm(f)>0.001 % check termination criterion
      g = -Alpha(layerA,layerB,r,w,dof)*f;                   % Eq. 4.19
      h = f + g;                                             % Eq. 4.10
      laYers(n).amplitude = laYers(n).amplitude + h;

      % reflected wave
      Transmit(n-dirn,-dirn,r,g,w,dof);

      % refracted wave
      if isequal(layerB.type,'L')
         Transmit(n+dirn,dirn,r,h,w,dof);
      end
   end
end
```

Listing 4.1 Recursive function forming heart of computer program

```
function S = DiskStiffness(layers, n, w, dof)
   global laYers % a globally visible and changeable copy of the layer data
   laYers = layers;

   % initial radius of cone
   r0 = laYers(n).disk;
   % initial amplitudes of displacement
   u0 = 1.0
   for k=1:size(laYers,2), laYers(k).amplitude = 0.0; end
   laYers(n).amplitude = u0;

   if laYers(n).type=='L' % transmit wave up across layer above the disk
      Transmit(n-1,-1,r0,u0,w,dof);
   end

   if laYers(n+1).type=='L' % transmit wave down across layer beneath the disk
      Transmit(n+1,1,r0,u0,w,dof);
   end

   % determine amplitude of force applied to disk, considering both initial cones
   P = ConeStiffness(laYers(n),r0,w,dof) + ConeStiffness(laYers(n+1),r0,w,dof);
   % divide by final amplitude of disk displacement to determine dynamic stiffness
   S = P/laYers(n).amplitude;
```

Listing 4.2 Driver function for determining dynamic-stiffness coefficient of single disk

required. A copy of the layers array is created as *laYers* and made available globally, so that this information does not have to be passed repeatedly between different instances of the recursive procedure *Transmit()*. The amplitude of the displacement of the disk is initialised to 1, and the amplitudes of the displacements of all other interfaces initialised to zero. If the layer above the disk is of finite thickness, the wave propagating upwards is transmitted through the corresponding cone segment, and the subsequent reflections and refractions processed using *Transmit()*. If the layer below the disk is of finite thickness, the wave propagating downwards is processed in the same way.

The amplitude of the force applied to the disk is determined by summing the dynamic-stiffness coefficient of the initial upper cone and that of the initial lower cone (since the initial wave amplitude is unity). The dynamic-stiffness coefficients of the cones are determined by the properties of the layer, the frequency and the degree of freedom in a function called *ConeStiffness()*, and are specified by Eq. 3.25 with Eq. 3.26 or Eq. 3.89, or Eq. 3.50 with Eq. 3.51 or Eq. 3.91, together with the information in Table 3.1. The ratio of the amplitude of this force and the final amplitude of the disk displacement determines the dynamic-stiffness coefficient. (The *ConeStiffness()* function is listed in Appendix F.)

As an illustration the example shown in Fig. 4.10a is processed for the vertical degree of freedom. The layer data is listed in Table 4.1a. For convenience G_0, ρ_0 and r_0 are assigned unit values and ζ is taken to be 0. The data structure array containing this information is entered manually at the MATLAB command line, or is entered into a text file (one row on each line with the columns separated by tabs) and read into the MATLAB environment, as described in Appendix F. Alternatively, the stand-alone executable program CONAN can be used to read and process the text file (Appendix E). The computed dynamic-stiffness coefficient is decomposed as indicated by Eq. 4.64, where K is the static-stiffness coefficient of the disk on a homogeneous half-space with the same material properties as the upper layer. The resulting spring and damping coefficients, $k(a_0)$ and $c(a_0)$, are plotted in Fig. 4.12. The analysis is repeated with the wave pattern followed to a magnitude of 0.01 of the initial wave magnitude, and again to a magnitude of 0.0001 of the initial wave magnitude. Figure 4.12 shows that the results for this example are not sensitive to this parameter.

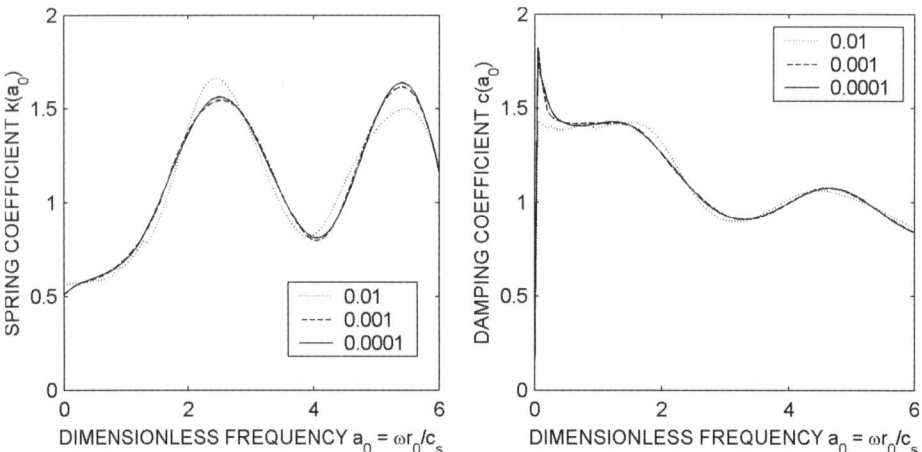

Figure 4.12 Dynamic-stiffness coefficient of disk on two layers overlying half-space computed with magnitude criterion, illustrating influence of termination magnitude

4.6 Termination criteria

An important aspect of the recursive computer procedure is the termination criterion. A simple approach is to specify a minimum value for the magnitude of the reflected and refracted waves below which further reflections and refractions are neglected. This criterion will be referred to as the *magnitude criterion*, and the minimum value for the magnitude of the waves (as a proportion of the initial disk displacement magnitude) will be referred to as the *termination magnitude*. In Ref. [42] a value of 0.001 for the termination magnitude has been recommended. For situations with a small number of layers overlying a flexible half-space, this value has proved to be adequate, as demonstrated in Fig. 4.12. However, there are two types of problems where the magnitude criterion may fail.

The first is where a non-homogeneous soil is modelled by a large number of layers, each of constant stiffness over its depth. The variation of stiffness between the layers is small (and becomes smaller the more layers are used to model the variation). The value of the reflection coefficient depends on the change in stiffness, and so the reflected waves at each interface are very small. Consequently, as the number of interfaces used to model the non-homogeneous half-space is increased, at a certain level all the reflections will be below the termination magnitude, and no reflections will be processed. Also, considering that the number of reflections to be processed increases dramatically with the number of layers present, it is prudent to limit the number of layers to a reasonable number, such as 10. This permits the magnitude criterion to be used.

The second type of problem where the magnitude criterion can cause significant difficulty is when layers overlie a rigid base. Waves reflected from the base reverse the sign of their amplitude, while waves reflected from the free surface preserve the sign of their amplitude. In both cases the magnitude of the wave is preserved. When the layers are thin, the geometric attenuation within the cones is small, and a large number of reflections must be processed before the magnitude falls to a sufficiently small level. Considering a disk on a single layer overlying a rigid base (Fig. 2.5), the amplitude of each wave reaching the disk is of the opposite sign to the proceeding one, so the total displacement amplitude oscillates around the final value as the amplitudes are cumulated. This problem has long been recognised, and Ref. [37] recommends forming two consecutive summations of reflections (i.e. for m reflections and for $m + 1$ reflections), and then averaging the two. The consequential error is far less than the current magnitude of the reflected waves.

As an illustration of the second type of problem, the half-space in the example of Fig. 4.10a is truncated with a rigid base, as shown in Fig. 4.13. The analysis is carried out using the magnitude criterion with termination magnitudes of 0.01, 0.001 and 0.0001. The corresponding dynamic-stiffness coefficients for the vertical degree of freedom are decomposed using Eq. 4.64 with the static-stiffness coefficient K for the disk on a homogeneous half-space with the same material properties as the upper layer. The resulting spring and damping coefficients, $k(a_0)$ and $c(a_0)$, are plotted in Fig. 4.14, along with an exact solution computed using the thin-layer method (Ref. [12]). In stark contrast to the example with layers overlying the homogeneous half-space (Fig. 4.12), as more reflections and refractions are processed the solution shows an increasing amount of numerical 'noise'.

Although the recursive algorithm described in Section 4.5 is simple and robust, as with all recursive algorithms it can cause a program to fail if the recursion does not terminate. Each recursive call to the *Transmit* procedure places a set of local variables in the computer memory 'stack'. If the procedure recurses too many times, the stack will overflow, and the computer program will behave unpredictably (possibly terminating the process with an 'out of stack space' error message).

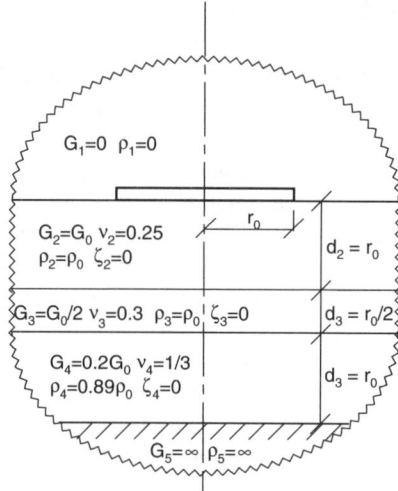

Figure 4.13 Disk on three layers overlying rigid base

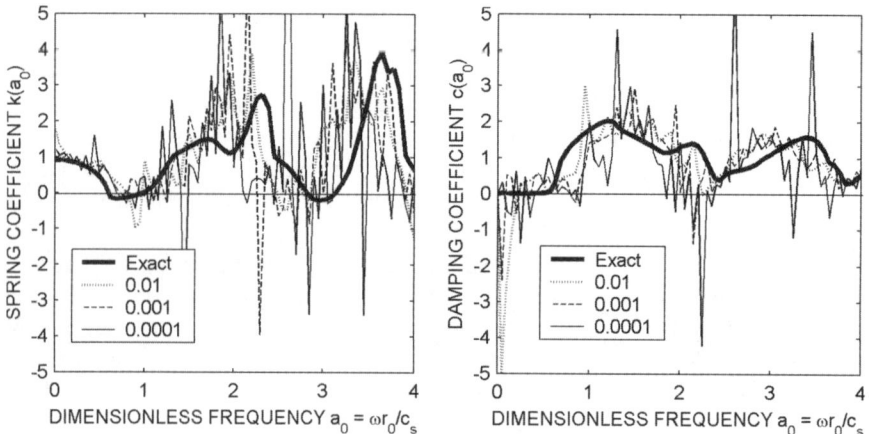

Figure 4.14 Dynamic-stiffness coefficient of disk on three layers overlying rigid base computed with magnitude criterion, illustrating effect of termination magnitude

A more sophisticated termination criterion should also count the number of cone segments through which the wave currently being processed has propagated (i.e. the depth down the wave tree) and commence termination when necessary to avoid numerical error and stack overflow. The final few reflections and refractions after termination is commenced should be averaged, in case their amplitudes are opposite in sign (improving the accuracy of systems overlying a rigid base). This termination criterion will be termed the *depth criterion*.

To count the number of generated waves, a global variable *depthCounter* is initialised to zero before the first call to *Transmit()*. Each nested call to *Transmit()* increments the counter on entry and decrements it on exit. Consequently the variable always contains the current depth of the traversal into the wave data tree. Once the *termination depth* is reached, the wave amplitude is attenuated by 10% in the following cone segment, 20% in the next cone segment, and so on, until its amplitude has been reduced to zero over a maximum of ten steps. The amplitudes over these

```
function Transmit(n, dirn, r, u0, w, dof)
    global laYers % global array of data structures describing soil layers
    global depthCounter % global variable for depth in wave tree
    depthCounter = depthCounter + 1;
    maxDepth = 20 + 2*(size(laYers,2) - 2);

    ... as in Listing 4-1 ...

    if norm(f)>0.0001
      if depthCounter > maxDepth
        f = f*(1.0 - 0.1*(depthCounter-maxDepth)); % kill off wave within 10
steps
      end

    ... as in Listing 4-1 ...

    end
    depthCounter = depthCounter - 1;
```

Listing 4.3 Changes required to implement depth criterion for termination

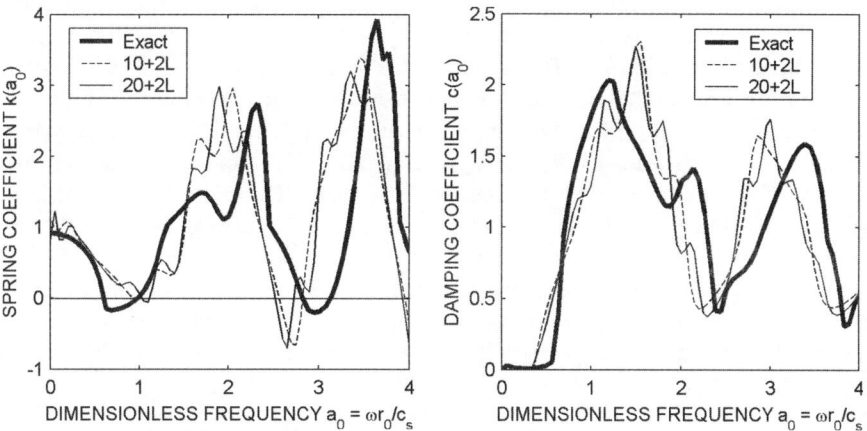

Figure 4.15 Dynamic-stiffness coefficient of disk on three layers overlying rigid base computed with depth criterion, illustrating effect of termination depth

last steps are effectively averaged, with higher weighting applied to the earlier steps than to the later steps. A termination depth of $20 + 2 \times L$ (where L is the number of layers of finite thickness present) has been found to provide accuracy of a similar level to the magnitude criterion with a termination magnitude of 0.001, but with increased stability.

The depth criterion also employs a termination magnitude of 0.0001 to avoid unnecessary computations when the wave amplitudes decrease rapidly. An implementation of this depth criterion is provided in Listing 4.3, which indicates the necessary changes to *Transmit()*.

Reanalysing the example illustrated in Fig. 4.13 using the depth criterion with termination depths of $10 + 2 \times L$ and $20 + 2 \times L$, and decomposing the dynamic-stiffness coefficients as described for Fig. 4.14, the spring and damping coefficients plotted in Fig. 4.15 are obtained. An exact solution from Ref. [12] obtained using the thin-layer method is also plotted. In this case, even when the depth is limited to $10 + 2 \times L$, adequate agreement with the exact solution is obtained. However, in other situations a depth of $20 + 2 \times L$ has been found to be necessary.

5

Foundation embedded in a layered half-space

To calculate the dynamic stiffness and the effective foundation input motion, the foundation is represented by a stack of disks in that part of the layered half-space which will be excavated, as already mentioned in Section 2.6. This concept of a primary dynamic system with redundants acting on the embedded disks is described in Section 5.1. As in the force method of structural analysis, the dynamic flexibility, i.e. the displacements of the disks caused by the redundants acting on the disks, is established for the free field in Section 5.2. Inversion of this relationship yields the dynamic stiffness of the free field discretised at the disks. The vertical wall of the embedded foundation is constrained to execute a rigid-body motion (considering the free-field motion of seismic waves, if present), and the trapped material of the excavation is eliminated from the formulation in Section 5.3, which leads to the dynamic stiffness of the foundation and the effective foundation input motion. Section 5.4 presents a computer implementation of these procedures, including MATLAB listings, and practical examples are addressed in Section 5.5.

5.1 Stack of embedded disks

The layered half-space without any excavation is called the free field. To be able to represent an embedded foundation, rigid massless disks are placed in the layered half-space in that region which will later be excavated. For instance, to model a cylindrical foundation with radius r_0 extending a depth e into the layered half-space (Fig. 5.1), m disks are placed. In other words, the region which will later be excavated is viewed as a sandwich of m disks separated by $m - 1$ layers. The first disk is on the surface of the layered half-space, and the mth coincides with the (rigid) base of the embedded foundation.

As modelling with cone segments addresses the discretisation on the axis in the vertical direction, a sufficient number of disks must be selected to be able to accurately represent the harmonic response for a specified frequency. The maximum vertical distance Δe between two neighbouring disks follows from the requirement that, for instance, 10 nodes (that is 10 disks) per wave length λ must be present. With $\lambda = c/(\omega/2\pi)$

$$\Delta e \leq \frac{\lambda}{10} = \frac{1}{10}\frac{2\pi}{\omega}c = \frac{\pi c}{5\omega} \tag{5.1a}$$

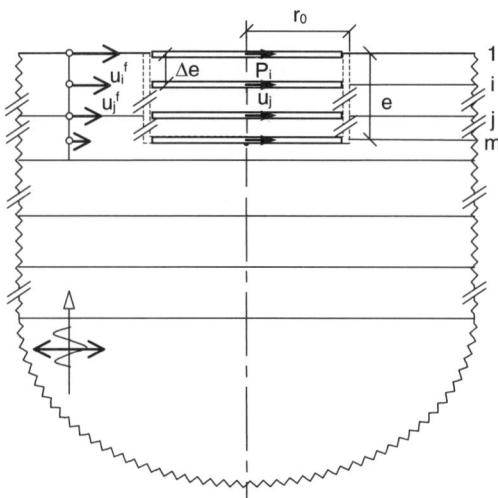

Figure 5.1 Stack of disks with redundants and free-field motion to represent embedded cylindrical foundation

follows. ω represents the highest frequency the dynamic model must accurately be able to represent, and c designates the appropriate wave velocity (in general the shear-wave velocity c_s). Expressed as a function of the dimensionless frequency $a_0 = \omega r_0/c_s$, this requirement (with $c = c_s$) is

$$\Delta e \leq \frac{\pi}{5} \frac{r_0}{a_0} \tag{5.1b}$$

In addition, disks are placed at all layer interfaces within the region of the half-space that will be excavated.

The force method of structural analysis is now applied. The dynamic system consisting of the m disks embedded in the layered half-space without excavation (free field) can be regarded as the primary system. In the centres of the disks, redundants for all degrees of freedom are present: forces acting in the horizontal and vertical directions, and bending and torsional moments. As an example the horizontal forces with amplitudes $P_i(\omega)$ ($i = 1, 2, \ldots, m$) are shown in Fig. 5.1. Each loaded embedded disk is modelled as a double cone with outward propagating waves and reflections and refractions occurring at the interfaces (including the free surface), while the disk on the surface is represented as a single cone. Each degree of freedom is modelled with corresponding cones, which are independent of those for the other degrees of freedom. The redundants will lead to corresponding displacements at the centres of the disks in the primary system. For instance, the horizontal forces will result in horizontal displacements with amplitudes $u_j(\omega)$ ($j = 1, 2, \ldots, m$), but will not lead to rotations or vertical displacements. Thus, no coupling between the degrees of freedom occurs in the primary system.

The analysis of the layered half-space leads, for seismic waves, to the free-field motion at the location of the disks (Appendix B.2). As only vertically propagating waves are addressed, the presence of the disks does not affect the free-field analysis. For a horizontal earthquake, horizontal displacements at the disks with amplitudes $u_j^f(\omega)$ ($j = 1, 2, \ldots, m$) follow (Figs 5.1 and 1.2a), and for a vertical earthquake, vertical displacements occur (Fig. 1.2b).

A non-cylindrical axi-symmetric foundation is represented analogously (Fig. 5.2). As in this case the walls will not be vertical, disks of variable radius are introduced. For illustration,

86 Foundation Vibration Analysis: A Strength-of-Materials Approach

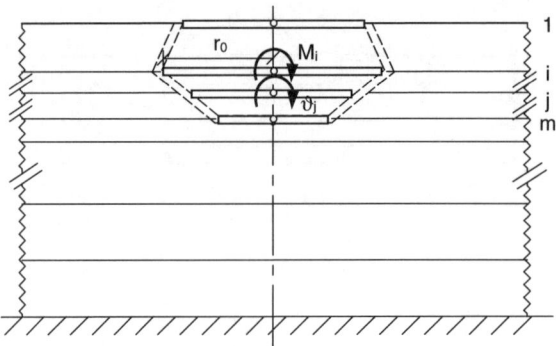

Figure 5.2 Stack of disks of variable radius with redundants to represent embedded axi-symmetric foundation

Fig. 5.2 shows bending moment redundants with amplitudes $M_i(\omega)$ leading to rocking rotations with amplitudes $\vartheta_j(\omega)$ in the primary system.

5.2 Dynamic flexibility of the free field

The primary system used for the dynamic analysis based on the force method consists of the disks embedded in the layered half-space without excavation (Section 5.1). The redundants act on the embedded disks. This system is called the *free field* for conciseness.

The relationship between the redundant forces and moments acting on the disks and the displacements and rotations of the disks is established. For harmonic excitation this leads to the dynamic-flexibility matrix of the free field.

The four components of the disk motions in the free field (horizontal displacement, vertical displacement, rocking rotation and torsional rotation) are independent when modelled with cones. Thus, the corresponding dynamic-flexibility matrices are uncoupled. For instance, the redundant horizontal forces acting on the embedded disks in Fig. 5.1 result in horizontal displacements of the embedded disks only, and the redundant bending moments in Fig. 5.2 lead to rocking motions only.

The horizontal motion (Fig. 5.1) is addressed first. The force amplitudes $P_i(\omega)$ ($i = 1, 2, \ldots, m$) form the elements of the vector $\{P(\omega)\}$, and the corresponding displacement amplitudes $u_j(\omega)$ ($j = 1, 2, \ldots, m$) form the elements of the vector $\{u(\omega)\}$. The displacement-force relationship is then expressed as

$$\{u(\omega)\} = [G(\omega)]\{P(\omega)\} \tag{5.2}$$

with the horizontal dynamic-flexibility matrix $[G(\omega)]$. For a specific element

$$u_j(\omega) = G_{ji}(\omega) P_i(\omega) \tag{5.3}$$

applies. The horizontal force with amplitude $P_i(\omega)$ is applied to the ith disk embedded in the layered half-space. The disk is modelled with a translational double cone consisting of initial cones with outward propagating waves (Section 3.6) which will impinge as incident waves at interfaces of layers yielding reflections and refractions (Section 4.1). The two building blocks to construct the wave pattern are thus applied. By tracking the reflection and refraction of each incident wave sequentially, the superimposed displacement wave pattern can be established. Such a wave pattern is illustrated in Fig. 2.12. This leads at the jth embedded disk to the horizontal displacement with amplitude $u_j(\omega)$, which defines the dynamic-flexibility coefficient $G_{ji}(\omega)$ in Eq. 5.3.

An analogous relationship to that formulated in Eq. 5.3 also applies for vertical motion.

For each of the two rotational motions

$$\{\vartheta(\omega)\} = [G_\vartheta(\omega)]\{M(\omega)\} \tag{5.4}$$

can be formulated. Figure 5.2 illustrates the rocking degree of freedom. Each term of the rotational dynamic-flexibility matrix $G_{\vartheta ji}(\omega)$ is constructed using the two building blocks for rotational cones and superimposing the rotational wave patterns.

Note that the dynamic-flexibility matrices are complex and frequency dependent. They are regular and can thus be inverted, yielding the dynamic-stiffness matrices. For the translational and rotational degrees of freedom, inverting Eqs 5.2 and 5.4 results in

$$\{P(\omega)\} = [S^f(\omega)]\{u(\omega)\} \tag{5.5}$$

with

$$[S^f(\omega)] = [G(\omega)]^{-1} \tag{5.6}$$

and

$$\{M(\omega)\} = [S^f_\vartheta(\omega)]\{u(\omega)\} \tag{5.7}$$

with

$$[S^f_\vartheta(\omega)] = [G_\vartheta(\omega)]^{-1} \tag{5.8}$$

The superscript f is added to denote that the system free field is referred to, or, more accurately, the primary system (specifically the disks embedded in the layered half-space without excavation).

The dynamic-flexibility matrices are not, in general, symmetric. However, the asymmetry is small and the cross-coupling terms can be averaged.

5.3 Dynamic stiffness and effective foundation input motion

For an axi-symmetric embedded foundation, the vertical and torsional degrees of freedom are uncoupled, while, because of the embedment, the horizontal and rocking degrees of freedom are coupled.

The computational procedure to determine the dynamic-stiffness coefficient and the effective foundation input motion for the vertical degree of freedom is now discussed (Fig. 5.3). The force-displacement relationship of the disks embedded in the layered half-space without excavation is described in Eq. 5.5, with the corresponding dynamic-stiffness matrix with respect to the disks of the free field denoted $[S^f(\omega)]$. In the free field the components of the vertical displacements with the amplitudes $\{u(\omega)\}$ are independent of one another. In a rigid embedded foundation, the displacements must be equal to the vertical displacement of the base, with amplitude $u_0(\omega)$. This

88 Foundation Vibration Analysis: A Strength-of-Materials Approach

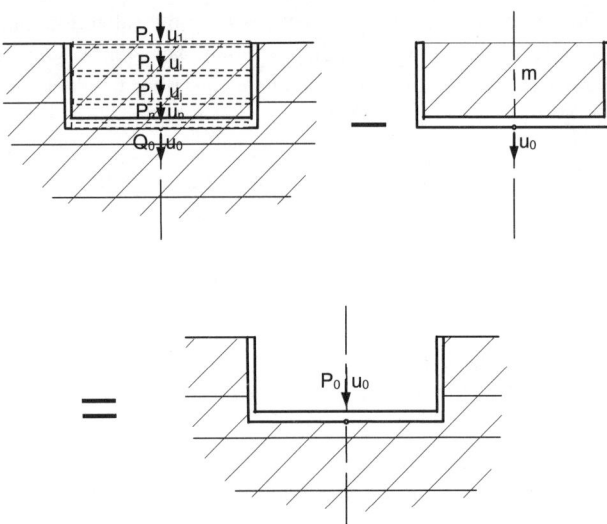

Figure 5.3 Enforcement of rigid-body displacement and excavation of trapped mass for vertical degree of freedom

rigid-body constraint is expressed as

$$\{u(\omega)\} = \{A\}u_0(\omega) \tag{5.9}$$

with the kinematic-constraint vector $\{A\} = [1 \; 1 \; 1 \; \ldots \; 1]^T$. The principle of virtual work, expressing equilibrium, implies that the force with amplitude $Q_0(\omega)$ is equal to the sum of the loads with amplitudes $P_i(\omega)$, or

$$Q_0(\omega) = \{A\}^T\{P(\omega)\} \tag{5.10}$$

Substituting Eq. 5.9 in Eq. 5.5 and then in Eq. 5.10 results in

$$Q_0(\omega) = \{A\}^T[S^f(\omega)]\{A\}u_0(\omega) \tag{5.11}$$

The triple product on the right-hand side of Eq. 5.11 represents the vertical dynamic-stiffness coefficient of the dynamic system shown as a shaded volume on the upper left of Fig. 5.3. The disks and the trapped domain between them are constrained to execute a vertical rigid-body translation. Therefore the domain between the disks may be analytically excavated from Eq. 5.11 simply by subtracting the mass times acceleration of the rigid interior domain (upper right of Fig. 5.3). The modified interaction force with amplitude $P_0(\omega)$ then becomes

$$P_0(\omega) = \{A\}^T[S^f(\omega)]\{A\}u_0(\omega) - m\ddot{u}_0(\omega) \tag{5.12}$$

with m representing the mass of the excavated domain. For harmonic motion $\ddot{u}_0(\omega) = -\omega^2 u_0(\omega)$, which, when substituted in Eq. 5.12, yields

$$P_0(\omega) = S^g_{00}(\omega)u_0(\omega) \tag{5.13}$$

where $S_{00}^g(\omega)$ is the vertical dynamic-stiffness coefficient of the rigid embedded foundation computed as

$$S_{00}^g(\omega) = \{A\}^T[S^f(\omega)]\{A\} + \omega^2 m \qquad (5.14)$$

This coefficient appears as the contribution of the unbounded soil in the basic equation of motion in a soil-structure-interaction analysis (Eq. B.4).

For a vertical earthquake, the vertical free-field motion with amplitude $\{u^f(\omega)\}$ on the level of the disks can be calculated (Fig. 1.2b, Appendix B.2). As under the free-field motion the disks are not loaded, the vertical displacement-force relationship (generalising Eq. 5.2) is

$$\{u^t(\omega)\} - \{u^f(\omega)\} = [G(\omega)]\{P(\omega)\} \qquad (5.15)$$

The forces depend on the motion relative to the free-field motion. $\{u^t(\omega)\}$ denotes the amplitudes of the total motion. Proceeding as above yields

$$P_0(\omega) = S_{00}^g(\omega) u_0^t(\omega) - \{A\}^T[S^f(\omega)]\{u^f(\omega)\} \qquad (5.16)$$

As the effective foundation input motion is equal to the displacement for vanishing interaction force, setting $P_0(\omega) = 0$ in Eq. 5.16 and replacing $u_0^t(\omega)$ by $u_0^g(\omega)$ leads to the amplitude of the effective foundation input motion, an 'averaged' vertical displacement,

$$u_0^g(\omega) = (S_{00}^g(\omega))^{-1}\{A\}^T[S^f(\omega)]\{u^f(\omega)\} \qquad (5.17)$$

This relationship is also expressed in Eq. B.5.

The torsional degree of freedom is processed analogously. The dynamic-stiffness coefficient of the embedded foundation is

$$S_{\vartheta 00}^g(\omega) = \{A\}^T[S_\vartheta^f(\omega)]\{A\} + \omega^2 m_\vartheta \qquad (5.18)$$

with the polar mass moment of inertia m_ϑ of the excavated domain and $\{A\}$ defined as before. For vertically propagating waves, the torsional effective foundation input motion vanishes.

The coupled horizontal and rocking degrees of freedom are now examined (Fig. 5.4). The horizontal displacements with amplitudes $\{u(\omega)\}$ and the rotations with amplitudes $\{\vartheta(\omega)\}$ of the embedded disks are related to the rigid-body motion of the base, a horizontal displacement with amplitude $u_0(\omega)$ and a rotation with amplitude $\vartheta_0(\omega)$, as

$$\begin{Bmatrix} \{u(\omega)\} \\ \{\vartheta(\omega)\} \end{Bmatrix} = [A] \begin{Bmatrix} u_0(\omega) \\ \vartheta_0(\omega) \end{Bmatrix} \qquad (5.19)$$

As, for instance, $u_i(\omega) = u_0(\omega) + h_i \vartheta_0(\omega)$ and $\vartheta_i(\omega) = \vartheta_0(\omega)$, the coefficients of the first half of the first column and the second half of the second column of the kinematic-constraint matrix $[A]$ are equal to 1 and the distances h_i of the base to the disks i make up the first half of the second column. The force amplitude vector at the base corresponding to $u_0(\omega)$ and

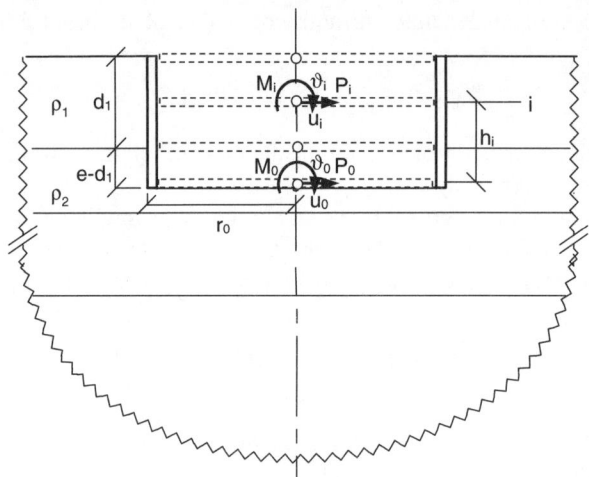

Figure 5.4 Enforcement of rigid-body motion for horizontal and rocking degrees of freedom

$\vartheta_0(\omega)$ is

$$\begin{Bmatrix} Q_0(\omega) \\ R_0(\omega) \end{Bmatrix} = [A]^T \begin{Bmatrix} \{P(\omega)\} \\ \{M(\omega)\} \end{Bmatrix} \tag{5.20}$$

Substituting Eq. 5.19 in Eqs 5.5 and 5.7, which are then substituted in Eq. 5.20, yields

$$\begin{Bmatrix} Q_0(\omega) \\ R_0(\omega) \end{Bmatrix} = [A]^T \begin{bmatrix} [S^f(\omega)] & \\ & [S^f_\vartheta(\omega)] \end{bmatrix} [A] \begin{Bmatrix} u_0(\omega) \\ \vartheta_0(\omega) \end{Bmatrix} \tag{5.21}$$

The coefficient matrix in Eq. 5.21 represents the dynamic-stiffness matrix of the layered half-space with the rigid-body motion of the walls of the embedded foundation enforced. The disks and the domain trapped between them are constrained to execute a rigid-body motion. This domain may be analytically excavated from Eq. 5.21 by simply subtracting the mass times the acceleration of the rigid interior with the rigid-body mass matrix $[M]$ corresponding to $[u_0(\omega)\vartheta_0(\omega)]^T$. For the problem shown in Fig. 5.4

$$[M] = \pi r_0^2 \begin{bmatrix} d_1\rho_1 + (e-d_1)\rho_2 & d_1\left(e-\frac{d_1}{2}\right)\rho_1 + \frac{(e-d_1)^2}{2}\rho_2 \\ d_1\left(e-\frac{d_1}{2}\right)\rho_1 + \frac{(e-d_1)^2}{2}\rho_2 & d_1\rho_1\left(e^2 - ed_1 + \frac{d_1^2}{3} + \frac{r_0^2}{4}\right) + (e-d_1)\rho_2\left(\frac{(e-d_1)^2}{3} + \frac{r_0^2}{4}\right) \end{bmatrix} \tag{5.22}$$

results, and $-[\ddot{u}_0(\omega)\,\ddot{\vartheta}_0(\omega)]^T = \omega^2[u_0(\omega)\,\vartheta_0(\omega)]^T$. The modified generalised interaction force-displacement relationship is

$$\begin{Bmatrix} P_0(\omega) \\ M_0(\omega) \end{Bmatrix} = [S^g_{00}(\omega)] \begin{Bmatrix} u_0(\omega) \\ \vartheta_0(\omega) \end{Bmatrix} \tag{5.23}$$

with the dynamic-stiffness matrix of the embedded foundation

$$[S^g_{00}(\omega)] = [A]^T \begin{bmatrix} [S^f(\omega)] & \\ & [S^f_\vartheta(\omega)] \end{bmatrix} [A] + \omega^2[M] \tag{5.24}$$

For a horizontal earthquake the horizontal free-field motion with amplitudes $\{u^f(\omega)\}$ on the level of the disks can be calculated (Fig. 1.2a, Appendix B.2). For vertically propagating waves no rotational free-field motion occurs. Equation 5.15, formulated for the horizontal direction, still applies. Proceeding as for the vertical direction, the effective foundation input motion consisting of an averaged horizontal component and a rocking component with amplitudes (Eq. 5.17)

$$\begin{Bmatrix} u_0^g(\omega) \\ \vartheta_0^g(\omega) \end{Bmatrix} = [S_{00}^g(\omega)]^{-1} [A]^T \begin{Bmatrix} [S^f(\omega)]\{u^f(\omega)\} \\ \{0\} \end{Bmatrix} \tag{5.25}$$

applies.

In the case of nearly-incompressible or incompressible soil ($1/3 < \nu \leq 1/2$, as discussed in Section 3.4) with vertical and rocking motion, besides limiting the wave velocity to $2c_s$, a trapped mass ΔM (Eq. 3.84) and a trapped mass moment of inertia ΔM_ϑ (Eq. 3.85) occur. The intermediate disks of the stack are separated by thin slices of soil. In the course of the analysis, however, the mass of the slices is subtracted out (excavated). As a result, trapped mass need only be included for the bottom-most disk to represent the cone of soil under the base of the embedded foundation, and, in the case of a fully embedded foundation, for the uppermost disk to represent the cone of soil above the foundation. Equation 5.14 becomes

$$S_{00}^g(\omega) = \{A\}^T [S^f(\omega)]\{A\} + \omega^2(m - \Delta M) \tag{5.26}$$

while, since only the rocking degree of freedom is affected, Eq. 5.24 becomes

$$[S_{00}^g(\omega)] = [A]^T \begin{bmatrix} [S^f(\omega)] & \\ & [S_\vartheta^f(\omega)] \end{bmatrix} [A] + \omega^2 \left([M] - \begin{bmatrix} 0 & 0 \\ 0 & \Delta M_\vartheta \end{bmatrix} \right) \tag{5.27}$$

Certain extensions are easily incorporated. For instance, for a wall with no stiffness the rigid-body constraint determining the $[A]$ matrix is not enforced. In this case the displacement and rotation amplitudes $\{u(\omega)\}$ and $\{\vartheta(\omega)\}$ are kept as the unknowns, and the dynamic-stiffness matrix of the embedded foundation is determined with respect to these degrees of freedom. The domain between the disks is still analytically excavated by subtracting the dynamic-stiffness matrix of the interior domain. The latter is (with superscript e for excavated)

$$[S^e(\omega)] = [K] - \omega^2[M] \tag{5.28}$$

with the static-stiffness matrix $[K]$ and mass matrix $[M]$ of the excavated part corresponding to $\{u(\omega)\}$ and $\{\vartheta(\omega)\}$. Assuming, for instance, that the interior domain can be modelled as a bar or beam with a vertical axis, $[K]$ and $[M]$ follow straightforwardly. The modified generalised interaction force-displacement relationship is

$$\begin{Bmatrix} \{P(\omega)\} \\ \{M(\omega)\} \end{Bmatrix} = [S^g(\omega)] \begin{Bmatrix} \{u(\omega)\} \\ \{\vartheta(\omega)\} \end{Bmatrix} \tag{5.29}$$

with the dynamic-stiffness matrix of the embedded foundation determined as the difference of the dynamic-stiffness matrices of the free field and the excavated part

$$[S^g(\omega)] = \begin{bmatrix} [S^f(\omega)] & \\ & [S_\vartheta^f(\omega)] \end{bmatrix} - [K] + \omega^2[M] \tag{5.30}$$

The case where the contact between the soil and wall exists only on part of the height can also be calculated. The rigid-body constraint applies only to that part where contact exists. Another case is discussed in Section 7.3 where, for a suction caisson, the interior domain is not analytically excavated and the location of the reference point O is changed.

It is of interest to study the behaviour for large embedment ratios. From the results presented, the dynamic-stiffness coefficients are accurate for all degrees of freedom of a cylinder (Section 5.5) up to $e/r_0 = 2$, and for the rotational degree of freedom of a hemi-ellipsoid (Section 6.6) up to $e/r_0 = 2.5$. This range covers most practical cases. For significantly larger embedment ratios, the double cone model leads to dynamic-flexibility coefficients at disks at large distances from the loaded disk which are too large for the translational degrees of freedom, especially for the vertical motion. The results are thus too flexible.

This can be verified by addressing the static case for simplicity. The vertical displacement at depth z follows for an embedded disk with load P acting vertically as (Eq. 4.1)

$$u(z) = \frac{1}{1 + z/z_0} u_0 \tag{5.31}$$

with, for the double cone,

$$u_0 = \frac{P}{2K} \tag{5.32}$$

Substituting Eq. 5.32 in Eq. 5.31, and noting that for large depths $(z \to \infty) z/z_0 \gg 1$,

$$g(z \to \infty) = \frac{z_0}{z} \frac{1}{2K} \tag{5.33}$$

results, where the displacement for a unit load is denoted as $g(z)$ to provide the link to the flexibilities in Section 5.2. Substituting K and z_0/r_0 specified in Table 3.1 for the vertical degree of freedom yields

$$g_v(z \to \infty) = \frac{\pi}{16} \frac{(1-v)^3}{1-2v} \frac{1}{G} \frac{1}{z} \tag{5.34}$$

Analogously, the horizontal displacement at depth $z \to \infty$ caused by a horizontal load acting on an embedded disk is

$$g_h(z \to \infty) = \frac{\pi}{128} (2-v)^2 \frac{1}{G} \frac{1}{z} \tag{5.35}$$

For this limit these flexibilities should be equal to the displacements caused by unit point loads acting in the corresponding directions in a full-space

$$u_v = \frac{1}{4\pi} \frac{1}{G} \frac{1}{z} \tag{5.36}$$

$$u_h = \frac{1}{16\pi} \frac{3-4v}{1-v} \frac{1}{G} \frac{1}{z} \tag{5.37}$$

The ratios are

$$\frac{g_v(z \to \infty)}{u_v} = \frac{\pi^2}{4} \frac{(1-v)^3}{1-2v} \tag{5.38}$$

and

$$\frac{g_h(z \to \infty)}{u_h} = \frac{\pi^2}{8} \frac{(1-v)(2-v)^2}{3-4v} \tag{5.39}$$

For $\nu = 1/3$, $g_v(z \to \infty)/u_v = 2.19$ for the vertical direction and $g_h(z \to \infty)/u_h = 1.37$ for the horizontal degree of freedom result. The double cone models (especially for the vertical degree of freedom) are too flexible at large distances from the loaded disk. This means that for an embedment that is an order of magnitude larger than the radius r_0 as encountered in pile foundations, modifications for the flexibilities have to be performed to use cone models. This aspect is, however, outside the scope of this book [Ref. 37].

The rotational degrees of freedom do not have to be addressed, as, for the static case, the rotations decay strongly (proportional to the inverse of the cube of the distance to the apex, $(z_0 + z)^3$, or, for $z \gg z_0$, z^3, Eq. 4.25).

5.4 Computer implementation

The general layered full-space description detailed in Section 4.5 is also employed for embedded foundations. Whereas previously only one interface was specified with a non-zero disk radius, now the radius of each stacked disk is specified at the appropriate interface. Some interfaces (those below or above the foundation) may have a zero disk radius.

The columns of the dynamic-flexibility matrix of the free field $[G(\omega)]$ (Eq. 5.2) are generated column by column by placing an excitation at each disk in turn. This is done using a procedure very similar to the *DiskStiffness()* procedure described in Section 4.5. However, rather than calculating the stiffness of a single disk, a vector containing the displacement of each disk due to an excitation force of unit magnitude at a single disk is generated. The function is called *DiskGreens()*, as it calculates values of the excited disk's Green's function at each disk, and is presented in Listing 5.1.

The first section of the *DiskGreens()* function is the same as *DiskStiffness()*. After initialising the displacement amplitude to unity at disk n and zero at all other disks, the wave propagating upwards across the layer above the disk and all the consequential reflections and refractions are computed using the *Transmit()* function. The wave propagating below the disk is then treated in the same way. The *Transmit()* function cumulates the amplitudes of all the waves at each disk, as described in Section 4.5. The force amplitude required on disk n to cause these displacements is computed based on the upper and lower cones, and the displacement amplitude at each disk divided by this force amplitude to yield the displacement amplitude due to a force of unit amplitude. The procedure returns the resulting vector of normalised displacement amplitudes (i.e. the nth column of the dynamic-flexibility matrix of the free field).

The dynamic-stiffness matrix of the free field is found by simply assembling the columns of the dynamic-flexibility matrix and then inverting the resulting matrix (Eq. 5.6). This is accomplished by the function *FreeField()*, specified in Listing 5.2.

Analytical excavation of the soil is accomplished for the vertical degree of freedom by using Eq. 5.14 and for the torsional degree of freedom by using Eq. 5.18. Equation 5.14 may also be applied for the horizontal degree of freedom when rotation of the foundation is prevented. (The restraining moment will not be computed if this approach is used.) However, the rocking dynamic stiffness cannot be computed in isolation from the horizontal dynamic stiffness.

Since the vector $\{A\}$ in Eqs 5.14 and 5.18 consists only of ones, the triple product $\{A\}^\mathrm{T}[S^f(\omega)]\{A\}$ is just a summation of all the elements of $[S^f(\omega)]$. Analytical excavation of the soil can therefore be carried out on one program line (in the function *DynamicStiffness()*, Listing 5.3). The mass of the excavated soil is computed in a separate function *Mass()*, which determines the soil volume and hence the mass for each cone frustum between two adjacent disks

```
function Gn = DiskGreens(layers, n, w, dof)
  global laYers     % a globally visible and changeable copy of the layer data
  laYers = layers;
  global depthCounter % global depth counter to allow Transmit to terminate by depth
  depthCounter = 0;

  % initial radius of cone
  r0 = laYers(n).disk;
  % initial amplitudes of displacement
  u0 = 1.0;
  for k=1:size(laYers,2), laYers(k).amplitude = 0.0; end
  laYers(n).amplitude = u0;

  if laYers(n).type=='L' % transmit wave up across the layer above the disk
     Transmit(n-1,-1,r0,u0,w,dof);
  end

  if laYers(n+1).type=='L' % transmit wave down across the layer below the disk
     Transmit(n+1,1,r0,u0,w,dof);
  end

  % determine amplitude of force applied to disk, considering both initial cones
  P = ConeStiffness(laYers(n),r,w,dof) + ConeStiffness(laYers(n+1),r,w,dof);

  % determine value of Green's function at each disk
  numInterfaces = size(laYers,2) - 1;   % number of interfaces between layers
  k1 = 1;
  for k=1:numInterfaces
    if laYers(k).disk>0.0 % only require value at disks
       Gn(k1) = laYers(k).amplitude/P; % displacement due to force of unit amplitude
       k1 = k1 + 1;
    end
  end
end
```

Listing 5.1 Function for determining Green's function of single disk

```
function Sf = FreeField(layers, w, dof)
  % assemble dynamic-flexibility matrix column by column
  numInterfaces = size(layers,2) - 1;
  k = 1;
  for n=1:numInterfaces
     if layers(n).disk>0.0
        G(:,k) = DiskGreens(layers,n,w,dof);
        k = k + 1;
     end
  end
  % invert dynamic-flexibility to obtain dynamic-stiffness matrix of free field
  Sf = inv(G);         % Eq. 5.6
```

Listing 5.2 Function for determining dynamic-stiffness matrix of free field

```
function Sg = DynamicStiffness(layers, w, dof)
  if isequal(dof,'R')
     error('This procedure does not compute rocking stiffness')
  else
     Sg = sum(sum(FreeField(layers,w,dof))) + w^2*Mass(layers,dof); % Eq. 5.14
  end

function M = Mass(layers, dof)
  % compute mass of soil in excavation
  M = 0.0;
  numInterfaces = size(layers,2) - 1;
  for n=1:numInterfaces
     if layers(n).disk>0.0 & layers(n+1).disk>0.0
        r = layers(n+1).disk;
        rr = layers(n).disk/r; % ratio of first disk radius to second
        dm = PI*layers(n+1).thickness*layers(n+1).rho*r^2;
        if isequal(dof,'T')
           M = M + dm*r^2*(1+rr+rr^2+rr^3+rr^4)/10;
        else
           M = M + dm*(1+rr+rr^2)/3;
        end
     end
  end
```

Listing 5.3 Functions for determining dynamic-stiffness coefficient of embedded foundation (except for rocking degree of freedom)

```
function ug = VerticalInputMotion(layers, w, uf)
  Sf = FreeField(layers,w,'V');    % find free field dynamic-stiffness matrix
  Sg = sum(sum(Sf)) + w^2*Mass(layers,'V'); % Eq. 5.14
  ug = 1/Sg * sum(Sf*uf);          % Eq. 5.17
```

Listing 5.4 Function for determining effective foundation input motion for vertical earthquake

making up the foundation and returns the sum of the masses. If the degree of freedom is torsional rotation, *Mass()* returns the sum of the polar mass moments of inertia of the cone frustums. The function *DynamicStiffness()* allows the vertical, horizontal and torsional dynamic-stiffness coefficients of the embedded foundation to be computed. Note that, for clarity, trapped mass terms resulting from the presence of nearly-incompressible or incompressible soil are not included. The version of this function presented in Appendix F (Listing F.11) includes these trapped mass terms (Eq. 5.26).

The effective foundation input motion for a vertical earthquake is obtained by implementing Eq. 5.17. Since $\{A\}$ is a vector of ones, premultiplication of the vector $[S^f(\omega)]\{u^f\}$ by $\{A\}^T$ is equivalent to summing the elements of the vector. Consequently the implementation is very simple, as demonstrated in the function *VerticalInputMotion()*, presented in Listing 5.4. On entry *uf* is a vector containing the vertical free-field motion amplitudes at the levels of the disks. For efficiency the dynamic-stiffness coefficient of the embedded foundation is computed directly from Eq. 5.14. In this way the free-field dynamic-stiffness matrix need only be determined once. (If $S_{00}^g(\omega)$ had been evaluated within the function by calling *DynamicStiffness()*, $[S^f(\omega)]$ would be

```
function Sg = HorzAndRockStiffness(layers, w, zRef)
  SfH = FreeField(layers,w,'H'); %free field dynamic-stiffness matrix horz dof
  SfR = FreeField(layers,w,'R'); %free field dynamic-stiffness matrix rocking dof

  numDisks = size(SfH,1);
  AH = zeros(numDisks,2);%upper portion of constraint matrix relating to horz dof
  AR = zeros(numDisks,2);%lower portion of constraint matrix relating to rocking dof

  % set up constraint matrix - first column horizontal translation
  %                          - second column rotation around zRef

  numInterfaces = size(layers,2) - 1;
  z = 0.0; % depth of interface
  k = 1;
  for n=1:numInterfaces
    if layers(n).disk>0.0
      AH(k,1) = 1.0;
      AH(k,2) = zRef - z;
      AR(k,2) = 1.0;
      k = k + 1;
    end
    z = z + layers(n+1).thickness;
  end

  Sg = AH'*SfH*AH + AR'*SfR*AR + w^2*HorzAndRockMass(layers, zRef); % Eq. 5.24
function M = HorzAndRockMass(layers, zRef)
  M = zeros(2,2);
  z = 0.0;
  numInterfaces = size(layers,2) - 1;
  for n=1:numInterfaces
    if layers(n).disk>0.0 && layers(n+1).disk>0.0
      r = layers(n+1).disk;
      rr = layers(n).disk/r; % ratio of first disk radius to second
      t = layers(n+1).thickness;
      h = zRef - z - t;
      dm = pi*layers(n+1).rho*r^2*t;
      a = (1+rr+rr^2)/3;
      b = (1+2*rr+3*rr^2)/6;
      c = r^2*(1+rr+rr^2+rr^3+rr^4)/20 + t^2*(1+3*rr+6*rr^2)/30;
      M(1,1) = M(1,1) + dm*a;
      M(1,2) = M(1,2) + dm*(a*h + b*t/2);
      M(2,2) = M(2,2) + dm*(c + a*h^2 + b*h*t);
    end
    z = z + layers(n+1).thickness;
  end
  M(2,1) = M(1,2);
```

Listing 5.5 Functions for determining coupled horizontal and rocking dynamic-stiffness coefficients of embedded foundation

evaluated on two separate occasions.) Again, the trapped masses for the nearly-incompressible and incompressible cases have not been included for clarity. Listing F.14 in Appendix F includes these terms.

To determine the coupled horizontal and rocking dynamic-stiffness coefficients, Eqs 5.19 to 5.24 are implemented. The result of the function *HorzAndRockStiffness()*, presented in Listing 5.5, is the two-by-two dynamic-stiffness matrix $[S_{00}^g(\omega)]$. The coefficients depend on the reference

```
function ug = HorizontalInputMotion(layers, w, zRef, uf)
  SfH = FreeField(layers,w,'H'); %free field dynamic-stiffness matrix horz dof
  SfR = FreeField(layers,w,'R'); %free field dynamic-stiffness matrix rocking dof


  numDisks = size(SfH,1);
  AH = zeros(numDisks,2);%upper portion of constraint matrix relating to horz dof
  AR = zeros(numDisks,2);%lower portion of constraint matrix relating to rocking dof

  % set up constraint matrix - first column horizontal translation
  %                          - second column rotation around zRef
  numInterfaces = size(layers,2) - 1;
  z = 0.0; % depth of interface
  k = 1;
  for n=1:numInterfaces
    if layers(n).disk>0.0
      AH(k,1) = 1.0;
      AH(k,2) = zRef - z;
      AR(k,2) = 1.0;
      k = k + 1;
    end
    z = z + layers(n+1).thickness;
  end

  Sg = AH'*SfH*AH + AR'*SfR*AR + w^2*HorzAndRockMass(layers, zRef); % Eq. 5.24
  ug = inv(Sg)*AH'*SfH*uf; % Eq. 5.25
```

Listing 5.6 Function for determining effective foundation input motion for horizontal earthquake

point, that is the height of the point on the central axis of the foundation where the degrees of freedom and the corresponding horizontal force and rocking moment are defined. For example, if the horizontal force and rocking moment are applied at the base of the foundation, the dynamic-stiffness coefficients will be different from the case where the force and rocking moment are applied at the top of the foundation. For this reason the depth of the reference point is specified on entry to the function (as *zRef*). (The description of the kinematic-constraint matrix [A] in Section 5.3 assumes *zRef* is the depth of the base of the foundation.) The rigid-body mass matrix is determined separately in the function *HorzAndRockMass()* to improve the readability of the *HorzAndRockStiffness()* function.

The effective foundation input motion can be found in a similar way, implementing Eq. 5.25. A function *HorizontalInputMotion()* is presented in Listing 5.6 which accomplishes this efficiently. On entry to the procedure the vector *uf* contains the horizontal free-field motion amplitudes at the levels of the disks. The function returns a vector with two elements, the horizontal input motion and the rocking input motion for the embedded foundation.

5.5 Examples

As the first example, the dynamic-stiffness coefficients of a rigid cylindrical foundation of radius r_0 embedded with a depth e in a homogeneous half-space with Poisson's ratio $\nu = 0.25$ (Fig. 5.5) are calculated for all degrees of freedom. The embedment ratios e/r_0 of 0.5, 1 and 2 are addressed. The results are determined for the dimensionless frequency $a_0 = \omega r_0/c_s$ ranging from 0 to 4.

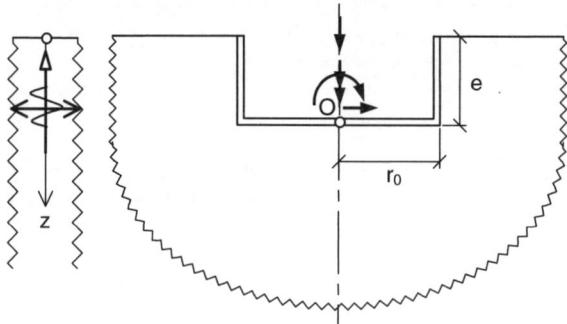

Figure 5.5 Cylindrical foundation embedded in homogeneous half-space with control motion specified at free surface

The maximum vertical distance Δe between two neighbouring disks follows from Eq. 5.1b with $a_0 = 4$ as $(\pi/20)r_0$. For the largest embedment ratio $e/r_0 = 2$, at least 13 layers (of finite thickness, i.e. slices) yielding 14 disks are required. However, for convenience of data preparation, 16 layers are used, so that $\Delta e = 0.125 r_0$. This vertical distance is preserved for each embedment ratio, leading to models with 8 layers for $e/r_0 = 1$ and 4 layers for $e/r_0 = 0.5$.

The dynamic-stiffness coefficients follow from Eqs 5.14, 5.18 and 5.24. They are then decomposed as specified in Eq. 1.5, where the static-stiffness coefficient K is used.

The static-stiffness coefficients of the embedded foundation are calculated using cones applying the same equations for $\omega = 0$. They are divided by the corresponding static-stiffness coefficients of a surface footing for the same motion and compared with the rigorous results of Ref. [2] in Fig. 5.6. The dashed curves correspond to the empirical formulas of Ref. [25], which are applicable for $e/r_0 \leq 2$, and are

$$\text{horizontal} \quad K_h = \frac{8Gr_0}{2-\nu}\left(1 + \frac{e}{r_0}\right) \quad (5.40a)$$

$$\text{vertical} \quad K_v = \frac{4Gr_0}{1-\nu}\left(1 + 0.54\frac{e}{r_0}\right) \quad (5.40b)$$

$$\text{rocking} \quad K_r = \frac{8Gr_0^3}{3(1-\nu)}\left(1 + 2.3\frac{e}{r_0} + 0.58\left(\frac{e}{r_0}\right)^3\right) \quad (5.40c)$$

$$\text{torsional} \quad K_t = \frac{16Gr_0^3}{3}\left(1 + 2.67\frac{e}{r_0}\right) \quad (5.40d)$$

$$\text{coupling} \quad K_{hr} = \frac{e}{3}K_h \quad (5.40e)$$

The first terms on the right-hand sides of Eq. 5.40 are the static-stiffness coefficients of a surface foundation. The accuracy of the cone models is excellent for all degrees of freedom.

The dynamic-stiffness coefficients are presented for the horizontal, vertical, rocking and torsional motions in Figs 5.7 to 5.10. The deviation from the rigorous results of Ref. [2], denoted

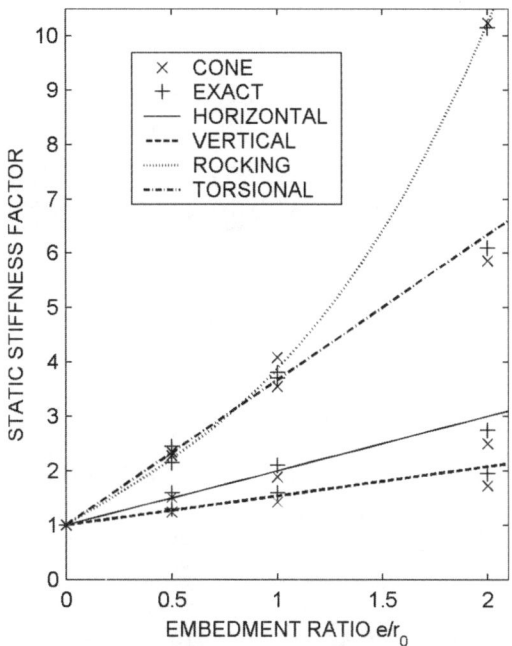

Figure 5.6 Static-stiffness factors for cylindrical foundation embedded in homogeneous half-space

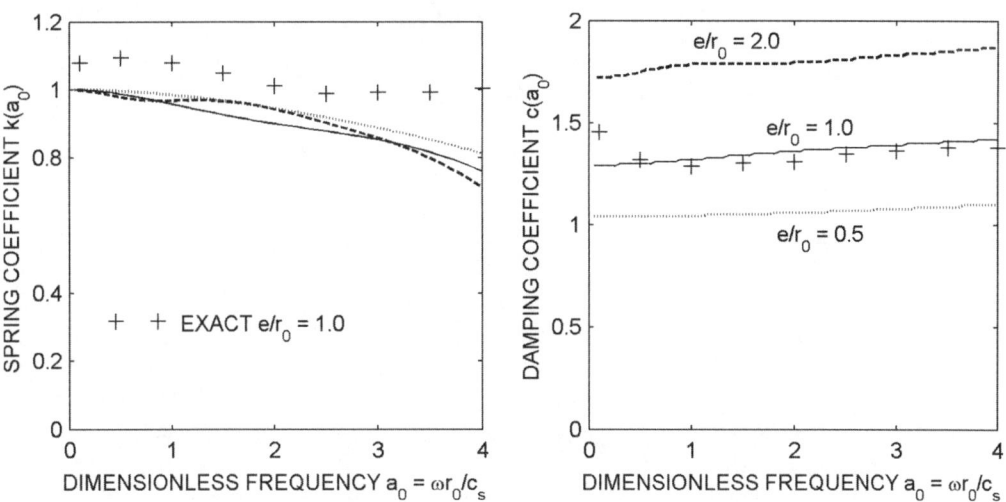

Figure 5.7 Dynamic-stiffness coefficient of cylindrical foundation embedded in homogeneous half-space in horizontal motion ($\nu = 0.25$)

as exact (shown here for the embedment ratio $e/r_0 = 1$, but comparable for the other embedment ratios) is generally less than 15%. It is observed that for horizontal and vertical translations (Figs 5.7 and 5.8) the spring coefficient k and the damping coefficient c are almost constant. The corresponding dynamic-stiffness can thus be represented by an ordinary spring and dashpot in parallel with frequency-independent coefficients, similar to the surface foundation (Fig. 2.2b).

100 Foundation Vibration Analysis: A Strength-of-Materials Approach

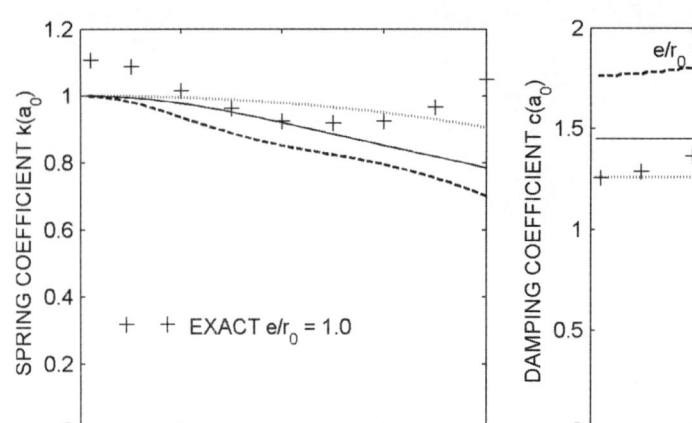

Figure 5.8 Dynamic-stiffness coefficient of cylindrical foundation embedded in homogeneous half-space in vertical motion ($\nu = 0.25$)

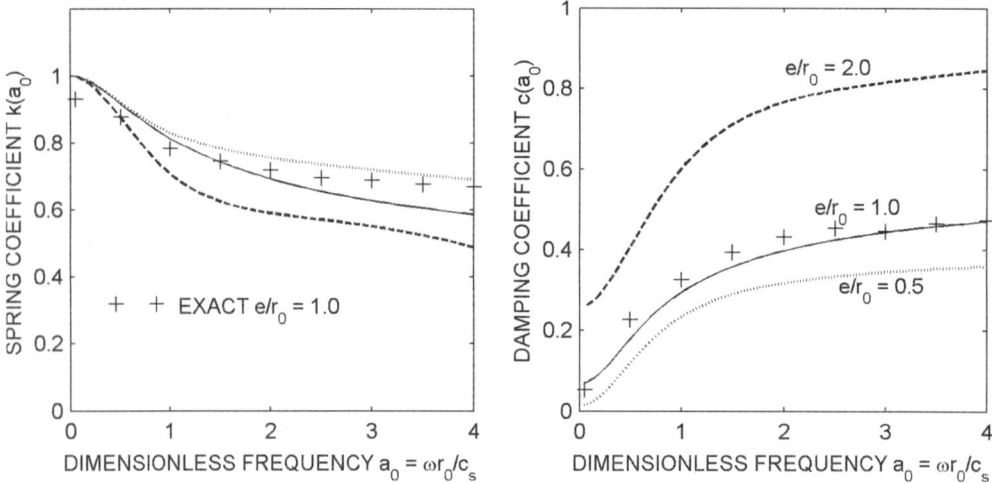

Figure 5.9 Dynamic-stiffness coefficient of cylindrical foundation embedded in homogeneous half-space in rocking motion ($\nu = 0.25$)

The effective foundation input motion is now calculated for vertically propagating S-waves with the control motion specified at the free surface (Fig. 5.5). The free-field motion with horizontal particle displacement with amplitude $u^f(z, \omega)$ is described as

$$u^f(z, \omega) = u^f(\omega) \cos \frac{\omega}{c_s} z \tag{5.41}$$

where the depth z is measured downwards from the free surface. The amplitude at the free surface is $u^f(\omega)$. $u^f(z, \omega)$ in Eq. 5.41 determines $\{u^f(\omega)\}$ on the level of the disks. The effective foundation

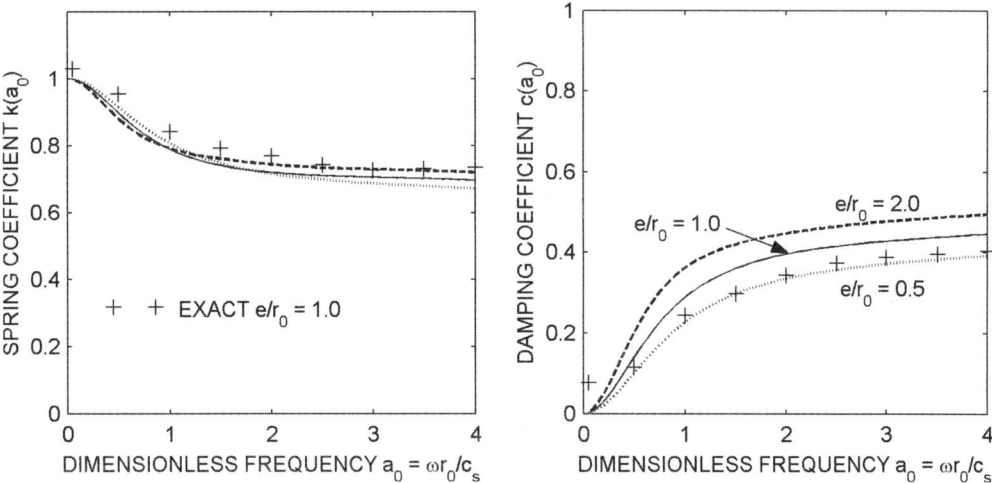

Figure 5.10 Dynamic-stiffness coefficient of cylindrical foundation embedded in homogeneous half-space in torsional motion

input motion, consisting of the horizontal component with amplitude $u_0^g(\omega)$ defined at the centre O of the base and the rocking component with amplitude $\vartheta_0^g(\omega)$, follows from Eq. 5.25. The real and imaginary parts are plotted as a function of a_0 in Fig. 5.11. The results using embedded disks with the corresponding cone models are in excellent agreement with the rigorous solutions of Ref. [16] for all embedment ratios.

As a second example, the dynamic-stiffness coefficients of a rigid cylindrical foundation of radius r_0 with an embedment ratio e/r_0 of 1 in an incompressible homogeneous half-space (Poisson's ratio $\nu = 1/2$) are calculated for vertical and rocking motion. A hysteretic damping ratio ζ of 5% is present. A vertical disk spacing of $\Delta e = 0.125 r_0$ is again employed. Dynamic-stiffness coefficients are computed for the dimensionless frequency range $0 \leq a_0 \leq 4$, and decomposed using the static-stiffness coefficient computed by the cone model. The computed dynamic-stiffness coefficients for the vertical and rocking motions are plotted in Figs 5.12 and 5.13, respectively. Also plotted in these figures are rigorous solutions presented in Ref. [31], which are denoted as exact and plotted as discrete points. The cone models agree impressively with the exact results, especially for the damping coefficient at higher dimensionless frequencies.

Finally, a cylindrical foundation of radius r_0 embedded with depth e in a soil overlying a flexible rock half-space with the geometry and material properties (shear modulus G, Poisson's ratio ν, mass density ρ, hysteretic damping ratio ζ) specified in Fig. 5.14 is analysed. The dynamic-stiffness coefficients are calculated for the vertical and torsional degrees of freedom in the dimensionless frequency range of $a_0 = \omega r_0/c_s$ from 0 to 3, with the shear-wave velocity of the layer $c_s = \sqrt{G_0/\rho_0}$. Results are also computed for the case where the half-space is regarded as rigid, i.e. the layer is fixed at its base at depth $3r_0$.

The maximum vertical distance Δe between two neighbouring disks follows for the torsional motion from Eq. 5.1b with $a_0 = 3$ as $(\pi/15)r_0$, and for the vertical motion with $c = c_p = 2c_s$ from Eq. 5.1a as $(2\pi/15)r_0$. For the torsional and vertical degrees of freedom at least 6 and 4 disks respectively are required, but for simplicity of data preparation a single model with 8 layers is employed.

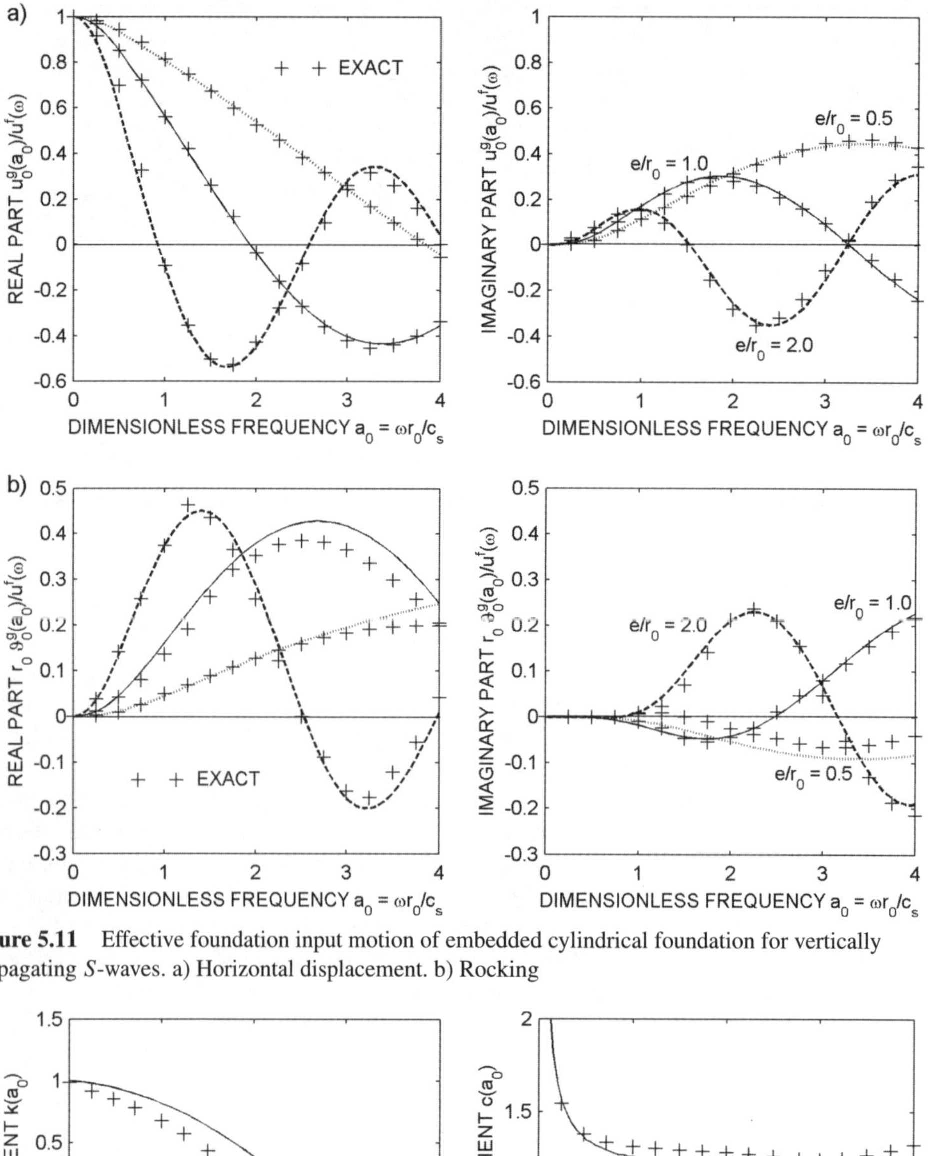

Figure 5.11 Effective foundation input motion of embedded cylindrical foundation for vertically propagating S-waves. a) Horizontal displacement. b) Rocking

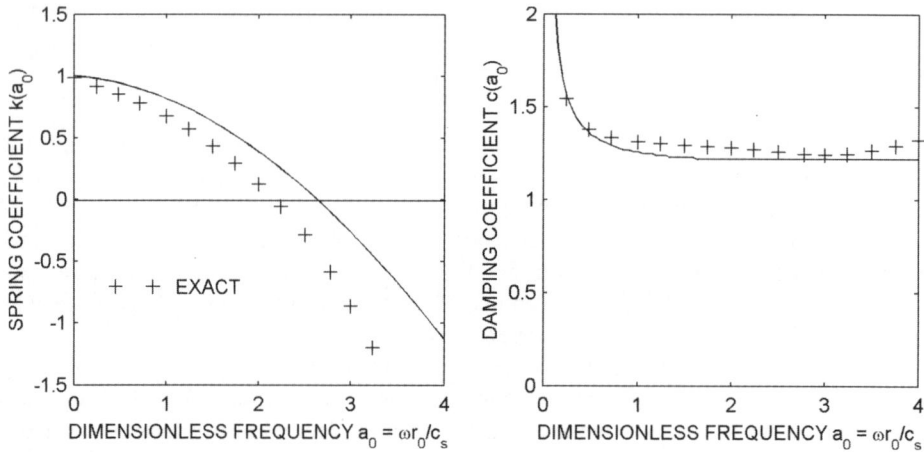

Figure 5.12 Dynamic-stiffness coefficient of cylindrical foundation embedded in homogeneous half-space in vertical motion ($\nu = 0.5$).

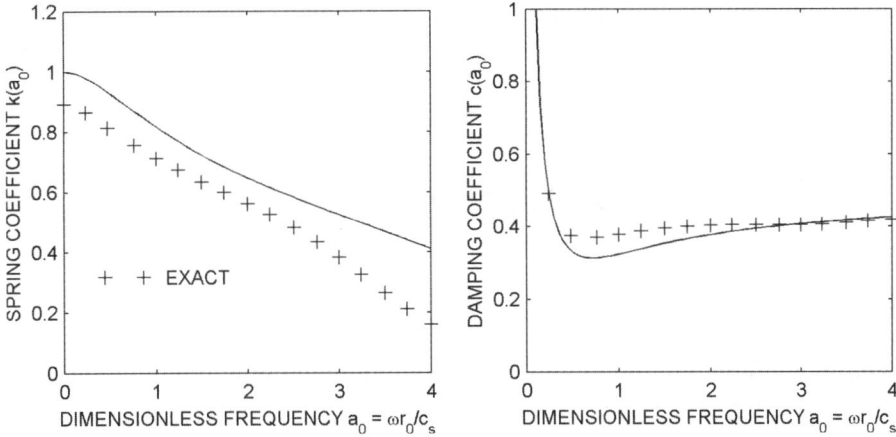

Figure 5.13 Dynamic-stiffness coefficient of cylindrical foundation embedded in homogeneous half-space in rocking motion ($\nu = 0.5$)

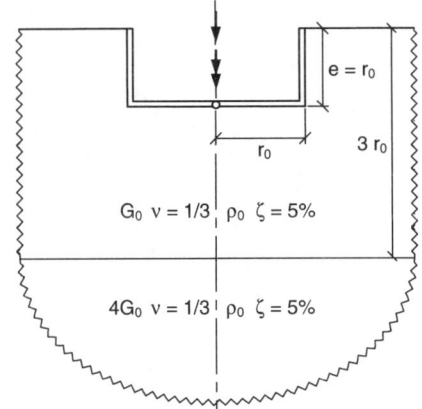

Figure 5.14 Cylindrical foundation embedded in layer overlying flexible half-space

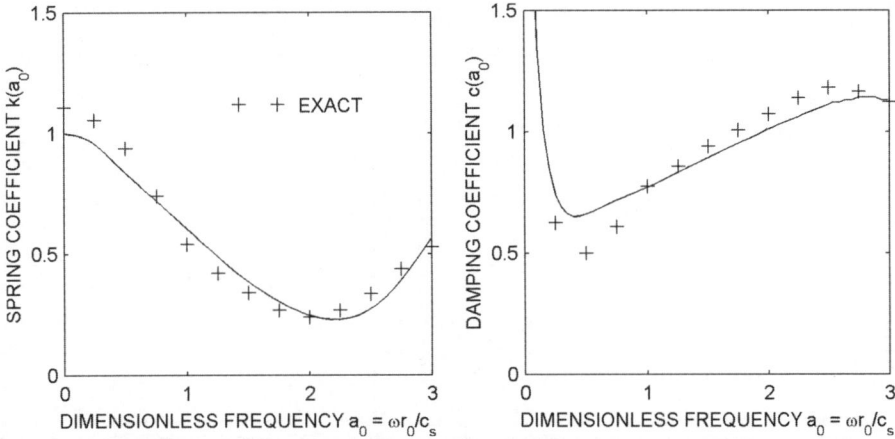

Figure 5.15 Dynamic-stiffness coefficient of cylindrical foundation in layer overlying flexible half-space in vertical motion

104 Foundation Vibration Analysis: A Strength-of-Materials Approach

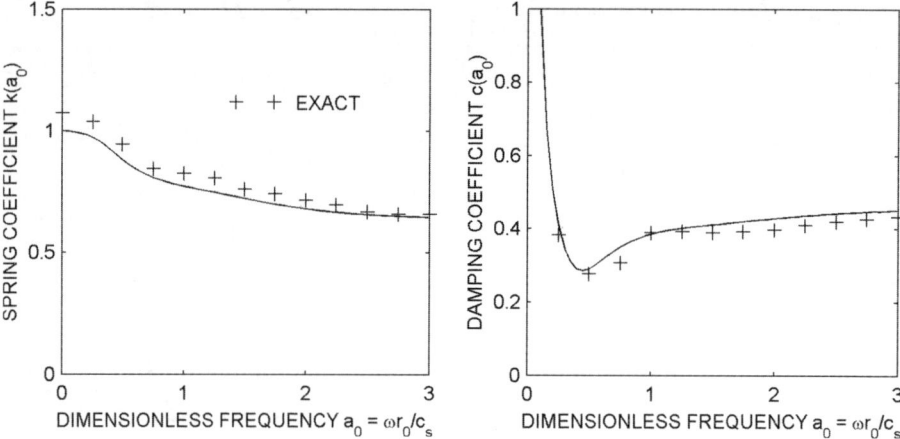

Figure 5.16 Dynamic-stiffness coefficient of cylindrical foundation in layer overlying flexible half-space in torsional motion

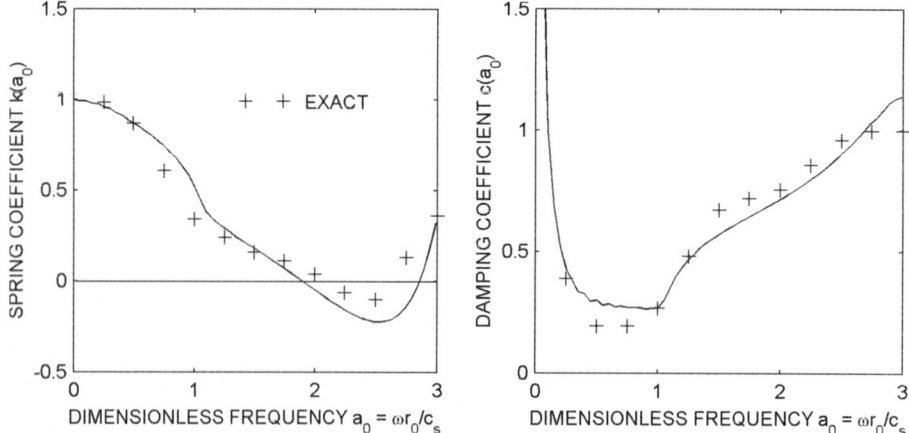

Figure 5.17 Dynamic-stiffness coefficient of cylindrical foundation in layer overlying rigid half-space in vertical motion

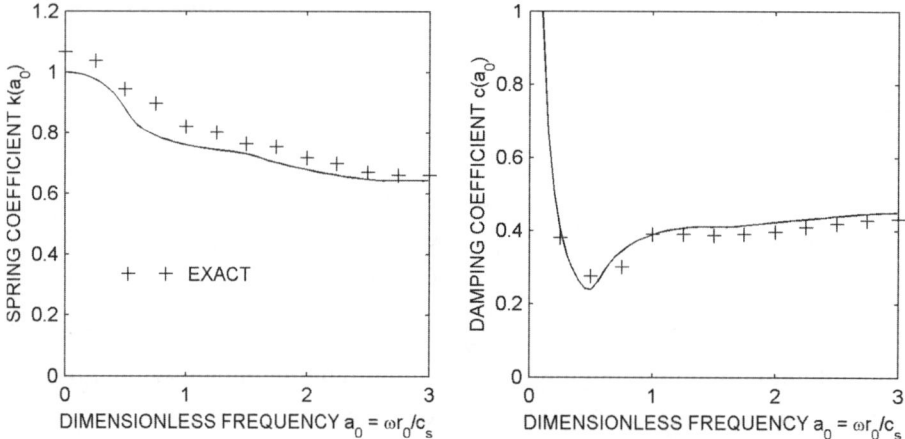

Figure 5.18 Dynamic-stiffness coefficient of cylindrical foundation in layer overlying rigid half-space in torsional motion

For comparison, a very fine boundary-element solution, denoted as exact, is available in Ref. [9].

The dynamic-stiffness coefficients follow from Eqs 5.14 and 5.18. In the decomposition of Eq. 1.5 K represents the static-stiffness coefficient computed by the cone model.

As can be seen from the comparison of the cone results with the exact solution for the dynamic system with a flexible underlying half-space (Figs 5.15 and 5.16), good agreement is achieved for both degrees of freedom. The same applies for the dynamic system with a fixed base (Figs 5.17 and 5.18).

6

Evaluation of accuracy

In the preceding chapters, the accuracy of academic examples only is evaluated. A homogeneous half-space and a homogeneous layer fixed at its base are addressed, both for surface and embedded cylindrical foundations. In addition, mainly to demonstrate the wave propagation in cone segments, a disk embedded in a full-space and a half-space is examined. To discuss the termination criteria, a disk on the surface of a layered half-space is considered, but only for one degree of freedom.

To gain confidence, a systematic evaluation of the accuracy for multiple-layered half-spaces is essential. The underlying half-space can either be flexible or rigid, in the latter case preventing wave propagation in the vertical direction towards infinity and thus radiation damping vertically from occurring. Besides the standard case where the underlying half-space is stiffer than the layers, the opposite situation must also be addressed. Compressible and incompressible half-spaces have to be discussed. Surface and embedded foundations must be examined. All degrees of freedom are to be investigated. A sufficiently large frequency range has to be considered. In particular, the behaviour in the low frequency range must be studied.

Besides evaluating the accuracy of cylindrical foundations, general axi-symmetric situations are important. In the latter case, the radii of the embedded disks will vary. For instance, hemi-spherical foundations are important. Fully-embedded foundations, such as embedded spheres, are also to be addressed.

The evaluation of the accuracy of the strength-of-materials approach using cones is the main goal of this chapter. In addition, certain limitations of the procedure are established, concerning the applicable range of the embedment ratio, and also the shape of the axi-symmetric embedded foundation.

Section 6.1 addresses surface foundations on a multi-layered half-space and Section 6.2 embedded cylindrical foundations. Section 6.3 examines modelling aspects of a site with gradually varying material properties where a large number of cone segments are required. Section 6.4 discusses the adequate representation of the dynamic behaviour below and above the so-called cutoff frequency, where an abrupt change in response occurs. Section 6.5 addresses a cylindrical foundation embedded in an incompressible multi-layered half-space. In Section 6.6, by varying the ratio of the radii of a hemi-ellipsoid embedded in a homogeneous half-space, the accuracy for axi-symmetric foundations modelled with disks of varying radii is studied. Finally, Section 6.7 examines a sphere embedded in a homogeneous full-space.

Sections 6.1 to 6.4 are based on Ref. [42].

Evaluation of accuracy 107

6.1 Foundation on the surface of a layered half-space

To evaluate the accuracy of the strength-of-materials approach using cones, a vast parametric study for a circular foundation of radius r_0 is performed (Figs 6.1a and 6.1b). Two rather extreme sites are investigated with the geometry and characteristics specified in the figure. The site in Fig. 6.1a consists of two layers overlying a flexible half-space, where waves will propagate vertically towards infinity, yielding radiation damping in the vertical direction, as for a site of a homogeneous half-space. Note that the shear modulus decreases with depth, which could correspond to a typical landfill situation. For the site in Fig. 6.1b where a fixed base exists, leading to total reflection of all waves, no radiation damping in the vertical direction occurs. Waves will still propagate towards infinity, but in the horizontal direction only, which, however, as will be discussed in Section 6.4, does not occur for low frequencies for vanishing material damping. The dynamic-stiffness coefficients are calculated up to $a_0 = 2\pi$.

The depth termination criterion described in Section 4.6 is used, with a termination depth of $20 + 2L$ (where L is the number of distinct soil layers) and a termination magnitude of 0.0001.

To be able to evaluate the accuracy, results obtained using the thin-layer method with a very fine discretisation with depth (Ref. [12]), denoted as exact, are considered.

The results of the strength-of-materials approach using cones are plotted in Figs 6.2 and 6.3 for the sites described in Figs 6.1a and 6.1b. The horizontal, vertical, rocking and torsional degrees of freedom are addressed. For each dynamic-stiffness coefficient $S(a_0)$ the standard decomposition and normalisation is performed, yielding

$$S(a_0) = K[k(a_0) + i\, a_0\, c(a_0)] \tag{6.1}$$

with the dimensionless frequency $a_0 = \omega r_0/c_{s1}$ referred to the shear-wave velocity of the first layer, $c_{s1} = \sqrt{G_0/\rho_0}$. For K, the static-stiffness coefficient of a disk on the surface of a homogeneous half-space with the material properties of the first layer, specified in Eqs 3.30, 3.27, 3.61 and

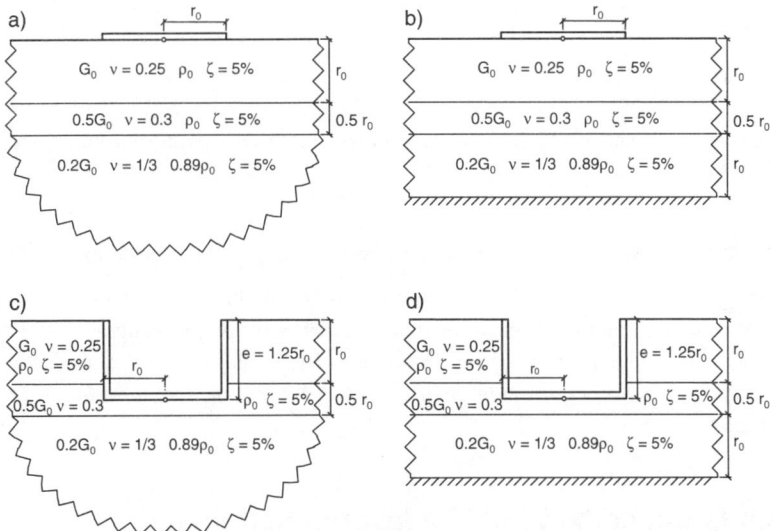

Figure 6.1 Multiple layered sites. a) Disk on two layers overlying flexible half-space. b) Disk on three layers fixed at base. c) Cylinder embedded in site consisting of two layers overlying flexible half-space. d) Cylinder embedded in site consisting of three layers fixed at base

108 Foundation Vibration Analysis: A Strength-of-Materials Approach

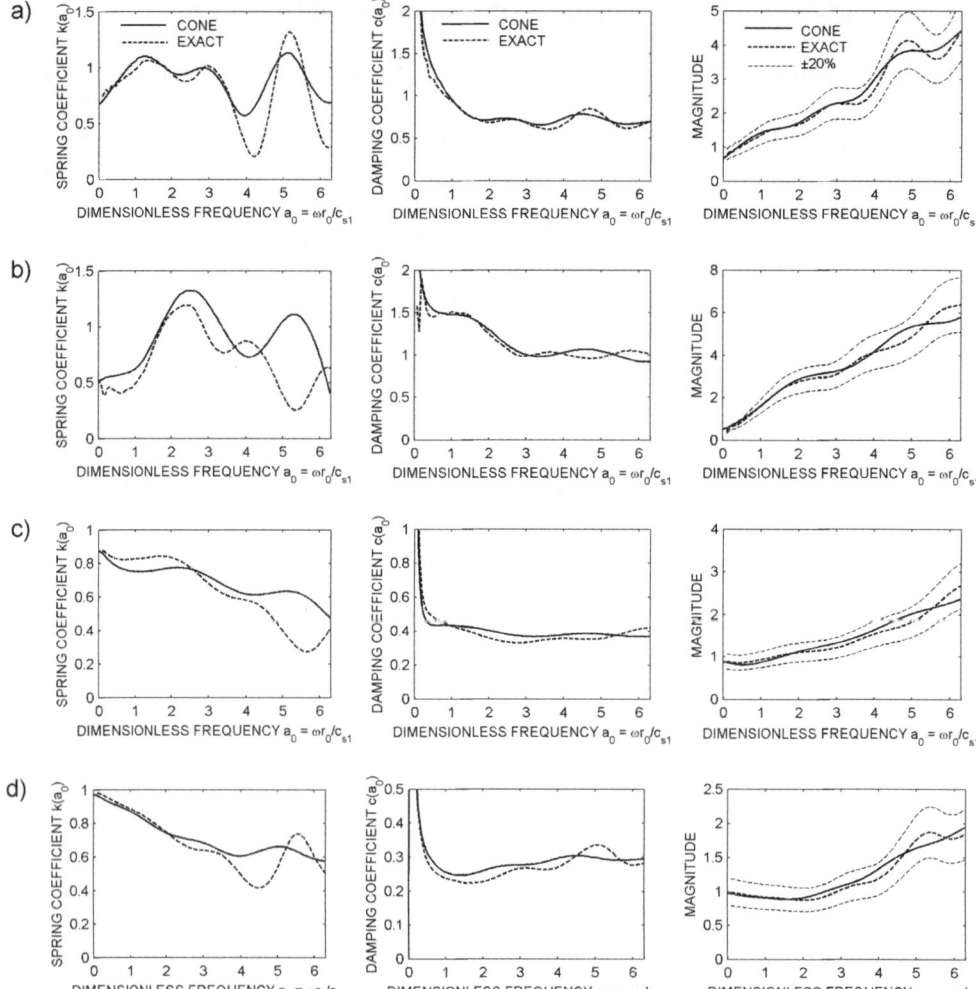

Figure 6.2 Dynamic-stiffness coefficients of disk on two layers overlying flexible half-space.
a) Horizontal. b) Vertical. c) Rocking. d) Torsional

3.54, is used. Besides the dimensionless spring coefficient $k(a_0)$ and damping coefficient $c(a_0)$, the magnitude $\sqrt{k^2(a_0) + a_0^2 c^2(a_0)}$ is also addressed. In the higher frequency range significant deviations exist for $k(a_0)$ and the same appears in the lower frequency range ($a_0 < 1$) for certain $c(a_0)$. However, these deviations strongly diminish in the magnitude. In general, the deviations are smaller than the *typical engineering accuracy of* ±20%, as is visible in the third column of the plots in Figs 6.2 and 6.3.

6.2 Foundation embedded in a layered half-space

For the same two sites addressed in Section 6.1 for a surface foundation, a cylindrical foundation of radius r_0 embedded with a depth e is examined (Figs 6.1c and 6.1d).

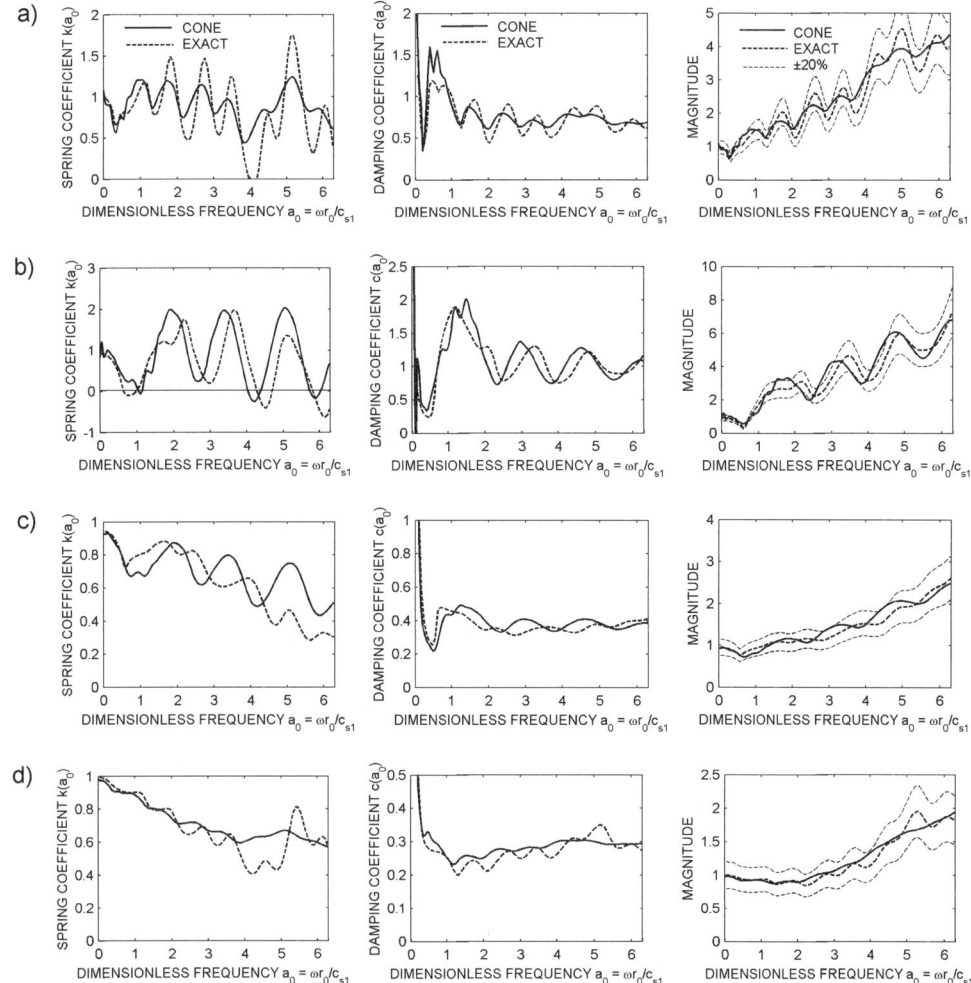

Figure 6.3 Dynamic-stiffness coefficients of disk on three layers fixed at base. a) Horizontal. b) Vertical. c) Rocking. d) Torsional

The number of disks in the stack in that part of the half-space which will be excavated is determined first. Applying Eq. 5.1b with $a_0 = 2\pi$, $\Delta e = r_0/10$, yielding 10 slices in the first layer. The shear-wave velocity in the second layer is lower, and Eq. 5.1a yields $\Delta e = r_0/(10\sqrt{2})$ and 3.5 slices in the second layer (rounded up to 4), which results in $10 + 4 = 14$ slices and 15 disks in total.

The depth criterion described in Section 4.6 is also used in this case, again with a termination depth of $20 + 2L$ and a termination magnitude of 0.0001.

For the comparison, the results of Ref. [12] are used once more. For the non-dimensionalisation (Eq. 6.1), K still represents the static-stiffness coefficient of a disk on the surface of a homogeneous half-space with the material properties of the first layer, with $K = 8G_0 r_0^2/(2-\nu)$ used for the coupling term between the horizontal and rocking degrees of freedom.

The accuracy of the dynamic-stiffness coefficients (vertical, horizontal, rocking, coupling and torsional degrees of freedom) for the embedded foundation is, for both sites (Figs 6.4 and 6.5),

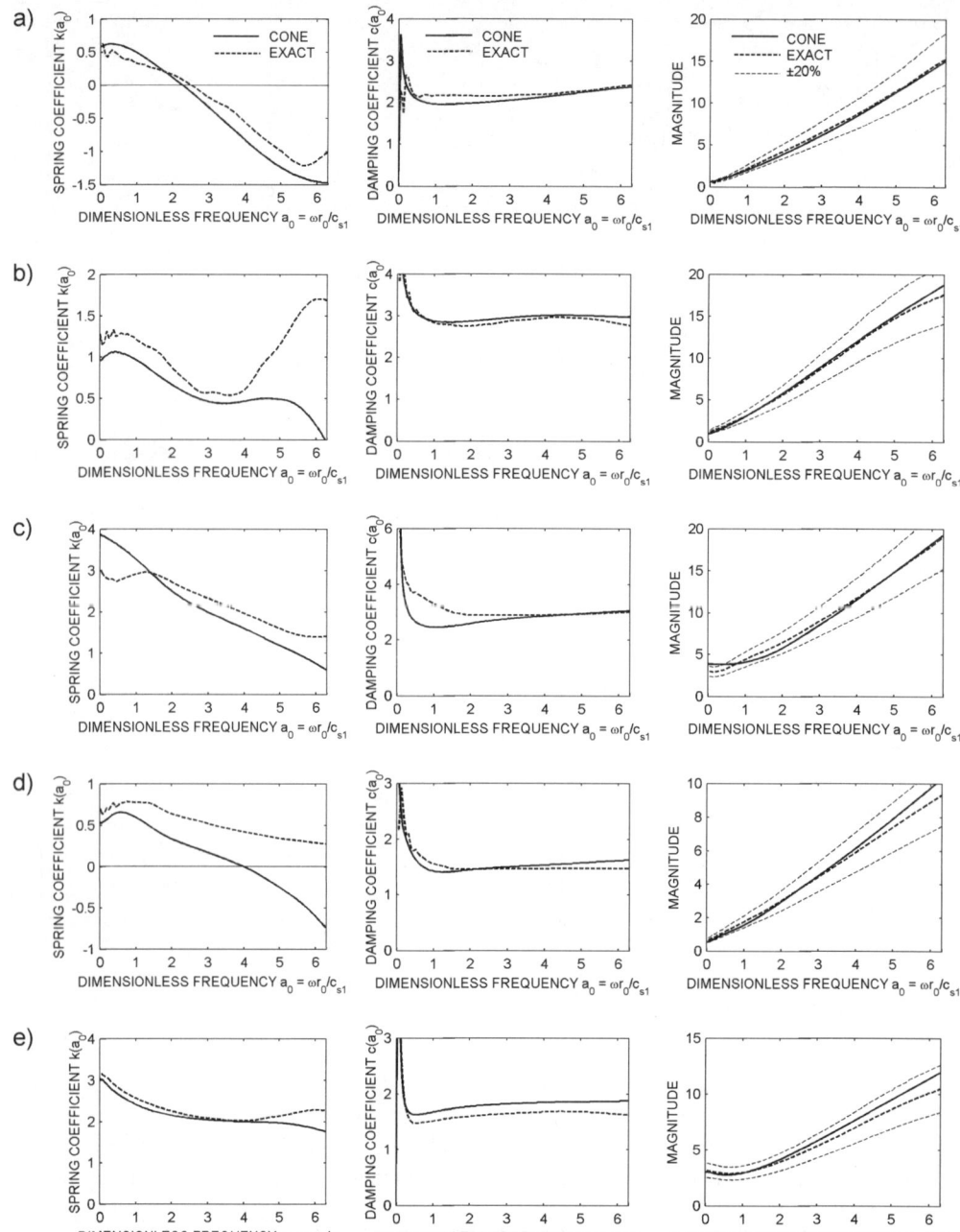

Figure 6.4 Dynamic-stiffness coefficients of cylinder embedded in site consisting of two layers overlying flexible half-space. a) Vertical. b) Horizontal. c) Rocking. d) Coupling. e) Torsional

Evaluation of accuracy 111

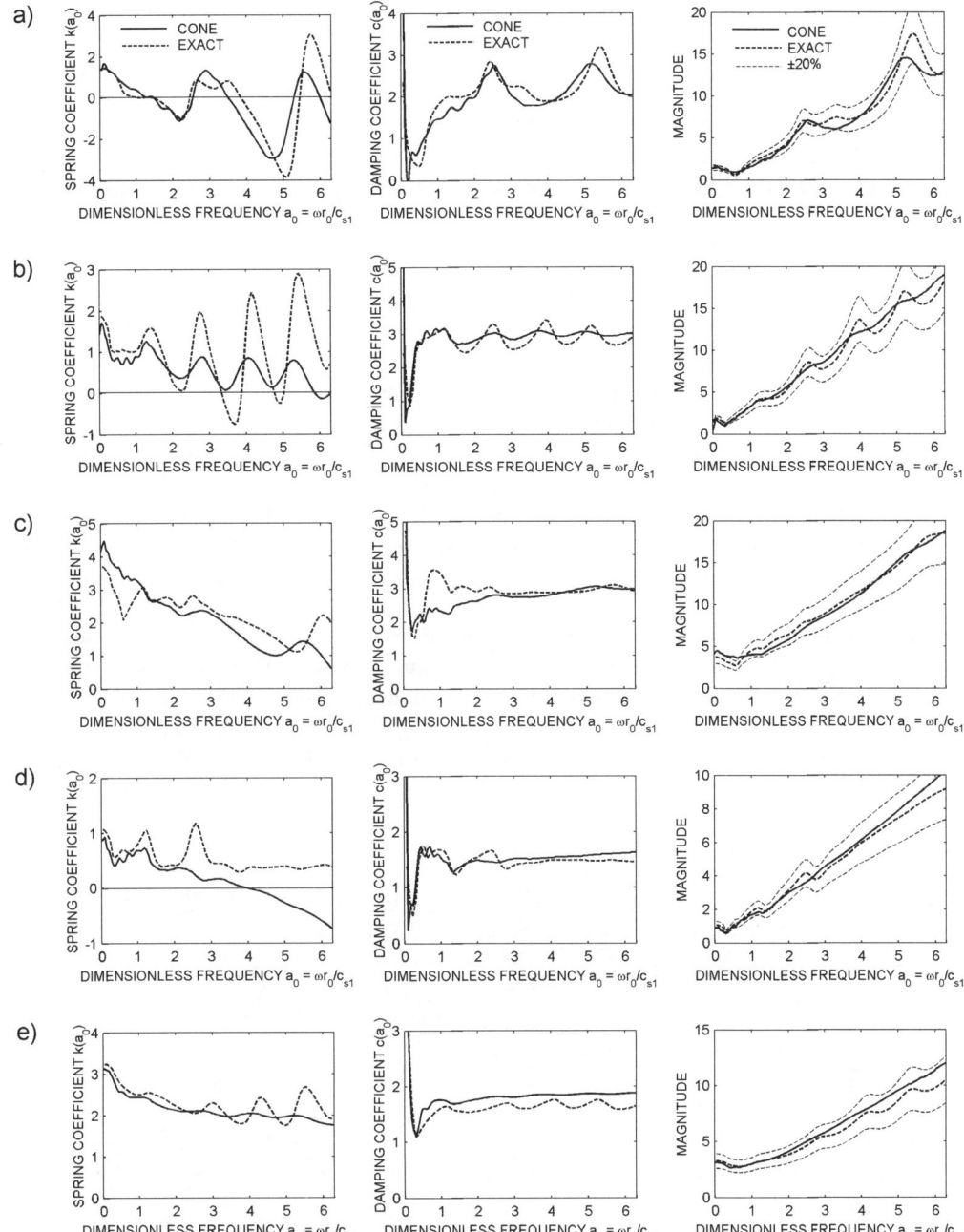

Figure 6.5 Dynamic-stiffness coefficients of cylinder embedded in site consisting of three layers fixed at base. a) Vertical. b) Horizontal. c) Rocking. d) Coupling. e) Torsional

112 Foundation Vibration Analysis: A Strength-of-Materials Approach

similar to that for the surface foundation. Engineering accuracy is thus achieved for all degrees of freedom throughout the frequency range.

6.3 Large number of cone segments

To demonstrate that a large number of layers can be processed using the strength-of-materials approach with cones, a surface disk of radius r_0 on the inhomogeneous half-space of Fig. 6.6 is investigated. The shear modulus increases with depth z as

$$G(z) = G_0 \left(1 + \frac{z}{r_0}\right) \tag{6.2}$$

with the shear modulus at the surface G_0. The mass density ρ is constant. Poisson's ratio equals $\nu = 1/3$. No hysteretic damping is present. The selected dynamic system consists of 20 layers with increasing G and thickness $d = r_0/4$ extending to a depth $5r_0$, where a homogeneous half-space with the $G(=6G_0)$ corresponding to this depth is placed. The dynamic-stiffness coefficient for the horizontal degree of freedom of a disk on this inhomogeneous half-space is addressed.

The termination criterion for a half-space with many layers is discussed in Section 4.6. In this example the depth criterion is used, again with a termination depth of $20 + 2L$ and a termination magnitude of 0.0001.

For the non-dimensionalisation of the dynamic-stiffness coefficient K in Eq. 6.1 is selected as the corresponding static-stiffness coefficient of a homogeneous half-space with the material properties at the free surface ($K = 8G_0 r_0/(2-\nu)$). The dimensionless frequency a_0 is also defined with the shear-wave velocity at the suface $c_{s0} = \sqrt{G_0/\rho}$. The dynamic-stiffness coefficient plotted in Fig. 6.7 agrees well with the result of the thin-layer method with a very fine discretisation in the vertical direction, denoted as exact (Ref. [36]). In particular, the results for low frequencies

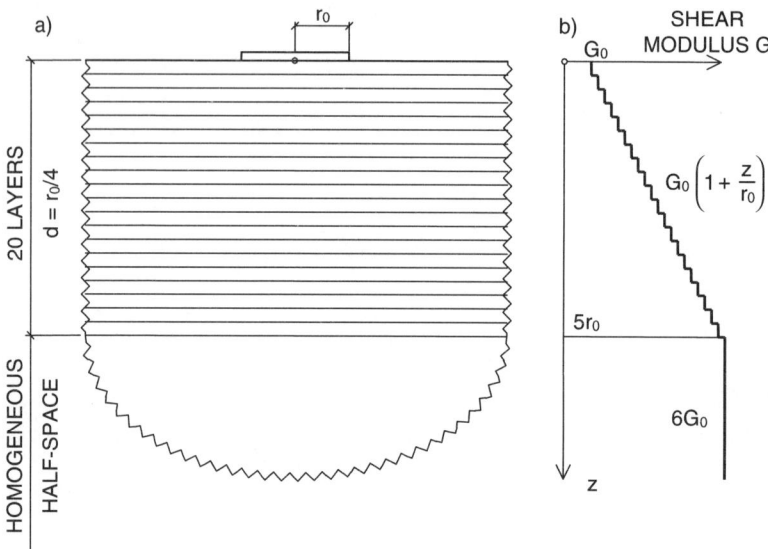

Figure 6.6 Disk on inhomogenous half-space. a) Discretisation with 20 layers resting on homogeneous half-space. b) Shear modulus linearly increasing with depth

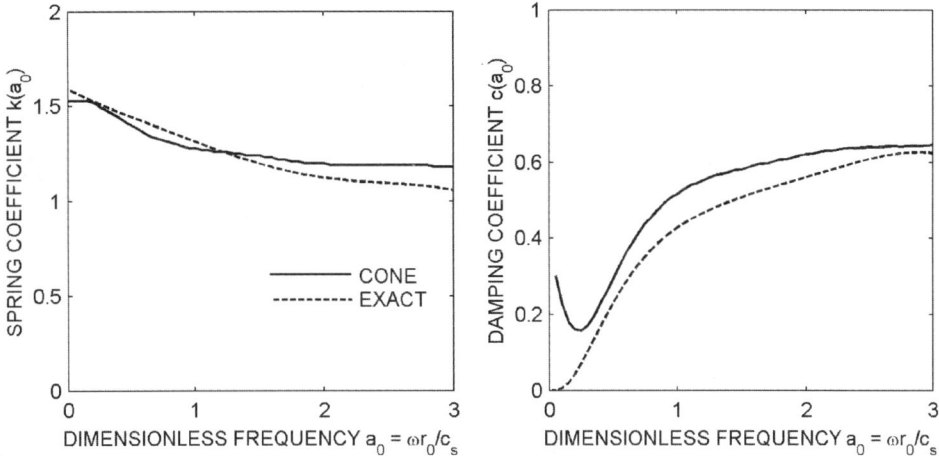

Figure 6.7 Dynamic-stiffness coefficient of disk on inhomogeneous half-space for horizontal degree of freedom

are as accurate as those for higher values. The termination criterion used is adequate in this case. However, if the number of slices were increased significantly, reducing the reflection coefficients (in magnitude) between each adjacent cone segment, a reduction of the termination magnitude would become necessary.

6.4 Cutoff frequency

For a site without material damping and fixed at its base, a so-called *cutoff frequency* exists, which separates the two completely different dynamic behaviours below and above this value. Below the cutoff frequency, the damping coefficient $c(a_0)$ representing radiation damping vanishes. It can be shown that the cutoff frequency is equal to the fundamental frequency of the soil column fixed at its base activating the corresponding deformations: for the horizontal and torsional degrees of freedom in shear and for the vertical and rocking degrees of freedom in dilatation (axial). The phenomenon of the cutoff frequency is also discussed at the end of Section 4.3 (Figs 4.4 and 4.5).

The ability of the cones to capture this effect for a layered half-space is investigated. The rocking and torsional degrees of freedom of a disk on three layers fixed at the base (Fig. 6.1b, with $\zeta = 0$) and the vertical and torsional degrees of freedom of a cylinder embedded in a site consisting of three layers fixed at the base (Fig. 6.1d, with $\zeta = 0$) are addressed in Figs 6.8 and 6.9. K in Eq. 6.1 is selected as described in Section 6.1. Again, the results of the thin-layer method (Ref. [12]) are regarded as exact. The agreement is good. In particular, below the cutoff frequency, the damping coefficients $c(a_0)$ calculated based on cones vanish, in general, from a practical point of view. The cutoff frequency for the torsional motions is also accurately predicted. For the torsional degree of freedom, the analytical value (horizontal fundamental frequency of a soil column deforming in shear) is $f = 0.049 c_{s1}/r_0$ which corresponds to $a_0 = 0.31$ (Figs 6.8b and 6.9b). The same applies for the rocking degree of freedom (Fig. 6.8a) governed predominantly by the dilatational-wave velocity c_p. A certain discrepancy exists for the vertical degree of freedom for the embedded cylinder also governed by c_p (Fig. 6.9a).

Thus, the strength-of-materials approach using cones can, through the superposition of the waves that are frequency dependent but governed by the same expressions for all frequencies,

114 Foundation Vibration Analysis: A Strength-of-Materials Approach

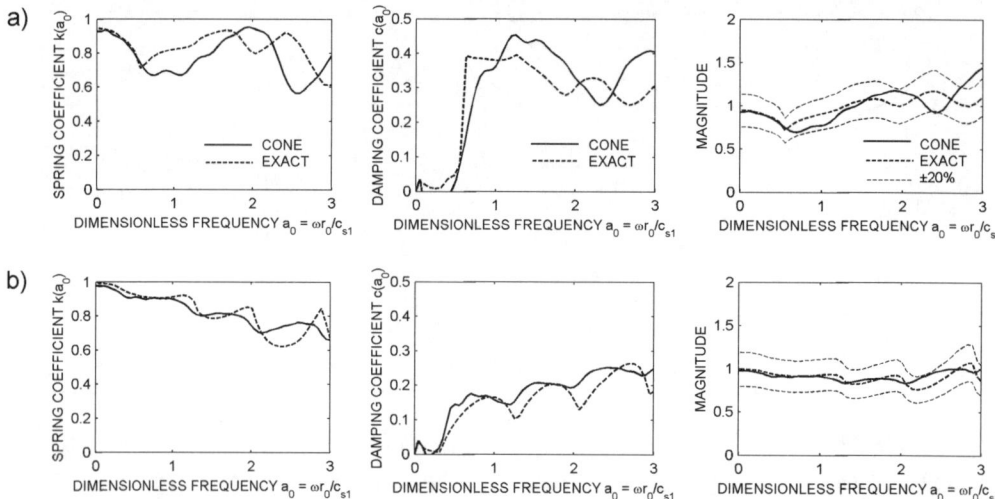

Figure 6.8 Dynamic-stiffness coefficients of disk on three layers fixed at base, no hysteretic damping. a) Rocking. b) Torsional

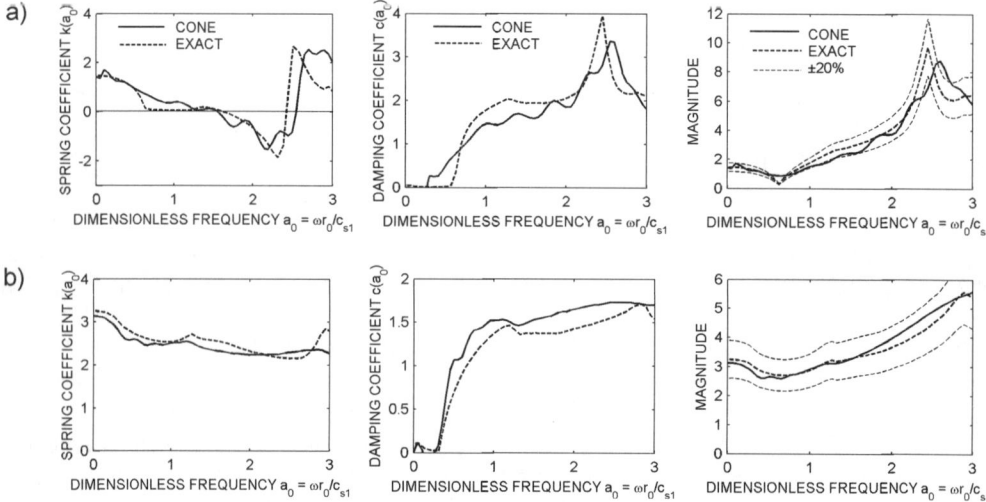

Figure 6.9 Dynamic-stiffness coefficients of cylinder embedded in site consisting of three layers fixed at base, no hysteretic damping. a) Vertical. b) Torsional

accurately predict the behaviour below the cutoff frequency, where the imaginary part of the resulting sum in the expression for the dynamic-stiffness coefficient vanishes approximately. This represents a very stringent test, passed successfully. The level of confidence is thus increased!

6.5 Incompressible case

The performance of the strength-of-materials approach using cones for embedded foundations in incompressible soil is now assessed. A cylinder embedded in three layers overlying a rigid base

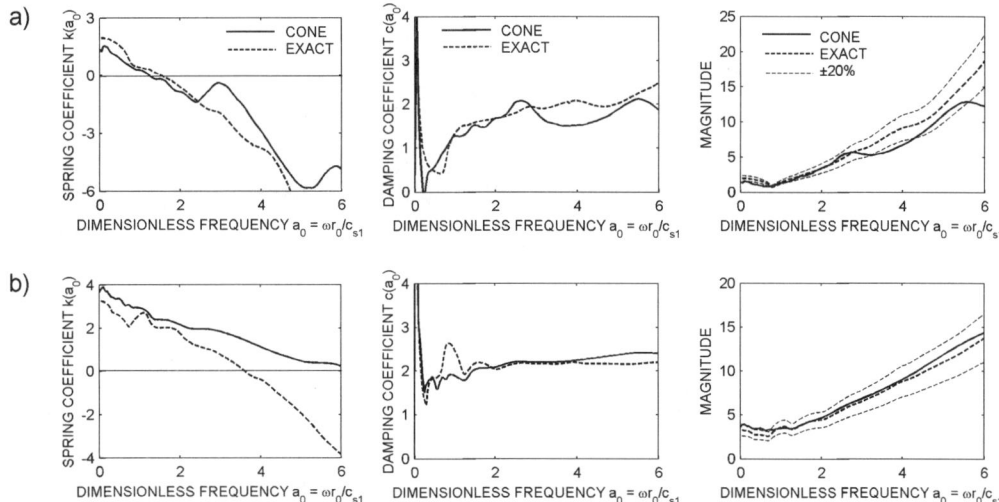

Figure 6.10 Dynamic-stiffness coefficients of cylinder embedded in site consisting of three incompressible layers fixed at base. a) Vertical. b) Rocking

(Fig. 6.1d) is analysed again, this time with the Poisson's ratio of each layer specified as 1/2. All other material properties and dimensions presented in Fig. 6.1d are retained. The accuracy of the computed dynamic-stiffness coefficients is determined by comparison with results obtained from Ref. [31]. These results were calculated rigorously using the thin-layer method, and will be denoted as exact. The vertical and rocking degrees of freedom are considered, since these are most severely affected by incompressibility. In each case the decomposition of Eq. 6.1 is applied, with the static-stiffness coefficient K of a disk of radius r_0 on the surface of a homogeneous half-space with the material properties of the first layer determined for the corresponding degree of freedom.

Figure 6.10a plots the dynamic-stiffness coefficient for vertical motion. Performance is similar to that achieved for compressible material (Fig. 6.5a), and, in the case of the magnitude, engineering accuracy is obtained over most of the frequency range.

Figure 6.10b plots the dynamic-stiffness coefficient for rocking motion. Although some discrepancy is apparent for the spring coefficient, particularly at high dimensionless frequencies, over the entire frequency range the magnitude of the dynamic-stiffness coefficient retains engineering accuracy.

6.6 Hemi-ellipsoid embedded in a homogeneous half-space

All the preceding examples involve disks or cylinders. When modelled using the technique described in Chapter 5, all the disks discretising such foundations have the same radius. However, the strength-of-materials approach using cones is sufficiently general to allow variation in the radii of the disks, permitting axi-symmetric foundations to be analysed. This section computes dynamic-stiffness coefficients for a range of hemi-ellipsoids embedded with depth e in a homogeneous half-space, as illustrated in Fig. 6.11, for torsional motion. Corresponding exact solutions are available in Ref. [1].

Some caution must be exercised when generating a cone model for a foundation of varying radius. It is implicit in the strength-of-materials approach that the radii of the disks modelling

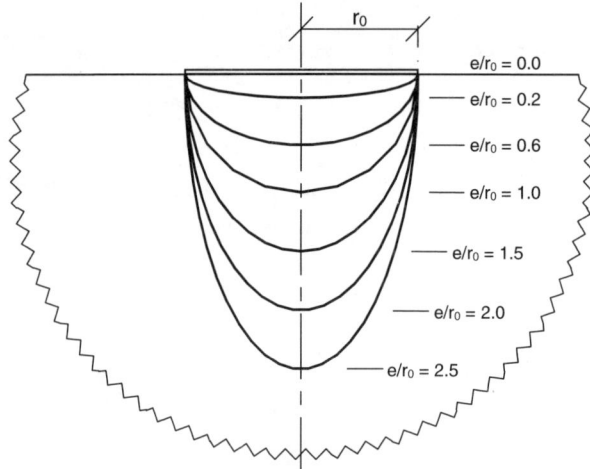

Figure 6.11 Hemi-ellipsoids embedded in homogeneous half-space

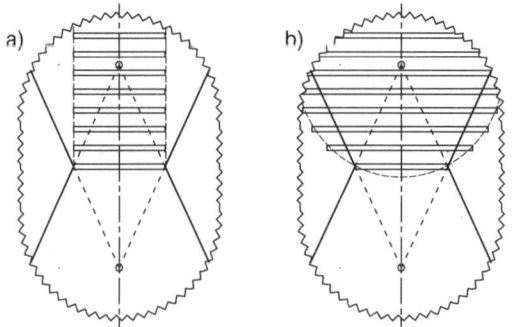

Figure 6.12 Cones of incident wave propagating from bottom-most disk. a) Disks of constant radius contained by upper cone. b) Disks of varying radius extending outside upper cone

the foundation are always less than the radius of the cone segment in which an incident wave propagates. When all disks have the same radius, this is guaranteed (Fig. 6.12a). However, if two adjacent disks are significantly different in radius but close together in the vertical direction, this assumption may be violated (Fig. 6.12b).

The radius of the hemi-ellipsoid (measured horizontally from the vertical axis of symmetry) varies continuously from the initial radius at the surface r_0 to zero at the depth of embedment e, which will be termed the toe. Near the toe the radius changes sharply with depth. To avoid generating a model that significantly violates the strength-of-materials assumptions, the following procedure is used. The minimum number of slices required in the vertical direction is determined in the usual way (Eq. 5.1). Here $\Delta e = 0.1 r_0$ is selected. The radius of the disk on each interface is determined by vertical projection of the intersection of the hemi-ellipsoid with the interface immediately *above* the current interface. This is illustrated on the left-hand side of Fig. 6.13 for a hemisphere. The resulting model boundary, shown on the right-hand side of Fig. 6.13 for a hemisphere, is slightly different from the original foundation near the toe. However, there is little variation at the top, and adequate results can be obtained.

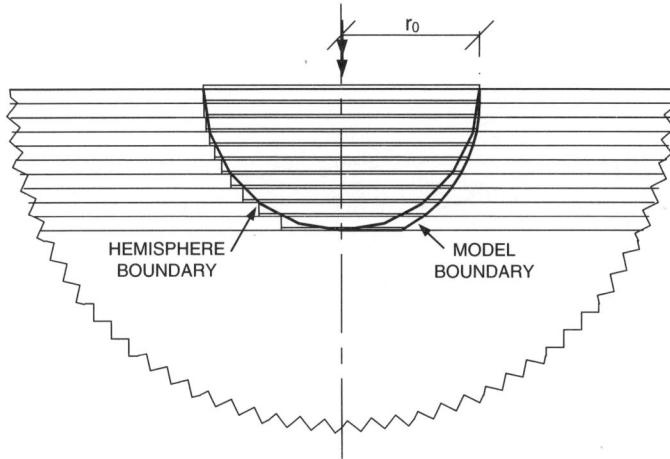

Figure 6.13 Modelling of embedded hemisphere with 11 disks (10 layers)

The dynamic-stiffness coefficients for the hemi-ellipsoids illustrated in Fig. 6.11 are plotted in Fig. 6.14, along with the exact solution of Ref. [1]. The decomposition of Eq. 6.1 is used, with the static-stiffness coefficient K taken as that of a hemisphere embedded in a half-space in torsional motion, $K = 4\pi G r_0^3$. Over the entire range of e/r_0 ratios calculated, the accuracy of the spring coefficient $k(a_0)$ is quite remarkable. The damping coefficient $c(a_0)$ is reasonably accurate, although showing a consistent error of between +15% and +20%. However, engineering accuracy is maintained across the frequency range, and the ratio of the damping coefficients between different hemi-ellipsoids in the sequence is predicted very well.

The overestimation of the damping coefficient is directly linked to the larger surface of the boundary of the model than that of the hemi-ellipsoid. In the high frequency limit $c(a_0)$ will be proportional to the surface area. Using 10 layers the model surface area for a hemisphere is 10% larger than $2\pi r_0^2$.

6.7 Sphere embedded in a homogeneous full-space

To evaluate the accuracy of the method using cones for a foundation of varying radius embedded in a homogeneous full-space, the case of a sphere is considered. Due to the point symmetry of the problem, only two types of motion exist, torsional and rectilinear. Exact solutions are available for each type of motion (Refs [4, 5]). The material properties of the full-space are selected as the shear modulus G, Poisson's ratio $\nu = 0.25$ and mass density ρ, with the radius of the sphere denoted as r_0.

The modelling procedure for the sphere is essentially identical to that described for hemi-ellipsoids in Section 6.6. A slice thickness of $\Delta e = 0.1 r_0$ is used, and the disk radii are calculated as illustrated in Fig. 6.15, resulting in a slightly different geometry from that of a true sphere.

Figure 6.16a plots the dynamic-stiffness coefficient calculated using the half-space calibration method (Table 3.1) for rectilinear motion. The decomposition of Eq. 6.1 is used with the exact static-stiffness coefficient for a sphere in a full-space in rectilinear motion, $K = 24\pi G r_0(1 - \nu)/(5 - 6\nu)$. Both the spring and damping coefficients are about 20% too small, causing the

118 Foundation Vibration Analysis: A Strength-of-Materials Approach

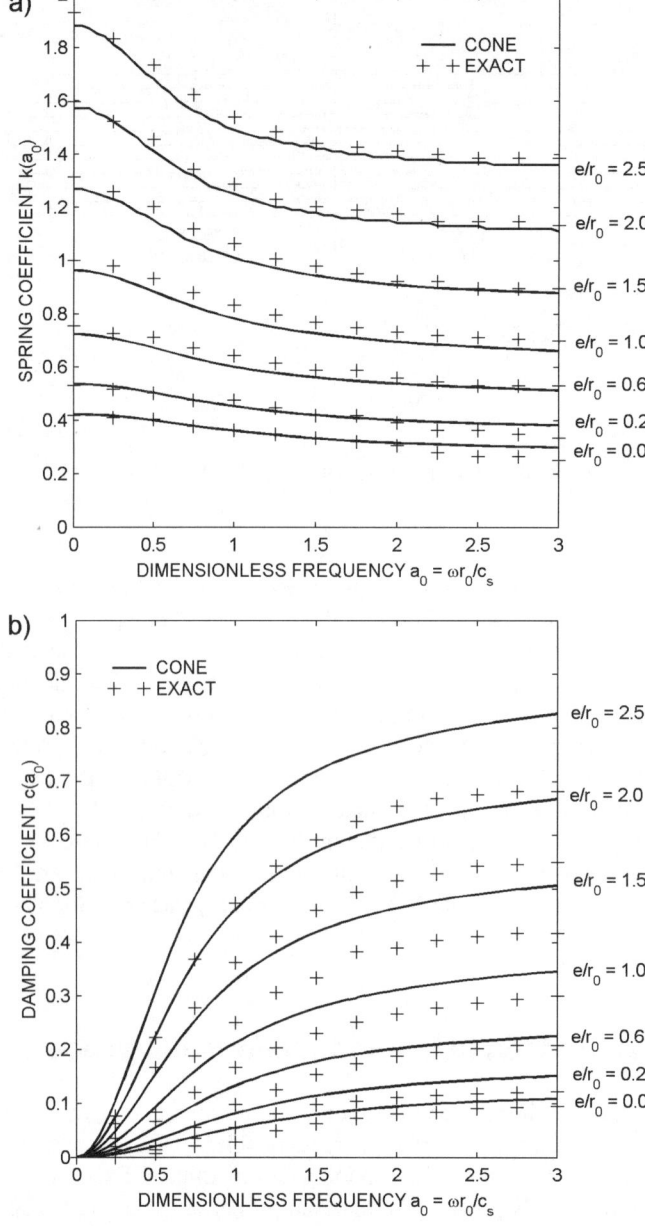

Figure 6.14 Dynamic-stiffness coefficients of hemi-ellipsoids embedded in homogeneous half-space for various radii ratios. a) Spring coefficient. b) Damping coefficient

magnitude to deviate by a similar amount. Although within engineering accuracy, these results are a little disappointing.

However, when the full-space calibration method is used (Section 3.6), the dynamic-stiffness coefficient plotted in Fig. 6.16b results. Here the performance is entirely satisfactory, illustrating the importance of the alternative calibration method in fully embedded problems.

Evaluation of accuracy 119

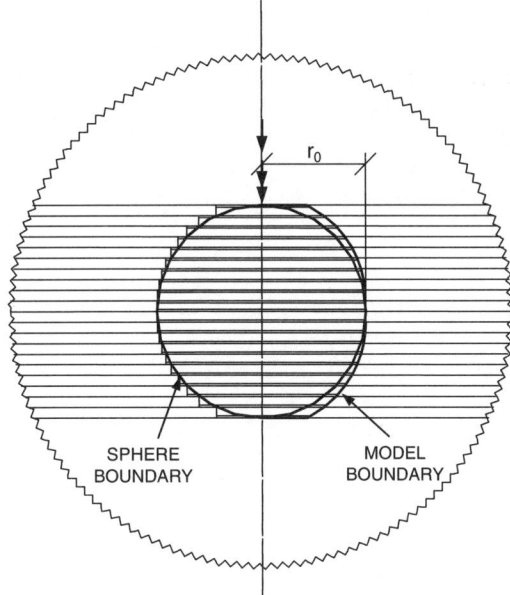

Figure 6.15 Modelling of sphere embedded in homogeneous full-space

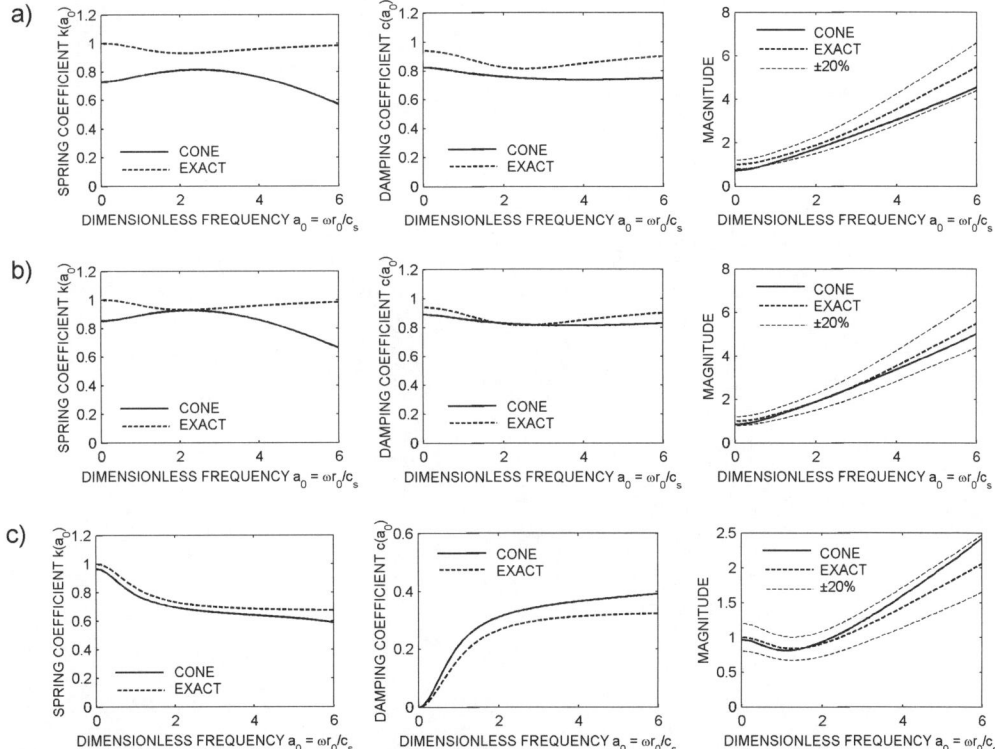

Figure 6.16 Dynamic-stiffness coefficients of sphere embedded in full-space. a) Rectilinear motion with half-space calibration. b) Rectilinear motion with full-space calibration. c) Torsional motion

For torsional motion both calibration methods yield identical models, and the dynamic-stiffness coefficient plotted in Fig. 6.16c is obtained. In the decomposition the exact static-stiffness coefficient for a sphere embedded in a full-space in torsional motion, $K = 8\pi\, G\, r_0^3$, is used. Excellent agreement with the exact solution is achieved, although the calculated damping is slightly too large. The same explanation as discussed in Section 6.6 applies.

7

Engineering applications

For the sake of illustration, three engineering applications are discussed. To be able to concentrate on the key aspects related to foundation vibration the problems are simplified, deleting all aspects that could distract from this goal. This permits a detailed discussion that the reader can follow step by step.

Section 7.1 analyses a machine foundation consisting of a block of concrete on the surface of a layered half-space excited by a reciprocating machine, a two-cylinder compressor with cranks at 90°. As will be shown, the periodic loading is equal to a harmonic vertical force, with the frequency of the machine, and two harmonic moments, the first with the frequency of the machine and the second with twice this frequency. Section 7.2 calculates a very simple case of seismic soil-structure interaction. The structure is modelled as a single degree of freedom system when fixed at its base. The structure's foundation is embedded in a layered half-space. The coupled structure-soil system is excited by a historic horizontal earthquake, a non-periodic (arbitrary) excitation. Section 7.3 analyses a wind turbine tower supported by a suction caisson foundation. Harmonic loading arising from the loss of lift on the blades as they pass the tower is addressed.

7.1 Machine foundation on the surface of a layered half-space

A block foundation made of concrete on the surface of a layered half-space supporting a reciprocating machine is addressed in this section (Fig. 7.1). As will be demonstrated, the two cylinders of the machine (separated by distance d) with cranks at 90° lead to a vertical force P and a moment M acting on the prism block foundation with dimensions $2a \times 2b \times h$. The site corresponds to the layered half-space addressed in Section 6.1 (Fig. 6.1a).

7.1.1 Dynamic load of single cylinder machine

First, a reciprocating machine consisting of a single cylinder is examined. The basic crank mechanism (Fig. 7.2) consists of a piston moving vertically within a guiding cylinder, a connecting rod of length l, fixed to the piston and to the crank, and a crank rod of radius r which rotates about the crankshaft with frequency ω. The mass of the connecting rod can be replaced by two lumped masses, one moving vertically together with the piston and the other rotating together with the

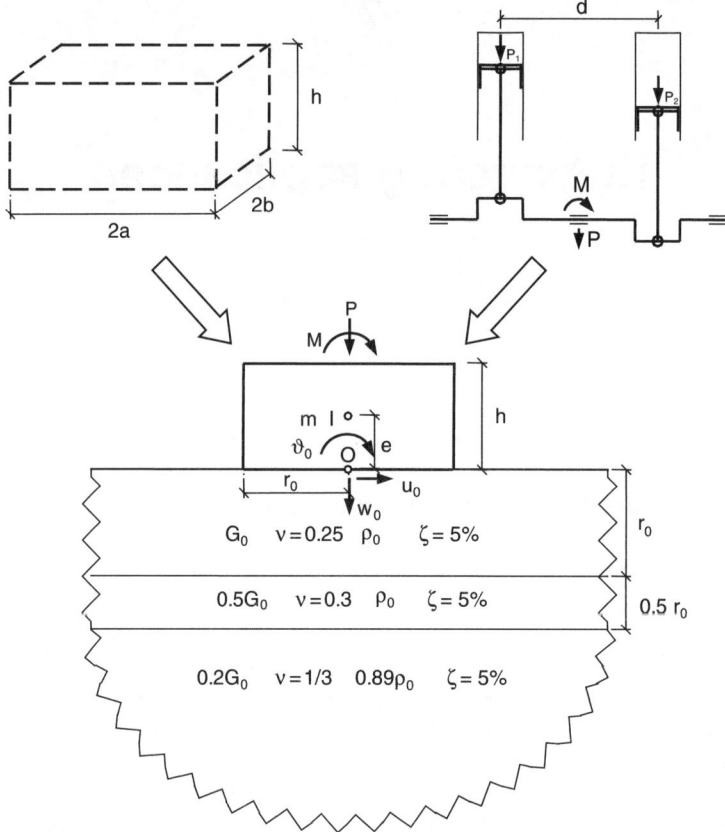

Figure 7.1 Block foundation for two-cylinder reciprocating machine on surface of multiple-layered site

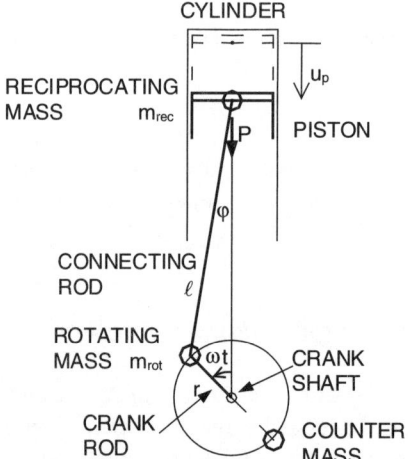

Figure 7.2 Motion of crank mechanism in single cylinder

crank rod. The total reciprocating mass m_{rec} and the total rotating mass m_{rot} follow by adding the contributions of the connecting rod to that of the piston and to that of the crank rod, respectively. The rotating mass will lead to a centrifugal load that is eliminated by installing a counter mass with the same m_{rot} at an angle of 180° (Fig. 7.2). Thus, only the vertical load P of the reciprocating mass m_{rec} remains.

The rotation of the crank with ω results in a vertical displacement $u_p(t)$ of the piston measured from its extreme outward position. The geometric relationships

$$u_p(t) + l \cos \varphi(t) + r \cos \omega t = l + r \tag{7.1a}$$

$$l \sin \varphi(t) = r \sin \omega t \tag{7.1b}$$

apply (Fig. 7.2). $\cos \varphi(t)$ follows from Eq. 7.1b as

$$\cos \varphi(t) = \sqrt{1 - \sin^2 \varphi(t)} = \sqrt{1 - \frac{r^2}{l^2} \sin^2 \omega t} \tag{7.2}$$

Since $r^2/l^2 \ll 1$, a series expansion with two terms is applied. Equation 7.2 yields

$$\cos \varphi(t) = 1 - \frac{r^2}{2l^2} \sin^2 \omega t \tag{7.3}$$

Substituting Eq. 7.3 into Eq. 7.1a results in

$$u_p(t) = r(1 - \cos \omega t) + \frac{r^2}{2l} \sin^2 \omega t$$

$$= r(1 - \cos \omega t) + \frac{r^2}{4l}(1 - \cos 2\omega t) \tag{7.4}$$

The corresponding acceleration follows from Eq. 7.4 as

$$\ddot{u}_p(t) = r\omega^2 \left(\cos \omega t + \frac{r}{l} \cos 2\omega t \right) \tag{7.5}$$

The inertial load acting vertically in the direction of the piston's motion

$$P(t) = m_{rec} \ddot{u}_p(t) \tag{7.6}$$

is

$$P(t) = m_{rec} r\omega^2 \cos \omega t + m_{rec} \frac{r^2}{l} \omega^2 \cos 2\omega t \tag{7.7}$$

The unbalanced dynamic load consists of a primary harmonic component with angular frequency ω and a secondary harmonic component with twice this frequency, 2ω.

124 Foundation Vibration Analysis: A Strength-of-Materials Approach

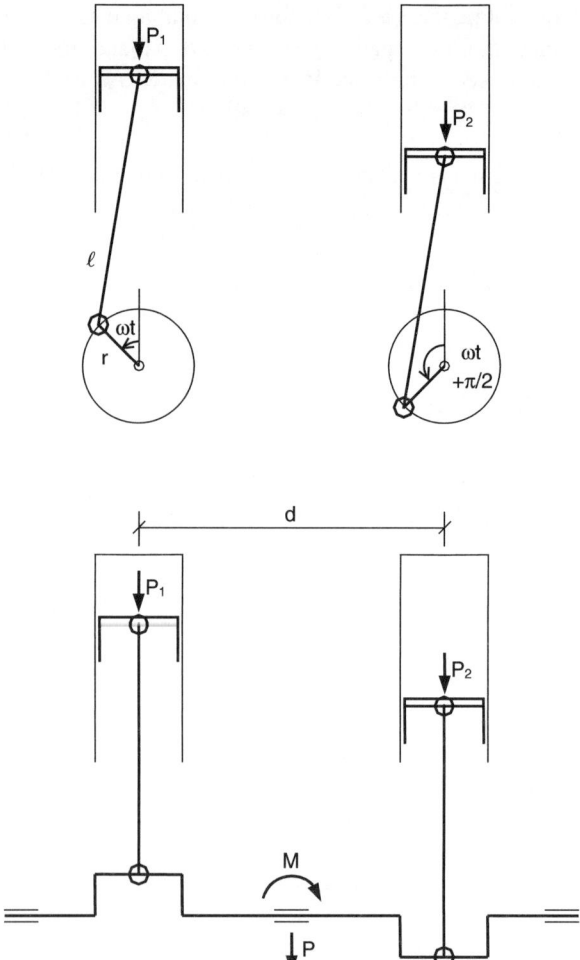

Figure 7.3 Two-cylinder reciprocating machine with cranks at 90°

7.1.2 Dynamic load of two-cylinder machine with cranks at 90°

Second, a reciprocating machine with two cylinders with pistons moving vertically mounted on a common horizontal crankshaft with a (relative) crank angle of 90° is addressed (Fig. 7.3). The crank angle defines the relative position of the pistons in the cylinders at any time. A relative crank angle of 90° implies that, for instance, when the piston in the first cylinder is in the extreme upward position, the piston in the second cylinder is half way towards the other extreme downward position.

The resulting unbalanced dynamic load is determined by superposition of the contributions of the two cylinders. The inertial load of the first cylinder with crank angle ωt is (Eq. 7.7)

$$P_1(t) = m_{rec} r \omega^2 \cos \omega t + m_{rec} \frac{r^2}{l} \omega^2 \cos 2\omega t \qquad (7.8)$$

and that of the second cylinder with crank angle $\omega t + \pi/2$

$$P_2(t) = m_{rec}\, r\omega^2 \cos(\omega t + \pi/2) + m_{rec}\frac{r^2}{l}\omega^2 \cos 2(\omega t + \pi/2) \tag{7.9}$$

$P_1(t)$ and $P_2(t)$ lead to the resultants (Fig. 7.3)

$$P(t) = P_1(t) + P_2(t) \tag{7.10a}$$

$$M(t) = -\frac{d}{2}P_1(t) + \frac{d}{2}P_2(t) \tag{7.10b}$$

with d the distance between the centrelines of the cylinders. Substituting Eqs 7.8 and 7.9 in Eq. 7.10 and using the well-known trigonometric identities for a cos-function with an argument consisting of a sum results in

$$\begin{aligned} P(t) &= m_{rec}\, r\omega^2 (\cos \omega t - \sin \omega t) \\ &= \sqrt{2}\, m_{rec}\, r\omega^2 \cos\left(\omega t + \frac{\pi}{4}\right) \end{aligned} \tag{7.11a}$$

$$\begin{aligned} M(t) &= -m_{rec}\, r\omega^2 d(\cos \omega t + \sin \omega t) - m_{rec}\frac{r^2}{l}\omega^2 d \cos 2\omega t \\ &= -\sqrt{2}\, m_{rec}\, r\omega^2 d \cos\left(\omega t - \frac{\pi}{4}\right) - m_{rec}\frac{r^2}{l}\omega^2 d \cos 2\omega t \end{aligned} \tag{7.11b}$$

The resultant dynamic load consists of a force with a primary component which is $\sqrt{2}$ as large as that of the single cylinder and a vanishing secondary component, and a moment with a primary component which equals $\sqrt{2}d$ times the primary component of the load of the single cylinder and a second component equal to d times the secondary component of the load of the single cylinder. Different phase angles arise. Thus, harmonic loads of two distinct frequencies $\omega_j (j = 1, 2)$, ω and 2ω, must be addressed.

In the example the mass of the piston is 12 kg, the mass of the connecting rod is 6 kg, the length of the connecting rod l is 0.5 m, the crank radius r is 0.2 m and the distance between the centrelines of the cylinders d is 1 m. The total reciprocating mass m_{rec} is therefore $12 + 0.5 \times 6 = 15\,\text{kg}$. The reciprocating machine including the driving motor, with a total mass of 4000 kg, operates at a frequency of 10 Hz ($\omega = 2\pi \times 10 = 62.83$ rad/s). The primary, secondary and total resulting dynamic loads consisting of the force $P(t)$ (Eq. 7.11a) and the moment $M(t)$ (Eq. 7.11b) are plotted in Fig. 7.4.

7.1.3 Dynamic system

Third, the dynamic system (Fig. 7.1) is discussed.

The block foundation, a prism, is modelled as a rigid body. It represents the structure in this dynamic soil-structure-interaction analysis. The inertial properties of the block foundation including the machine with respect to the centre of mass (gravity) are the mass m and the mass moment of inertia I. An equivalent radius r_0 of the contact area of the block foundation and the soil (structure-soil interface) can be determined.

The dynamic load consisting of the vertical force $P(t)$ and the moment $M(t)$ acts on the block foundation.

126 Foundation Vibration Analysis: A Strength-of-Materials Approach

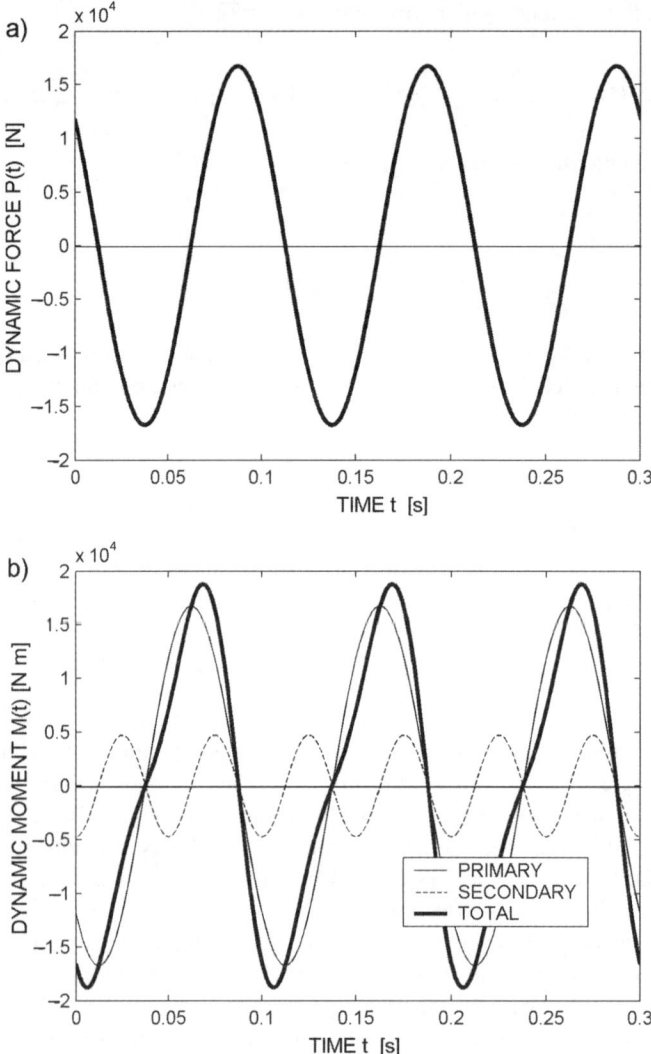

Figure 7.4 Dynamic load of two-cylinder reciprocating machine. a) Dynamic force. b) Dynamic moment

The dynamic properties of the soil, a layered half-space, are described by the dynamic-stiffness coefficients with respect to the centre O of the equivalent circular surface foundation. Besides the obvious vertical and rocking degrees of freedom $w_0(t)$ and $\vartheta_0(t)$, which dominate the response, the horizontal degree of freedom $u_0(t)$ is also excited, although to a lesser extent. This occurs as the centre of gravity of the block foundation is at a distance $e = 0.5h$ from O, measured in the vertical direction, yielding coupling of the rocking and horizontal degrees of freedom.

In the example, the prism with mass density 2.5×10^3 kg/m^3 has a length $2a$ of 3 m, a width $2b$ of 2.5 m and a height h of 1.5 m. Its mass m_f is therefore 28.125×10^3 kg. Adding the mass of the machine (4×10^3 kg) to this value yields $m = 32.125 \times 10^3$ kg. The mass moment of inertia of the block foundation with respect to the centre of gravity $I_f = m_f/12[(2a)^2 + h^2] = 26.367 \times 10^3$ kg m^2. Assuming the eccentricity of the machine on top of the block foundation is $h/2$, its contribution will be $0.75^2 \times 4 \times 10^3 = 2.25 \times 10^3$ kg m^2, leading to a total value $I = 28.617 \times 10^3$ kg m^2.

The equivalent radius r_0 is determined based on the area of the block foundation-soil interface $r_0 = \sqrt{4ab/\pi} = 1.545$ m. (Theoretically, a slightly different value could be determined for the rocking degree of freedom, $r_0 = \sqrt[4]{2b(2a)^3/(3\pi)} = 1.636$ m. However, the effect on the results is minimal and a single r_0 is adopted here for all degrees of freedom.)

The applied dynamic load is specified in Eq. 7.11. For the first frequency $\omega = 62.83$ rad/s (10 Hz) a harmonic vertical force $P(t) = \sqrt{2} m_{rec} r\omega^2 \cos(\omega t + \pi/4) = 1.675 \times 10^4 \cos(\omega t + \pi/4)$ N and a harmonic moment $-\sqrt{2} m_{rec} r\omega^2 d \cos(\omega t - \pi/4)$ act, and for the second frequency $2\omega = 125.66$ rad/s (20 Hz) the harmonic moment $-m_{rec}(r^2/l)\omega^2 d \cos 2\omega t$ is applied. These expressions correspond to the magnitude-phase angle description for harmonic motion with, for instance, for the vertical force, the magnitude $|P(\omega)| = 1.675 \times 10^4$ N and the phase angle $\varphi(\omega) = \pi/4$ (see Appendix A.1, Eq. A.22). The corresponding complex amplitude description is $P(\omega) = \text{Re } P(\omega) + i \text{Im} P(\omega)$ with (Eq. A.23) Re $P(\omega) = |P(\omega)| \cos \varphi(\omega) = 1.184 \times 10^4$ N and Im $P(\omega) = |P(\omega)| \sin \varphi(\omega) = 1.184 \times 10^4$ N.

The properties of the soil are described in Fig. 7.1 with the length $r_0 = 1.545$ m, shear modulus $G_0 = 112.5 \times 10^6$ N/m^2 and mass density $\rho_0 = 1800$ kg/m^3 of the first layer, resulting in a shear-wave velocity $c_{s1} = \sqrt{G_0/\rho_0} = 250$ m/s. The dynamic-stiffness coefficients for this site are calculated in Section 6.1. For the vertical, horizontal and rocking degrees of freedom the coefficients $S_v(a_0)$, $S_h(a_0)$ and $S_r(a_0)$ are plotted in Figs 6.2b, 6.2a and 6.2c with the decomposition specified in Eq. 6.1. They are calculated for the two dimensionless frequencies $\omega_j r_0/c_{s1}$ ($j = 1$: $\omega_1 = \omega$; $j = 2$: $\omega_2 = 2\omega$) with $\omega = 62.83$ rad/s of the harmonic load.

Figure 7.5 presents a physical interpretation of the dynamic system. Equation 6.1 permits the force-displacement relationship for each degree of freedom described by the corresponding dynamic-stiffness coefficient to be interpreted as a spring and dashpot in parallel with frequency-dependent coefficients, as discussed in Section 1.1 (Fig. 1.6). Such a dynamic model is thus established for each of the two frequencies ω_j of the dynamic load, which only affect the spring coefficients $k(\omega_j)$ and damping coefficients $c(\omega_j)$.

7.1.4 Equations of motion

The degrees of freedom are defined in the centre O of the block foundation-soil interface with amplitudes $w_0(\omega_j)$ ($j = 1$) in the vertical direction, $u_0(\omega_j)$ ($j = 1, 2$) in the horizontal direction and $\vartheta_0(\omega_j)$ ($j = 1, 2$) for rocking (Fig. 7.5). The amplitudes of the dynamic load are $P(\omega_j)$ ($j = 1$) for the vertical force and $M(\omega_j)$ ($j = 1, 2$) for the moment. (As the vertical force consists

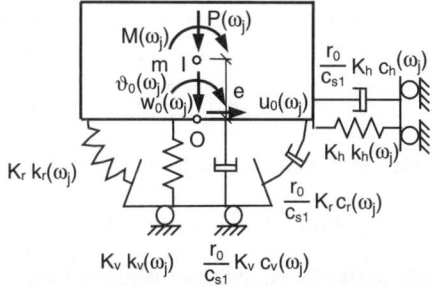

Figure 7.5 Dynamic system with degrees of freedom and frequency-dependent springs and dashpots

of a primary component only, this will also apply for the response, the vertical displacement.) The equilibrium equations are formulated for harmonic motion ω_j (in the frequency domain) at the centre of gravity of the block foundation. In the vertical direction

$$m\ddot{w}_0(\omega_j) + S_v(\omega)w_0(\omega_j) = P(\omega_j) \quad j = 1 \quad (7.12)$$

results, with $m\ddot{w}_0(\omega_j)$ representing the amplitude of the inertial load acting in the negative direction. The horizontal displacement amplitude at the centre of gravity of the block foundation is $u_0(\omega_j) + e\vartheta_0(\omega_j)$. The equilibrium equations in the horizontal direction and for rocking are

$$m(\ddot{u}_0(\omega_j) + e\ddot{\vartheta}_0(\omega_j)) + S_h(\omega_j)u_0(\omega_j) = 0 \quad j = 1, 2 \quad (7.13\text{a})$$

$$I\ddot{\vartheta}_0(\omega_j) - eS_h(\omega_j)u_0(\omega_j) + S_r(\omega_j)\vartheta_0(\omega_j) = M(\omega_j) \quad j = 1, 2 \quad (7.13\text{b})$$

To derive a symmetric system, Eq. 7.13b is replaced by the sum of Eq. 7.13b and Eq. 7.13a multiplied by e. This corresponds to formulating the rocking equilibrium equation with respect to O, where the degrees of freedom are defined, leading to

$$m\ddot{u}_0(\omega_j) + em\ddot{\vartheta}_0(\omega_j) + S_h(\omega_j)u_0(\omega_j) = 0 \quad j = 1, 2 \quad (7.14\text{a})$$

$$em\ddot{u}_0(\omega_j) + (I + e^2m)\ddot{\vartheta}_0(\omega_j) + S_r(\omega_j)\vartheta_0(\omega_j) = M(\omega_j) \quad j = 1, 2 \quad (7.14\text{b})$$

Substituting $\ddot{w}_0(\omega_j) = -\omega_j^2 w_0(\omega_j)$, etc. in Eqs 7.12 and 7.14 leads to the equations of motion in the (unknown) amplitudes of the degrees of freedom $w_0(\omega_j)$, $u_0(\omega_j)$ and $\vartheta_0(\omega_j)$,

$$(-\omega_j^2 m + S_v(\omega_j))w_0(\omega_j) = P(\omega_j) \quad j = 1 \quad (7.15)$$

$$(-\omega_j^2 m + S_h(\omega_j))u_0(\omega_j) - \omega_j^2 em\vartheta_0(\omega_j) = 0 \quad j = 1, 2 \quad (7.16\text{a})$$

$$-\omega_j^2 emu_0(\omega_j) + (-\omega_j^2(I + e^2m) + S_r(\omega_j))\vartheta_0(\omega_j) = M(\omega_j) \quad j = 1, 2 \quad (7.16\text{b})$$

Equation 7.15 is solved for $w_0(\omega_j)$ for $j = 1$, yielding the amplitude $w_0(\omega_1)$ ($= w_0(\omega)$). Analogously, the coupled Eq. 7.16 results in $u_0(\omega_j)$ and $\vartheta_0(\omega_j)$ for $j = 1, 2$, i.e. $u_0(\omega)$, $\vartheta_0(\omega)$, $u_0(2\omega)$, $\vartheta_0(2\omega)$.

Note that a new system of algebraic equations is solved for each ω_j.

To determine the displacements and rotations in the time domain for each harmonic j, the complex amplitudes are multiplied by $e^{i\omega t}$ and only the real parts of the products are considered (Eq. A.2). For instance, Eq. 7.16 leads, for example for $j = 2$, to $u_0(2\omega) = \text{Re}\, u_0(2\omega) + i\, \text{Im}\, u_0(2\omega) = 2.272 \times 10^{-6}$ m $+ i\, 4.463 \times 10^{-6}$ m. Multiplying $u_0(2\omega)$ by $e^{i2\omega t}$ yields $(2.272 \times 10^{-6} \cos 2\omega t - 4.463 \times 10^{-6} \sin 2\omega t)$ m $+ i\,(2.272 \times 10^{-6} \sin 2\omega t - 4.463 \times 10^{-6} \cos 2\omega t)$ m. Only the real part $(2.272 \times 10^{-6} \cos 2\omega t - 4.463 \times 10^{-6} \sin 2\omega t)$ m is considered, representing $u_0(t)$ for the second harmonic.

Finally, superposition of the response of the two harmonic excitations yields the final result. For instance, the vertical displacement of the left side face of the block foundation

$$w(t) = w_0(\omega)e^{i\omega t} - a\,\vartheta_0(\omega)e^{i\omega t} - a\vartheta_0(2\omega)e^{i2\omega t} \quad (7.17)$$

applies, and analogously for the horizontal displacement of the top side

$$u(t) = u_0(\omega)e^{i\omega t} + u_0(2\omega)e^{i2\omega t} + h\vartheta_0(\omega)e^{i\omega t} + h\vartheta_0(2\omega)e^{i2\omega t} \quad (7.18)$$

The time history of the vertical displacement $w(t)$ is plotted for the example in Fig. 7.6. The three terms on the right-hand side of Eq. 7.17 are also presented. The peak-to-peak displacement

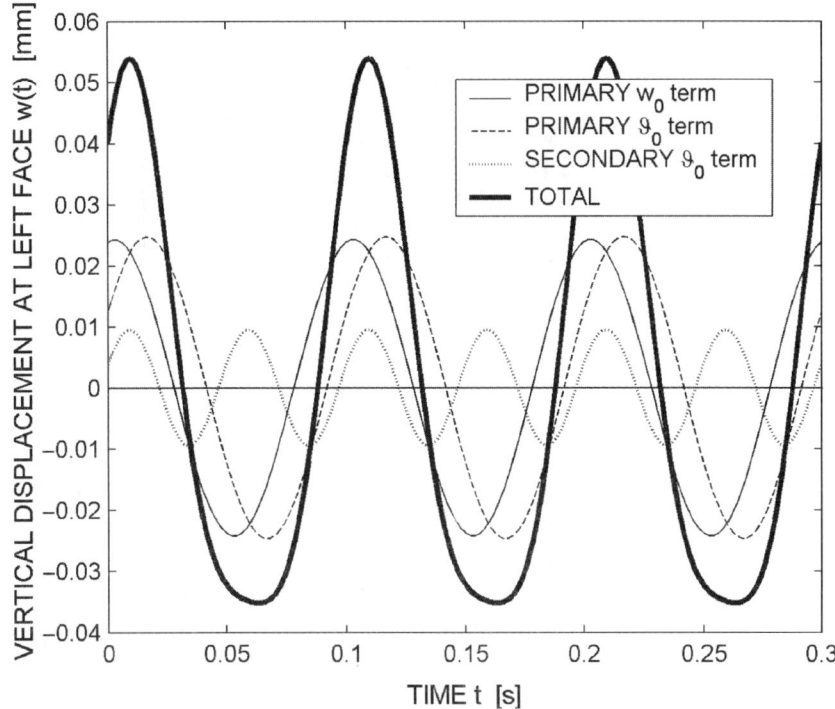

Figure 7.6 Vertical displacement of left side face of block foundation

equals 0.088 mm. This value satisfies a design criterion of 0.1 mm, which could be specified for a machine operating at 10 Hz with higher frequency components present.

7.2 Seismic analysis of a structure embedded in a layered half-space

The dynamic soil-structure interaction is analysed for three structures with different dynamic properties, ranging from a flexible configuration with small mass to a stiff one with large mass, subjected to a horizontal earthquake. A foundation embedded in a layered half-space is present in each case.

The three structures, shown schematically in Fig. 7.7, are a reactor building, a chimney stack and a frame. As this book does not address modelling aspects of the structure, the simplest possible model is selected. For a fixed base, the structure is modelled with a mass m and a static spring with coefficient k (representing the lateral stiffness) connected to a rigid bar of height h (Fig. 7.8). A hysteretic damping ratio ζ also applies. The parameters m, k and h either correspond to the fundamental frequency of the structure with a fixed base or are equivalent values determined approximately, taking the effect of higher modes into consideration by, for example, prescribing a certain lateral displacement pattern. They are specified, together with the equivalent r_0 of the base-mat, in Table 7.1. The natural frequency of the fixed-base structure is

$$\omega_s = \sqrt{\frac{k}{m}} \qquad (7.19)$$

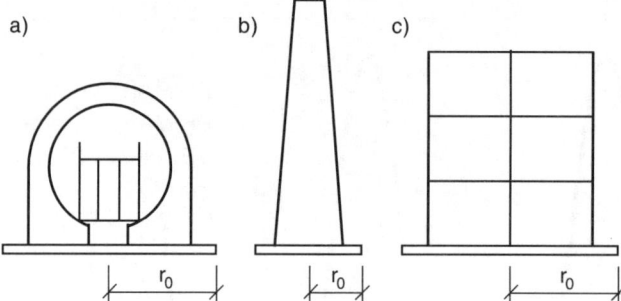

Figure 7.7 Structures. a) Reactor building. b) Chimney stack. c) Frame

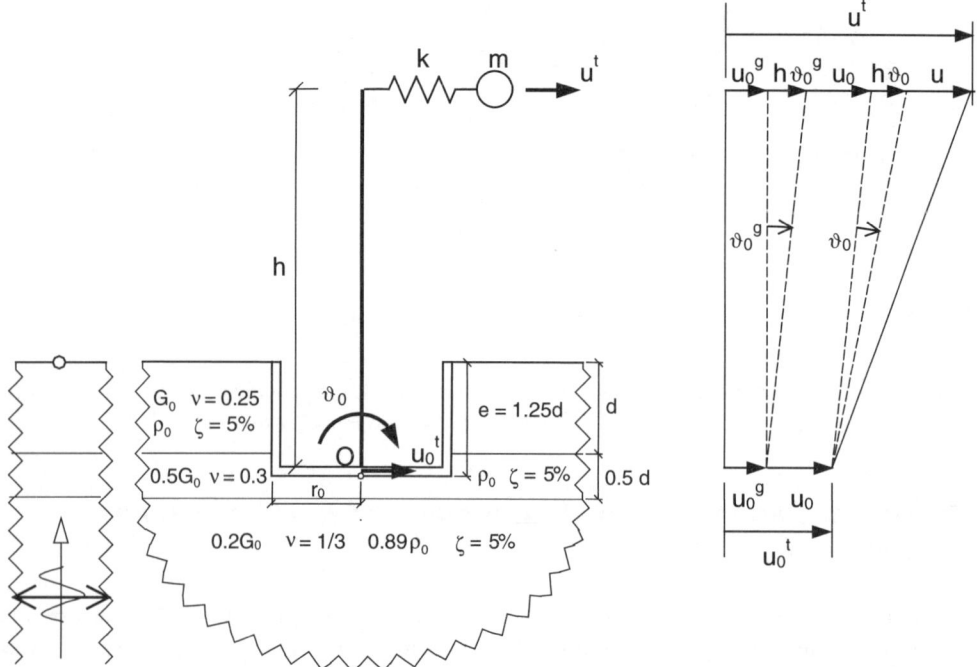

Figure 7.8 Coupled dynamic system of structure and embedded foundation for horizontal earthquake

with $f_s = \omega_s/2\pi$ also listed. The flexible chimney stack has a low frequency (0.5 Hz) and the stiff reactor building with large mass has a high value (4 Hz), while the stiff frame's frequency is even higher (12 Hz).

The cylindrical foundation embedded with depth e in a layered half-space is similar to the case addressed in Section 6.2 (Fig. 6.1c). However, to maintain realistic foundation depths, the embedded depth e is $1.25d$, with d (the thickness of the first layer) taken to be 5 m in all cases. The foundation radius r_0 is specified for the three structures in Table 7.1. The structure-soil interface is rigid. For the analysis the shear modulus of the first layer is chosen as $G_0 = 1124 \times 10^6$ N/m^2 and the mass density $\rho_0 = 1800$ kg/m^3, resulting in a shear-wave velocity $c_{s1} = \sqrt{G_0/\rho_0} = 790$ m/s. (The shear-wave velocity of the underlying flexible half-space is 375 m/s.)

Table 7.1 Parameters of structures fixed at base

	Reactor building	Chimney stack	Frame
Base-mat radius r_0 (m)	30	5	15
Height h (m)	20	50	17
Mass m (10^6 kg)	73	1.2	0.37
Spring coefficient k (10^9 N/m)	46	0.0118	2.1
Hysteretic damping ratio ζ	0.025	0.025	0.025
Frequency f_s (Hz)	4	0.5	12

The horizontal earthquake control motion is specified at the control point on the free surface of the site (left-hand side of Fig. 7.8). Simulated strong ground motion for the Newcastle (Australia) earthquake of 1989 (Refs [13, 14]) based on a soil site with a site period of approximately 0.5 s is processed. The peak ground acceleration is 0.2 g and the peak ground velocity is 0.14 m/s. The acceleration time-history is presented in Fig. 7.9a, and the displacement response spectrum for 2.5% damping in Fig. 7.9b. As is to be expected for any historic earthquake, peaks occur at certain frequencies in the response spectrum. The horizontal control motion is generated by vertically propagating S-waves with horizontal particle motion.

The coupled dynamic model is shown in Fig. 7.8. The foundation's degrees of freedom consist of the horizontal displacement with amplitude $u_0(\omega)$ and rocking with amplitude $\vartheta_0(\omega)$ with respect to the centre of the base O (Fig. 7.8). The fixed-base structure introduces the distortion in the spring with amplitude $u(\omega)$ as an additional degree of freedom.

To perform the analysis, that part of the layered half-space which will be excavated is modelled with a stack of disks. The highest circular frequency of the earthquake ω considered to determine the vertical distance Δe between two neighbouring disks (Eq. 5.1) is selected as $2\pi \times 20 = 125.7$ rad/s. This leads to a minimum $\Delta e = 0.62$ m for the first layer and to a minimum $\Delta e = 0.44$ m for the second layer. For convenience of data generation Δe is taken to be 0.5 m for the first layer and 0.3125 m for the second layer, generating 14 layers and 15 disks.

The analysis sequence is now described. (Appendix F.8 shows how this complete analysis may be performed in the MATLAB environment.) First, the acceleration (and displacement) time history is transformed to the frequency domain using a discrete Fourier transform. This corresponds to processing a periodic function. To ensure that the response has decayed to zero at the beginning of the loading a quiet zone of 10.48 seconds is added to the 10 s time-history. The augmented period T of 20.48 s is discretised in $N = 1024$ time steps of $\Delta t = 0.02$ s, which permits, with $\Delta \omega = 2\pi/(N\Delta t) = 0.3068$ rad/s, $\omega = 125.7$ rad/s to be represented adequately with 410 harmonics. The higher frequencies are neglected. The fast Fourier transformation leads to the amplitudes of horizontal displacements $u_c(\omega_j)$ ($j = 1, 2, \ldots, 410$) at the control point. Each harmonic is subsequently processed separately.

Second, the dynamic-stiffness matrix of the embedded foundation is calculated. As an intermediate step, the dynamic-stiffness matrices of the free field with respect to the disks for horizontal motion $[S^f(\omega_j)]$ (Eq. 5.6) and for rocking $[S^f_\vartheta(\omega_j)]$ (Eq. 5.8) are determined. The dynamic-stiffness matrix of the embedded foundation $[S^g_{00}(\omega_j)]$, consisting of the coefficients $S_h(\omega_j)$ and $S_r(\omega_j)$, and the coupling terms, $S_{hr}(\omega_j)$ and $S_{rh}(\omega_j)$ (which are averaged), follows from Eq. 5.24 (Fig. 6.4b, c and d in the case of the chimney stack, for which $r_0 = d$ applies).

Third, the effective foundation input motion is determined. The horizontal free-field motion $\{u^f(\omega_j)\}$ on the level of the disks is calculated (Appendix B.2), applying Eq. B.24. The effective

132 Foundation Vibration Analysis: A Strength-of-Materials Approach

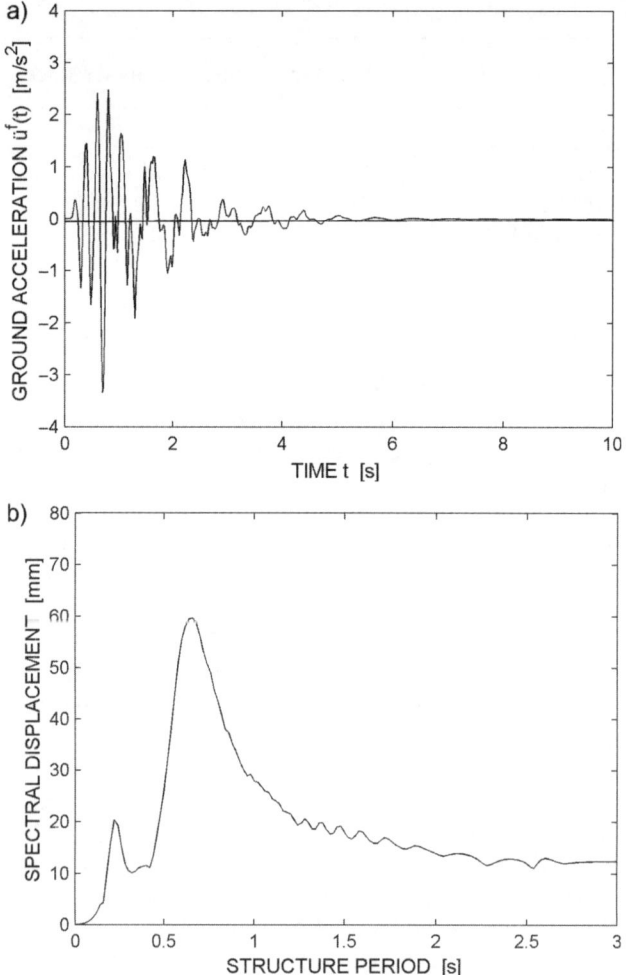

Figure 7.9 1989 Newcastle earthquake. a) Acceleration time history. b) Response spectrum (2.5% damping)

foundation input motion, consisting of the horizontal displacement with amplitude $u_0^g(\omega_j)$ and the rocking component $\vartheta_0^g(\omega_j)$ defined at O, then follows from Eq. 5.25. (As another example, a foundation embedded in a homogeneous half-space is analysed in Section 5.5 (Fig. 5.11 with Fig. 5.5).)

Fourth, the equations of motion of the structure-soil system (Eq. B.4) are formulated. These are expressed in total displacement amplitudes, which, for the two displacements, are (right-hand side of Fig. 7.8)

$$u_0^t(\omega_j) = u_0^g(\omega_j) + u_0(\omega_j) \tag{7.20a}$$

$$u^t(\omega_j) = u_0^g(\omega_j) + h\vartheta_0^g(\omega_j) + u_0(\omega_j) + h\vartheta_0(\omega_j) + u(\omega_j) \tag{7.20b}$$

The structural distortion with amplitude $u(\omega_j)$ is thus referred to a moving frame of reference (in translation and rocking) attached to the base. Formulating dynamic equilibrium of the node coinciding with the mass and the horizontal and rocking equilibrium equations of the total system yields

$$-\omega_j^2 m[u_0(\omega_j) + h\vartheta_0(\omega_j) + u(\omega_j)] + k(1 + 2\zeta\,\mathrm{i})u(\omega_j)$$
$$= \omega_j^2 m[u_0^g(\omega_j) + h\vartheta_0^g(\omega_j)] \qquad (7.21\mathrm{a})$$

$$-\omega_j^2\, m[u_0(\omega_j) + h\vartheta_0(\omega_j) + u(\omega_j)] + S_h(\omega_j)u_0(\omega_j) + S_{hr}(\omega_j)\vartheta_0(\omega_j)$$
$$= \omega_j^2\, m[u_0^g(\omega_j) + h\vartheta_0^g(\omega_j)] \qquad (7.21\mathrm{b})$$

$$-\omega_j^2\, hm[u_0(\omega_j) + h\vartheta_0(\omega_j) + u(\omega_j)] + S_{rh}(\omega_j)u_0(\omega_j) + S_r(\omega_j)\vartheta_0(\omega_j)$$
$$= \omega_j^2\, hm[u_0^g(\omega_j) + h\vartheta_0^g(\omega_j)] \qquad (7.21\mathrm{c})$$

Note that the inertial load acting in the negative direction is calculated with the total acceleration. Dividing Eqs 7.21a and 7.21b by $\omega_j^2\,m$ and Eq. 7.21c by $\omega_j^2\,h\,m$ and substituting Eq. 7.19 yields the equations of motion of the coupled system

$$\begin{bmatrix} \dfrac{\omega_s^2}{\omega_j^2}(1+2\zeta\,\mathrm{i}) - 1 & -1 & -1 \\ -1 & \dfrac{S_h(\omega_j)}{\omega_j^2\,m} - 1 & \dfrac{S_{hr}(\omega_j)}{\omega_j^2\,h\,m} - 1 \\ -1 & \dfrac{S_{rh}(\omega_j)}{\omega_j^2\,h\,m} - 1 & \dfrac{S_r(\omega_j)}{\omega_j^2\,h^2\,m} - 1 \end{bmatrix} \begin{Bmatrix} u(\omega_j) \\ u_0(\omega_j) \\ h\,\vartheta_0(\omega_j) \end{Bmatrix}$$
$$= \begin{Bmatrix} 1 \\ 1 \\ 1 \end{Bmatrix} u_0^g(\omega_j) + \begin{Bmatrix} 1 \\ 1 \\ 1 \end{Bmatrix} h\vartheta_0^g(\omega_j) \qquad (7.22)$$

with the unknowns $u(\omega_j)$, $u_0(\omega_j)$ and $\vartheta_0(\omega_j)$. Since $S_{hr}(\omega_j) = S_{rh}(\omega_j)$ is assumed, Eq. 7.22 is symmetric. Solving Eq. 7.22 for all the unknowns leads to $u(\omega_j)$, $u_0(\omega_j)$ and $\vartheta_0(\omega_j)$. (Equation 7.22 has to be solved for all js! This can be achieved readily in the MATLAB environment.)

The response in the time domain follows for each harmonic j as, for example, (Eq. A.2)

$$u_j(t) = u(\omega_j)\mathrm{e}^{\mathrm{i}\omega_j t} \qquad (7.23)$$

Fifth, after processing each harmonic j, the total time response is determined as the superposition of the time responses for all harmonics. For instance,

$$u(t) = \sum_j u_j(t) \qquad (7.24)$$

The structural distortion with amplitude $u(\omega_j)$ is also representative of the shear force in the structure with an amplitude $k\,u(\omega_j)$ and the overturning moment with amplitude $h\,k\,u(\omega_j)$. It is thus the most important response quantity.

For harmonic excitation with frequency ω_j the structural distortion $u(\omega_j)$ non-dimensionalised with the amplitude of the control motion $u_c(\omega_j)$ is calculated as a function of ω_j/ω_s. The reader is reminded that ω_s for each structure is different. The results are presented in Figs 7.10a, b and c

134   Foundation Vibration Analysis: A Strength-of-Materials Approach

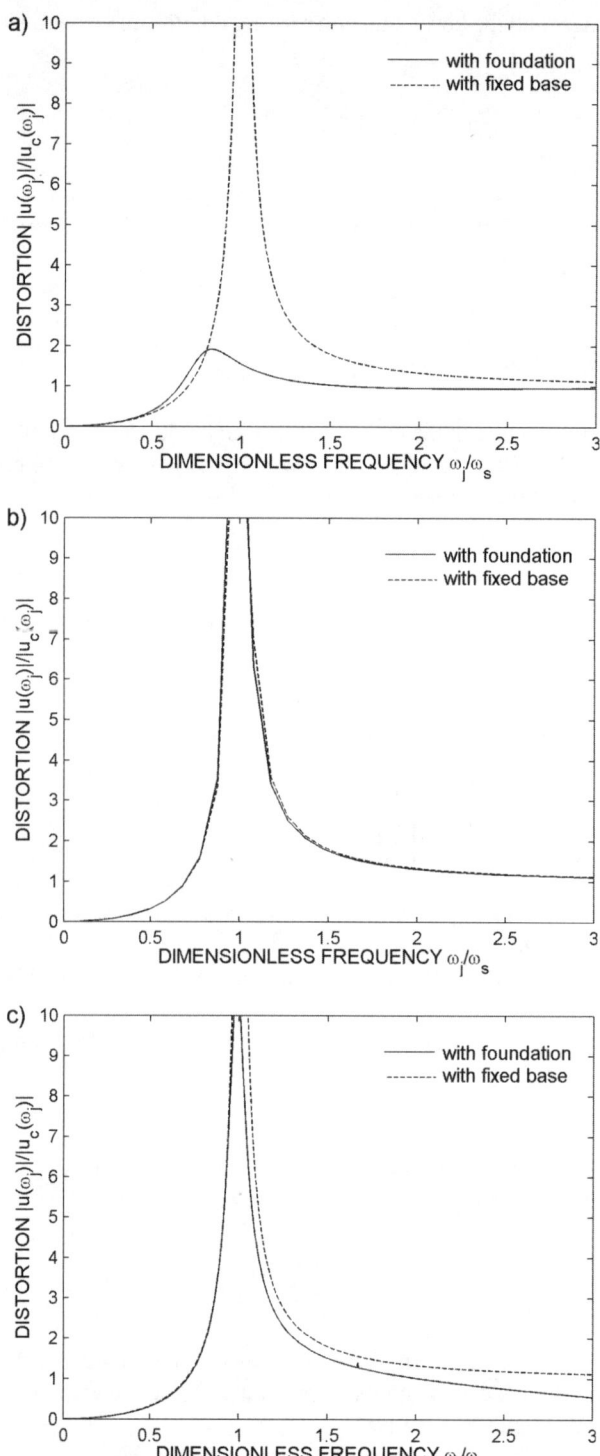

Figure 7.10 Structural distortion as a function of excitation frequency. a) Reactor building. b) Chimney stack. c) Frame

Table 7.2 Maximum structural distortion for 1989 Newcastle earthquake

	Reactor building	Chimney stack	Frame
Structural distortion u_{max} (mm)	5.7	13.6	0.70
(with base fixed)	(17.0)	(13.9)	(0.79)

for the three structures. The solutions disregarding the soil, i.e. for a fixed base, are also presented as dashed lines. These correspond to the well-known curve for a one-degree-of-freedom system with the peak of $1/(2\zeta)$ (i.e. $1/(2 \times 0.025) = 20$) occurring at $\omega_j/\omega_s = 1$. For the reactor building (Fig. 7.10a), for which the soil-structure interaction is important, the peak response of the structural distortion of the coupled structure-soil system is significantly smaller than that of the same structure on a rigid soil. The peak occurs at a smaller frequency, corresponding to a more flexible system, and is also broader, indicating that the damping ratio is larger when the soil is modelled. As for the reactor building the horizontal dimension is much larger than the height, the horizontal degree of freedom with large radiation damping contributes significantly to the global response. For a specific frequency, however, the response can be smaller or larger. For instance, for $\omega_j/\omega_s = 0.7$, taking soil-structure interaction into account increases the structural distortion. This even occurs for this structure with significant embedment yielding large radiation damping. Turning to the chimney (Fig. 7.10b), the effect of soil-structure interaction is small and can, in general, be neglected. For this tall structure, the rocking degree of freedom with small radiation damping dominates the global response. For the frame the effect of soil-structure interaction is noticeable but small.

Finally, processing the historical earthquake of Fig. 7.9, the maximum structural distortion u_{max} is listed in Table 7.2 for the three structures. As already mentioned, this value is also representative of the maximum shear force in the structure and the maximum overturning moment, both acting on the base at point O. The results for the structure fixed at its base are also specified. The maximum distortions reflect the observations made above. The stiff reactor building's response is strongly reduced by the presence of the soil. The response of the chimney stack is virtually unaltered, while the response of the frame is marginally reduced.

7.3 Offshore wind turbine tower with a suction caisson foundation

The final example concerns the response of a wind turbine tower to periodic loading generated by the varying lift exerted on the blades as they pass in front of the tower. The dynamics of a wind turbine system are rather complicated, as the response of the tower and foundation influences the dynamic response of the blades, and consequently the loads on the tower. Treatment of these aspects is beyond the scope of this book. Here a simplified system is analysed. However, treatment of the foundation by the method described could easily be included in a more comprehensive dynamic analysis of a complete rotor-turbine-tower-foundation system.

A novel option proposed recently for the foundations of the towers that support such turbines is a skirted circular foundation, or *caisson* (Ref. [3]). Such cylindrical caissons can be installed in offshore conditions through the use of suction, as illustrated in Fig. 7.11. With the caisson sitting on the seabed at the desired location, a valve at the top of the foundation is opened and

136   Foundation Vibration Analysis: A Strength-of-Materials Approach

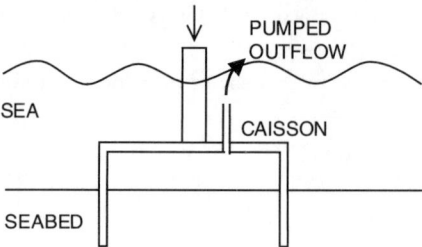

Figure 7.11   Suction caisson

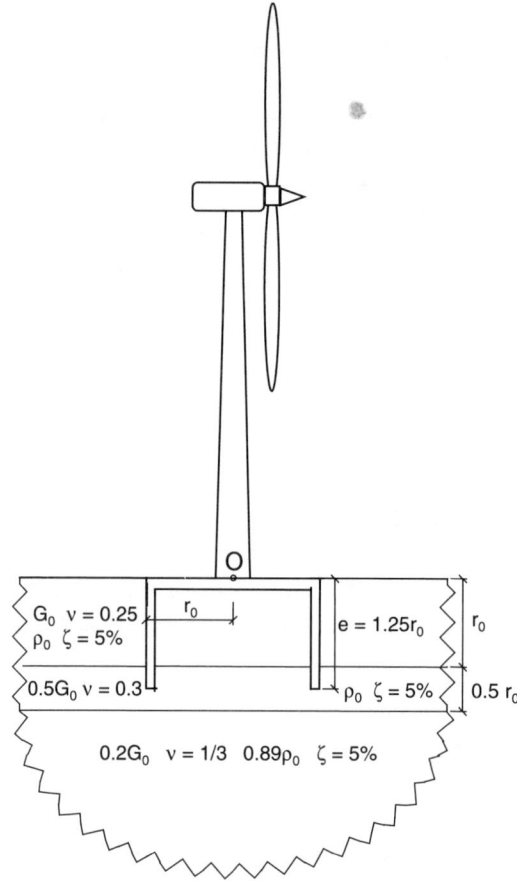

Figure 7.12   Offshore wind turbine with tower and suction caisson foundation

the water inside evacuated by pumping. The combination of the dead weight of the foundation and the differential hydrostatic pressure causes bearing failure at the base of the cylinder, and the foundation gradually self-buries. When the design depth is reached, pumping ceases and the valve is closed. Such caissons are termed *suction caissons*.

The example treated here is based on the two-bladed prototype wind turbine developed for offshore applications by Kvaerner Turbin (Ref. [10]) installed on a single suction caisson foundation of the type proposed in Ref. [3] (Fig. 7.12). Both the diameter of the rotor and the height of the turbine above the foundation are 80 m. The radius of the caisson r_0 is 10 m, and the depth of the skirt is 12.5 m. The layered half-space into which the caisson is installed corresponds to the case addressed in Section 6.2 (Fig. 6.1c). The caisson is assumed to be rigid. For the analysis the shear modulus of the first layer G_0 is selected as 28.1×10^6 N/m² and the mass density $\rho_0 = 1800$ kg/m³, resulting in a shear-wave velocity $c_{s1} = \sqrt{G_0/\rho_0} = 125$ m/s. (The shear-wave velocity of the underlying flexible half-space is 60 m/s.)

As stated above, there is complex interaction between the dynamics of the supporting system and the dynamics of the blades. For simplicity, the forces computed in Ref. [6] for the prototype wind turbine will be applied directly to the structure, and interaction effects will be neglected. The flap moment at the blade root for a uniform wind speed of 13 m/s will be employed. At this wind speed the rotor rotates at 0.37 Hz. The flap moment for a single blade is plotted in Fig. 7.13. Sharp reductions in lift occur as the blade passes in front of the tower. A single cycle of the flap moment is discretised at intervals of $\Delta t = 0.0575$ s and a fast Fourier transformation performed. The first eight computed Fourier terms are used. As indicated in Fig. 7.13, these terms represent the flap moment extremely well.

The horizontal force exerted by the blade on the tower structure is approximated by dividing the flap moment by 2/3 of the length of the blade. Since the second blade produces the same horizontal force as the first but 180° out of phase, each term in the Fourier series is multiplied by $(1 + e^{ij\pi})$, where j is the number of the Fourier term. The resulting force $V(t)$ is plotted against time in Fig. 7.14. As expected, the horizontal force oscillates at twice the rotation frequency of the rotor, and only the amplitudes for $j = 0, 2, 4$ and 6 are non-zero. The frequencies $\omega_0, \omega_2, \omega_4$ and ω_6 are listed in Table 7.3, together with the corresponding Fourier coefficients, which are the amplitudes $V(\omega_0), V(\omega_2), V(\omega_4)$ and $V(\omega_6)$.

The summed oscillating component of the flap moments of the two blades also generates an oscillating moment on the structure. However, it is assumed here that the rotor assembly is articulated such that no significant moment is transmitted to the tower. Depending on the electrical load on the turbine generator, a torque along the axis of the rotor will also be generated. This torque will not be treated here. Thus the simplified example consists of the wind turbine tower excited at the top by the horizontal force plotted in Fig. 7.14.

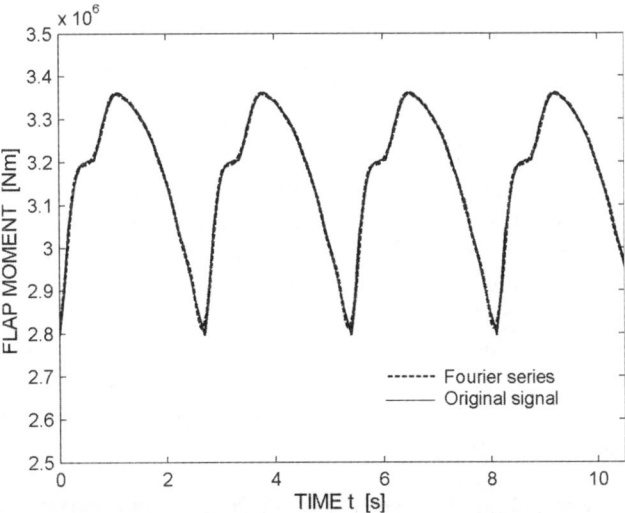

Figure 7.13 Flap moment for single blade

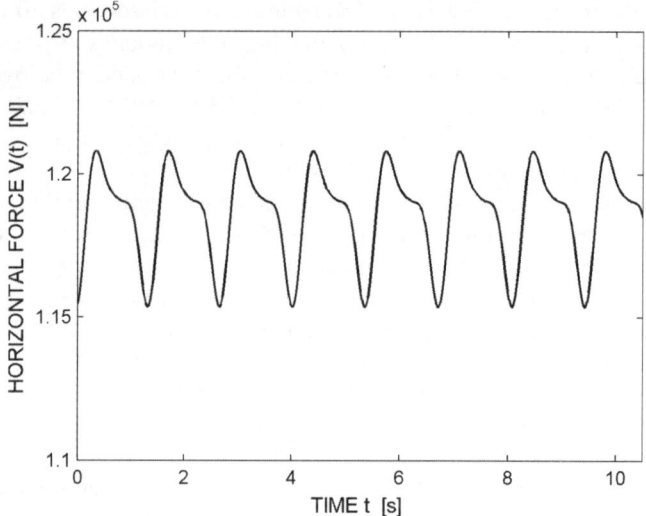

Figure 7.14 Net horizontal force acting on tower

Table 7.3 Force amplitudes and dynamic-stiffness coefficients at frequencies in Fourier series for horizontal loading on wind turbine

j	ω_j (rad/s)	$V(\omega_j)$ (10^3 N)	$S_h(\omega_j)$ (10^9 N/m)	$S_r(\omega_j)$ (10^9 Nm)	$S_{hr}(\omega_j)$ (10^9 N)
0	0	119	$1.224 + 0.122i$	$407.7 + 40.8i$	$-8.489 - 0.849i$
2	4.654	$-1.64 - 1.00i$	$1.202 + 1.545i$	$374.1 + 140.3i$	$-7.863 - 10.769i$
4	9.308	$-1.20 - 0.21i$	$0.695 + 2.796i$	$310.2 + 264.7i$	$-4.245 - 20.762i$
6	13.963	$-0.29 - 0.03i$	$-0.181 + 4.089i$	$227.5 + 395.5i$	$0.902 - 30.817i$

As a further simplification, only the first mode of the tower vibration is considered, reducing the tower and turbine to a single degree of freedom system relative to the vertical line extending from the foundation. The caisson foundation moves horizontally and rocks under the applied horizontal and moment loadings. The complete model indicated in Fig. 7.15 has three degrees of freedom. This model is essentially the same as addressed in Section 7.2 (Fig. 7.8). Computation of the dynamic stiffness of the caisson foundation requires two minor changes to the procedure described in Chapter 5. The soil within the caisson can still be considered to undergo rigid body motion with the foundation. However, the soil *remains in place and is not analytically excavated*. The point O is located at the top of the foundation rather than at the bottom, and so the constraint matrix $[A]$ described in Section 5.3 is modified accordingly.

The necessary dynamic-stiffness coefficients for the site ($S_h(\omega)$, $S_r(\omega)$ and $S_{hr}(\omega)$) computed taking into account these changes for the frequencies of interest are tabulated in Table 7.3, and are plotted in Fig. 7.16 against a_0 using the decomposition of Eq. 6.1, with the static-stiffness coefficients K being the real parts of $S_h(\omega = 0)$, $S_r(\omega = 0)$ and $S_{hr}(\omega = 0)$, respectively (first line of Table 7.3).

The mass of the tower and turbine assembly is 600 000 kg, with 400 000 kg being lumped at the top of the tower (m_1 in Fig. 7.15), and 200 000 kg being lumped at the bottom. The mass of

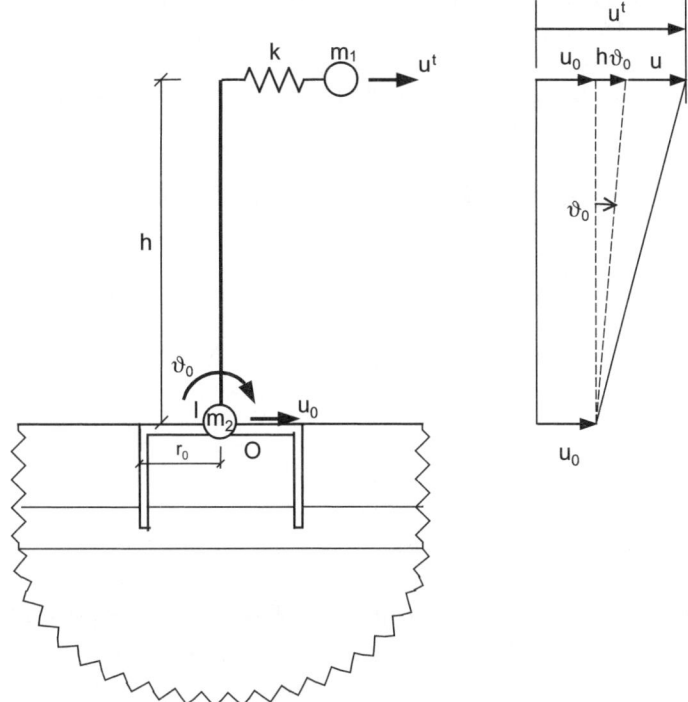

Figure 7.15 Coupled dynamic system of wind turbine tower and suction caisson foundation for horizontal loading on turbine blades

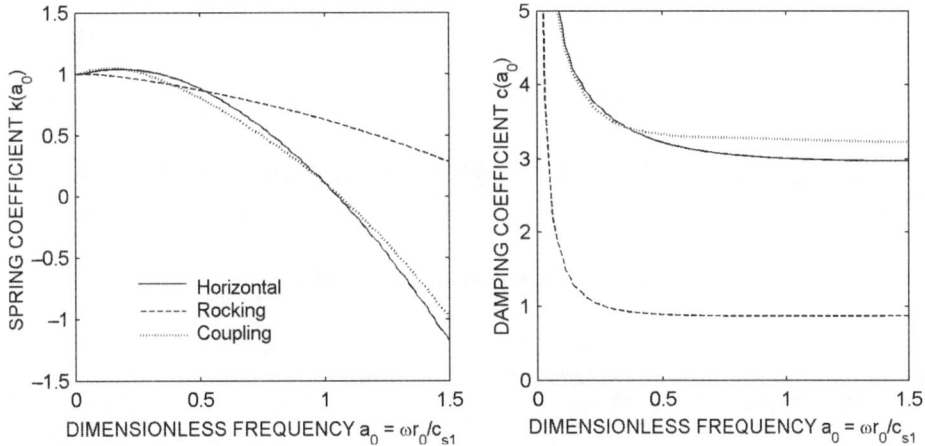

Figure 7.16 Horizontal, rocking and coupling dynamic-stiffness coefficients for suction caisson

the caisson foundation is 220 000 kg, and so the total mass lumped at the bottom of the tower (m_2 in Fig. 7.15) is 420 000 kg. The mass moment of inertia of the caisson about the bottom of the tower (I in Fig. 7.15) is 14.9×10^6 kg m². The spring representing the lateral stiffness of the first mode of vibration of the tower has a spring constant k of 9×10^6 N/m, indicating a fundamental natural frequency of 0.755 Hz for the fixed-base condition. Hysteretic damping within the tower and turbine is assumed to be 2% of critical (hysteretic damping ratio $\zeta = 0.02$).

The equations of motion for the three-degree-of-freedom system (Fig. 7.15) expressed in the frequency domain (formulating dynamic equilibrium of the mass m_1 and horizontal and rocking equilibrium of the total system at point O) are

$$-\omega_j^2 m_1[u_0(\omega_j) + h\vartheta_0(\omega_j) + u(\omega_j)] + k(1 + 2\zeta i)u(\omega_j) = V(\omega_j) \quad (7.25a)$$

$$-\omega_j^2 m_2 u_0(\omega_j) - \omega_j^2 m_1[u_0(\omega_j) + h\vartheta_0(\omega_j) + u(\omega_j)] + S_h(\omega_j)u_0(\omega_j) + S_{hr}(\omega_j)\vartheta_0(\omega_j)$$
$$= V(\omega_j) \quad (7.25b)$$

$$-\omega_j^2 I\vartheta_0(\omega_j) - \omega_j^2 m_1 h[u_0(\omega_j) + h\vartheta_0(\omega_j) + u(\omega_j)] + S_{rh}(\omega_j)u_0(\omega_j) + S_r(\omega_j)\vartheta_0(\omega_j)$$
$$= hV(\omega_j) \quad (7.25c)$$

or, written in matrix form,

$$\begin{bmatrix} k(1+2\zeta i) - \omega_j^2 m_1 & -\omega_j^2 m_1 & -\omega_j^2 m_1 h \\ -\omega_j^2 m_1 & S_h(\omega_j) - \omega_j^2(m_1 + m_2) & S_{hr}(\omega_j) - \omega_j^2 m_1 h \\ -\omega_j^2 m_1 h & S_{rh}(\omega_j) - \omega_j^2 m_1 h & S_r(\omega_j) - \omega_j^2(I + m_1 h^2) \end{bmatrix} \begin{Bmatrix} u(\omega_j) \\ u_0(\omega_j) \\ \vartheta_0(\omega_j) \end{Bmatrix}$$

$$= \begin{Bmatrix} 1 \\ 1 \\ h \end{Bmatrix} V(\omega_j) \quad (7.26)$$

Equation 7.26 is solved for $j = 0, 2, 4$ and 6, leading to the amplitudes $u(\omega_j)$, $u_0(\omega_j)$ and $\vartheta_0(\omega_j)$ tabulated in Table 7.4.

The total displacement at the top of the tower is related to the model degrees of freedom by

$$u^t(t) = u_0(t) + h\vartheta_0(t) + u(t) \quad (7.27)$$

Consequently, the amplitudes $u^t(\omega_j)$ are obtained by summing the contributions of the amplitudes $u(\omega_j)$, $u_0(\omega_j)$ and $\vartheta_0(\omega_j)$ as

$$u^t(\omega_j) = u_0(\omega_j) + h\vartheta_0(\omega_j) + u(\omega_j) \quad (7.28)$$

Table 7.4 Displacement and rotation amplitudes at frequencies in Fourier series for horizontal loading on wind turbine

j	ω_j (rad/s)	$u(\omega_j)$ (10^{-3} m)	$u_0(\omega_j)$ (10^{-3} m)	$\vartheta_0(\omega_j)$ (10^{-6} rad)	$u^t(\omega_j)$ (10^{-3} m)
0	0	13.200 − 0.500i	0.225 − 0.022i	18.425 − 1.843i	14.899 − 0.670i
2	4.654	−2.200 + 3.000i	−0.004 + 0.051i	−1.075 + 5.079i	−2.290 + 3.458i
4	9.308	0.046 + 0.010i	0.000 + 0.000i	0.045 − 0.034i	0.050 + 0.007i
6	13.963	0.004 + 0.000i	0.000 + 0.000i	0.001 − 0.004i	0.004 − 0.001i

Engineering applications

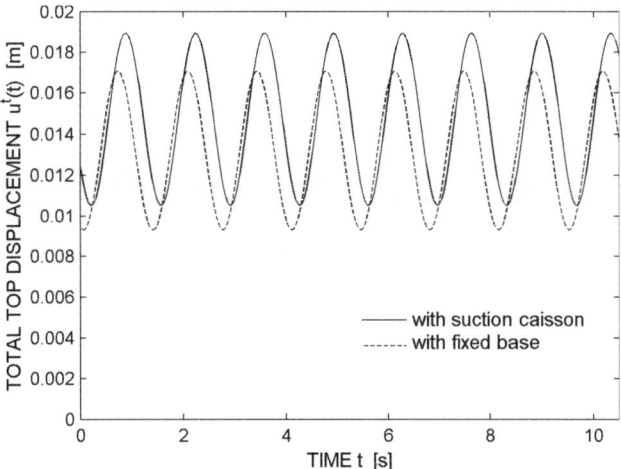

Figure 7.17 Total displacement at top of tower, with suction caisson foundation and with fixed base

These amplitudes are also tabulated in Table 7.4. The total displacement at the top of the tower is

$$u^t(t) = \sum_{j=0}^{6} u^t(\omega_j) e^{i\omega_j t} \qquad (7.29)$$

which is plotted in Fig. 7.17. Also shown is the total displacement at the top of the tower calculated for a fixed base. The effect of the interaction with the foundation is small but noticeable in this case. As the strength-of-materials approach using cones maintains accuracy of within 20%, and the effect of the interaction with the foundation on the final result is only about 10%, the effect of the strength-of-materials assumptions on the accuracy of the final result is less than 2%.

8

Concluding remarks

To analyse the vibrations of a foundation on the surface of or embedded in a layered half-space, an approach using conical bars and beams, called cones, is developed. The complicated exact formulation of three-dimensional elastodynamics is replaced by the simple one-dimensional description of the theory of the strength of materials, postulating the deformation behaviour ('plane sections remain plane').

The half-space with linear elastic behaviour and hysteretic material damping can consist of any number of horizontal layers either overlying a half-space or fixed at its base. Besides cylindrical foundations, axi-symmetric configurations of arbitrary embedment shape can be processed, with the wall and base of the embedded foundation assumed to be rigid. The dynamic-stiffness coefficients describing the interaction force-displacement relationship and the effective foundation input motion for vertically propagating S- and P-waves in seismic excitation are calculated for all frequencies.

Only approximations of the one-dimensional strength-of-materials approach based on wave propagation in cones apply. No other assumptions are made. For each degree of freedom only one type of body wave exists: for the horizontal and torsional motions S-waves propagating with the shear-wave velocity; and for the vertical and rocking motions P-waves propagating with the dilatational-wave velocity. The corresponding displacements can be formulated directly in closed form as a function of the depth of the site, without any spatial Fourier transformation into the wave number domain.

Two building blocks are required to construct the procedure to analyse the vibrations of a foundation in a layered half-space. The first addresses the outward wave propagation occurring from a disk embedded in a full-space modelled as a double cone. The sectional property of these initial cones increases in the direction of wave propagation, modelling the spreading of the disturbance in the medium. The cones are thus radiating. The opening angle of the cone is determined by equating the static-stiffness coefficient of the truncated semi-infinite cone to that of a disk on a half-space determined using the three-dimensional theory of elasticity. The opening angle depends only on Poisson's ratio (and the degree of freedom). Alternatively, the calibration can be performed addressing the static-stiffness coefficient of a disk embedded in a full-space. In the case of nearly-incompressible and compressible material (Poisson's ratio larger than 1/3), the wave velocity is limited to twice the shear-wave velocity and a trapped mass and mass moment of inertia are introduced for the vertical and rocking degrees of freedom, respectively.

The second building block addresses the wave mechanism generated at a material discontinuity corresponding to an interface between two layers. When the incident wave propagating in the

Concluding remarks

initial cone (described in the first building block) encounters a discontinuity, a reflected wave and a refracted wave, each propagating in its own cone, are created. Enforcement of compatibility of displacement and equilibrium of the interface permits the reflected and refracted waves to be expressed as a function of the incident wave. The reflection coefficient, defined as the ratio of the reflected wave to the incident wave, depends on the frequency and on the properties of the two materials present at the interface, in particular their impedances. (In more detail, the reflection coefficient is equal to the difference of the dynamic-stiffness coefficients of the cones with the reflected and refracted waves at the interface, divided by their sum.)

The reflected and refracted waves generated at an interface will also encounter material discontinuities as incident waves at a later stage, yielding additional reflections and refractions. By tracking the reflections and refractions sequentially, the superimposed wave pattern can be established for a layered site up to a certain stage. The termination criterion addresses the number of cone segments in which the waves have propagated and the magnitude of the created waves, and applies averaging over the last ten steps before termination to improve the accuracy in cases where the amplitudes of successive waves oscillate in sign.

The embedded foundation is modelled with a stack of disks in that part of the soil which will be excavated. This leads to a primary dynamic system with redundants acting on the embedded disks. As in the force method of structural analysis, the dynamic flexibility of the free field with respect to the displacements of the disks caused by the redundants is established addressing the wave pattern in the layered half-space. Inversion of this relationship, enforcement of the rigid-body motion of the foundation (considering the free-field motion of the seismic waves, if present) and excavation of the trapped material yield the dynamic-stiffness coefficients of the embedded foundation and the effective foundation input motion.

In a nutshell, the computational procedure can be characterised as follows. In this one-dimensional strength-of-materials approach, waves of one type for each degree of freedom propagate in cone segments outwards with reflections and refractions occurring at layer interfaces. The sectional property of the cone segments increases in the direction of wave propagation, downwards as well as upwards. The restrictions of tapered bar and beam theory only apply. No other assumptions are enforced. In particular, no curve fitting is used, with the exception of selecting meaningful parameters in the incompressible case for the vertical and rocking degrees of freedom. For a surface foundation the computational procedure is analytical. For an embedded foundation that part of the half-space which will be excavated is discretised with a stack of disks, and thus a numerical approximation is required.

Turning to the features of the cone models, the following requirements are met: conceptual clarity with physical insight, simplicity in mechanics and mathematical formulation permitting an exact mathematical formulation, sufficient generality concerning layering and embedment for all degrees of freedom and all frequencies, and sufficient engineering accuracy. These features permit cone models to be applied for everyday practical foundation vibration and dynamic soil-structure interaction analyses in a design office. The rigorous methods, which belong more to the field of computational mechanics than to civil engineering, should be used for large projects of critical facilities and in those cases that are not covered by the cone models. For instance, when the thickness of a layer is not constant in the horizontal direction, leading to an inclined interface, rigorous methods such as the boundary-element method can be appropriate.

The accuracy is evaluated not only for academic examples, most with analytical solutions available for comparison, but also for actual multi-layered sites with vastly varying characteristics. For the latter, results determined with a rigorous procedure, the thin-layer method with a very fine discretisation, which can be regarded as exact, are used. Foundations on the surface of and embedded in sites with material properties decreasing with depth, as well as sites with several layers fixed at the base with no radiation damping occurring in the vertical direction, are investigated for all degrees of freedom and for a large frequency range. Compressible and incompressible layered half-spaces are addressed. Besides cylindrical foundations, general axi-symmetric situations and fully embedded foundations are examined. In all these cases the systematic evaluation confirms that sufficient engineering accuracy is obtained, with deviations within the range ±20%. The cone models also work well for a site with material properties varying gradually with depth, where a large number of layers are required. The same applies for the very stringent test of an undamped site fixed at its base, where the damping coefficient representing the imaginary part of the dynamic stiffness calculated with cones almost vanishes for frequencies below the cutoff frequency.

A complete MATLAB implementation of the method is provided (Appendix F), together with a comprehensive description of the workings of each function. These can be used along with the built-in facilities of MATLAB, such as the fast Fourier transformation, to perform all necessary analyses in the MATLAB environment. A reader knowledgeable in computer programming can use these listings as a guide to constructing his own programs in other languages. An executable program (CONAN, Appendix E) is also provided, along with complete details of its use. This program permits the reader without access to MATLAB to duplicate the practical engineering examples provided in Chapter 7, and to apply the method to his own problems.

As already mentioned, cone models permit a vast class of reasonably complicated practical cases to be analysed for harmonic excitation. The dynamic-stiffness coefficients and the effective foundation input motion for vertically propagating seismic waves can be calculated for all degrees of freedom. The cone models work well for the static case, for the low and intermediate frequency ranges important for machine vibrations and earthquakes, and for the limit of very high frequencies as occur in impact loads. The following key aspects are adequately represented:

- *Shape of the foundation-soil interface* – Besides the embedded cylinder, axi-symmetric configurations with varying radii of the disks can be processed. More general foundations can be transformed to axi-symmetric cases. Of course, limits exist. For instance, determining an equivalent circular disk for an L-shaped surface foundation can be problematic.
- *Soil profile* – A layered half-space with any number of (horizontal) interfaces which overlies a half-space or is fixed at its base can be processed.
- *Embedment* – Surface and embedded foundations can be processed.

Certain limitations do exist, which are not repeated here for conciseness. They are discussed in the preceding chapters when the corresponding aspect of the computational procedure is addressed.

The accuracy has been evaluated systematically for harmonic excitation by comparing the results with those of rigorous methods in a parametric study covering all key aspects discussed above including extreme cases. In addition further results are available in Ref. [37] where, for example, a site consisting of a layer overlying a half-space which is stiffer or more flexible than the layer is also investigated. This information is valuable as no approximate method, including cone models, should be used for cases with parameters which lie outside the performed systematic evaluation. Besides the comparison with results of rigorous methods, confidence can also be built up based on theoretical considerations. These are discussed in Section 2.5 of Ref. [37], where the three main reservations against cone models (based on the theory of strength of materials, portion of half-space outside cone neglected, no representation of Rayleigh surface waves) are proved to be unfounded.

This book concentrates on analysis for harmonic excitation (in the frequency domain), which is appropriate for linear systems. However, time domain procedures using cone models also exist, which are described in Ref. [37], where further aspects such as representing frictional material damping and modelling a half-plane in contrast to a half-space are developed.

As discussed in Appendix G, when frequency-independent reflection coefficients are applied, the analysis can be performed directly in the time domain without any transformations to and from the frequency domain. In Appendix G the analysis is described in detail for a surface foundation on a layered half-space for translational motion, and is sketched for rotational motion. When the high frequency reflection coefficients, which are frequency independent, are used, the procedure is still governed by the one-dimensional strength-of-materials theory with the additional assumption that at the interface with a material discontinuity the cones are replaced locally by prismatic bars and beams when addressing the wave mechanism. Based on the same logic of the computational algorithm as for the frequency-dependent case, the displacement of the foundation as a function of the applied load is directly determined in the time domain. In addition, the interaction force-displacement relationship of the foundation is calculated, which permits dynamic soil-structure-interaction analysis even with non-linear behaviour of the structure to be performed in the time domain. Thus the cone models with frequency-independent reflection coefficients permit the modelling of a foundation on a multi-layered half-space in a dynamic analysis performed in the time domain in an efficient straightforward manner with physical insight.

It is important to stress that the cones model a three-dimensional situation, a foundation embedded in a half-space, although the representation is one dimensional. It is not a two-dimensional model, for which the characteristics of wave propagation are significantly different than in the three-dimensional case. For a practical engineering problem (which is nearly always three dimensional) it is much better to construct an approximate three-dimensional model than a two-dimensional one. If possible, to reduce the computational effort the model should be axi-symmetric, in which case cones represent an attractive method of analysis.

When analysing a foundation embedded in a layered half-space, the one-dimensional discretisation is in the vertical direction, coinciding with the axis of symmetry. The behaviour of the cones is fully described by the displacements and rotations defined on the axes (plane sections remain plane) and the corresponding sectional properties. The cones have no width measured horizontally. The ability of cone models to represent the behaviour of the half-space in the horizontal plane outside the cone is thus very limited. It is to be expected that cones are not well suited to address problems where this behaviour in the horizontal plane governs the response, such as the through-soil coupling of two adjacent embedded foundations. A notable exception is that the proportions of the cones determined by the opening angles are compatible with the proper horizontal wave-propagation velocities, which are based on physical reasoning, for all motions (Section 2.5.3 of Ref. [37]).

Summarising, using the cone models does lead to some loss of precision compared to applying the rigorous methods. However, this is more than compensated for, by the many advantages discussed. The achieved accuracy using cone models is more than sufficient. It cannot be the aim of the engineer to calculate the complex reality as closely as possible, as this is not required for a safe and economical design. The accuracy is limited anyway because of the many uncertainties that cannot be eliminated, such as the wide scatter of the dynamic material properties of the soil.

Finally, it is appropriate to recall the goal of this book. Starting from scratch, a treatise is written developing the one-dimensional strength-of-materials theory of conical bars and beams called cones, which are then applied to practical foundation vibration analyses. Confidence in cones is gained as the procedure to analyse foundations is the same as routinely used in structural analysis, and a systematic evaluation for a wide range of actual sites demonstrates sufficient engineering accuracy. A short computer program written in MATLAB forms an integral part of the book, and a user-friendly executable program is also provided.

Appendix A

Frequency-domain response analysis

This appendix discusses the frequency-domain analysis of linear dynamic systems. Alternative descriptions for harmonic motion with frequency ω are addressed in Appendix A.1. In particular, the representation using a complex amplitude of frequency ω is developed. In Appendix A.2 the complex frequency response function is discussed, which represents the steady-state response of the dynamic system to an excitation of frequency ω. In Appendix A.3 a periodic excitation in the time domain (for instance a load) is decomposed into a Fourier series with terms equal to the amplitudes of $\omega_j (j = 1, \ldots, n)$. Determining these amplitudes represents the Fourier transformation. Multiplying the amplitude of each term with the corresponding complex frequency response function yields the amplitude of the response (for instance a displacement) for that term. The definition of the Fourier series then yields the response in the time domain. In Appendix A.4 an arbitrary excitation is processed, leading to Fourier integrals.

A.1 Alternative descriptions of harmonic motion

Several alternative ways exist to describe a harmonic motion for a specific frequency ω, which repeats itself indefinitely. Two are discussed in the following for a displacement, but they also apply for a force.

In the form with *real quantities* only

$$u(t) = |u(\omega)| \cos(\omega t + \varphi(\omega)) \tag{A.1}$$

with the magnitude $|u(\omega)|$ (a positive value) and the phase angle $\varphi(\omega)$ (covering the range $-\pi$ to $+\pi$).

In the form with a *complex amplitude* $u(\omega) = \operatorname{Re} u(\omega) + i\operatorname{Im} u(\omega)$ and

$$u(t) = \operatorname{Re}(u(\omega)e^{i\omega t}) \tag{A.2}$$

which implies that the result is equal to the real part of the product of the complex amplitude $u(\omega)$ and the exponential function with an imaginary argument $e^{i\omega t}$. The imaginary part is disregarded at the end of the operation. To simplify the nomenclature, the symbol Re in Eq. A.2 is omitted.

To establish the relationship between the constants of the two forms represented by Eqs A.1 and A.2, Eq. A.1 is expanded as

$$u(t) = |u(\omega)| \cos \varphi(\omega) \cos \omega t - |u(\omega)| \sin \varphi(\omega) \sin \omega t \qquad (A.3)$$

and Eq. A.2 as

$$u(t) = \operatorname{Re} u(\omega) \cos \omega t - \operatorname{Im} u(\omega) \sin \omega t \qquad (A.4)$$

Setting the coefficients of $\cos \omega t$ and $\sin \omega t$ equal in Eqs A.3 and A.4 yields

$$\operatorname{Re} u(\omega) = |u(\omega)| \cos \varphi(\omega) \qquad (A.5a)$$
$$\operatorname{Im} u(\omega) = |u(\omega)| \sin \varphi(\omega) \qquad (A.5b)$$

which permits the constants of the complex-amplitude form (Eq. A.2) to be calculated from the real magnitude-phase angle form (Eq. A.1). Going the other way leads to

$$|u(\omega)| = \sqrt{\operatorname{Re} u(\omega)^2 + \operatorname{Im} u(\omega)^2} \qquad (A.6a)$$
$$\tan \varphi(\omega) = \arctan \frac{\operatorname{Im} u(\omega)}{\operatorname{Re} u(\omega)} \qquad (A.6b)$$

To be able to determine the quadrant of $\varphi(\omega)$, the signs of $\operatorname{Im} u(\omega)$ and $\operatorname{Re} u(\omega)$ must also be examined.

The complex amplitude form provides a powerful and compact way of processing harmonic motion. This is further discussed in Appendix A.2. Differentiation with respect to time is also easy to handle. With the displacement

$$u(t) = u(\omega) e^{i\omega t} \qquad (A.7)$$

the velocity $\dot{u}(t)$ follows as

$$\dot{u}(t) = i\omega u(\omega) e^{i\omega t} = \dot{u}(\omega) e^{i\omega t} \qquad (A.8)$$

from which

$$\dot{u}(\omega) = i\omega u(\omega) \qquad (A.9)$$

follows. Differentiation with respect to time thus amounts to multiplication of the amplitude by $i\omega$. For the acceleration

$$\ddot{u}(t) = -\omega^2 u(\omega) e^{i\omega t} \qquad (A.10)$$

applies.

It is worth stressing that, when working with the complex amplitude form, the result in the time domain is equal to the real part of the product of the complex amplitude and $e^{i\omega t}$ (Eq. A.2), and not just the real part of the amplitude (multiplied by $\cos \omega t$, for example).

A.2 Complex frequency response function

The complex frequency response function describes the response of a linear dynamic system to harmonic excitation of frequency ω, which is called the steady-state response.

First, the single-degree-of-freedom system consisting of a spring with constant K, a dashpot with constant C and a mass M illustrated in Fig. A.1a is addressed. The equation of motion in the time domain expresses the equilibrium of the inertial load $M\ddot{u}(t)$, the dashpot force $C\dot{u}(t)$, the spring force $Ku(t)$ and the exterior load $P(t)$, yielding

$$M\ddot{u}(t) + C\dot{u}(t) + Ku(t) = P(t) \tag{A.11}$$

For harmonic excitation with complex amplitudes, as in Eq. A.7,

$$P(t) = P(\omega)e^{i\omega t} \tag{A.12a}$$
$$u(t) = u(\omega)e^{i\omega t} \tag{A.12b}$$

apply. Substituting Eqs A.12, A.8 and A.10 in Eq. A.11 yields, after cancelling $e^{i\omega t}$, the force-displacement relationship expressed in amplitudes

$$P(\omega) = (-\omega^2 M + i\omega C + K)u(\omega) \tag{A.13}$$

or

$$P(\omega) = S(\omega)u(\omega) \tag{A.14}$$

with the dynamic-stiffness coefficient

$$S(\omega) = K\left(1 - \omega^2 \frac{M}{K} + i\omega \frac{C}{K}\right) \tag{A.15}$$

The *dynamic-stiffness coefficient* $S(\omega)$, which represents a *complex frequency response function*, represents the force amplitude for a unit displacement amplitude. The frequency-dependent $S(\omega)$ can be non-dimensionalised with the static-stiffness coefficient K (the spring constant), and a dimensionless frequency

$$a_0 = \omega\sqrt{\frac{M}{K}} \tag{A.16}$$

can be introduced, yielding

$$S(a_0) = K(k(a_0) + ia_0 c(a_0)) \tag{A.17}$$

with the dimensionless spring coefficient $k(a_0)$ and the dimensionless damping coefficient $c(a_0)$

$$k(a_0) = 1 - a_0^2 \tag{A.18a}$$
$$c(a_0) = \frac{C}{\sqrt{KM}} \tag{A.18b}$$

The dimensionless frequency a_0 has been introduced to prepare for the nomenclature used to denote the dynamic-stiffness coefficient of the unbounded soil. Note that for the single-degree-of-freedom system of Fig. A.1a with a constant dashpot constant, the dimensionless damping

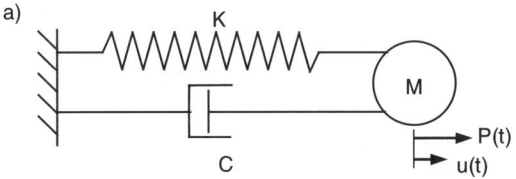
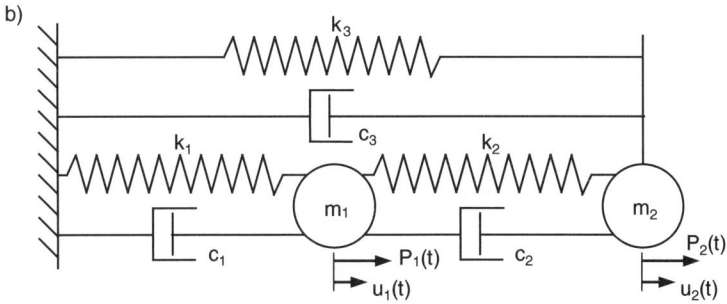

Figure A.1 Dynamic systems. a) Single-degree-of-freedom system. b) Two-degree-of-freedom system

coefficient $c(a_0)$ is a constant, independent of a_0. However, if the dashpot were to model linear hysteretic damping, with a dashpot constant inversely proportional to the frequency $2\zeta K/\omega$, the dashpot force as the product of the constant and the velocity for harmonic motion would be equal to $i2\zeta K$ in Eq. A.13 (hysteretic damping ratio ζ). The frequency ω would cancel, and $c(a_0)$ would become (Eq. A.17)

$$c(a_0) = 2\frac{\zeta}{a_0} \tag{A.19}$$

Solving for $u(\omega)$ in Eq. A.13 leads to

$$u(\omega) = G(\omega)P(\omega) \tag{A.20}$$

with the dynamic-flexibility coefficient $G(\omega)$ (Green's function), another complex frequency response function,

$$G(\omega) = \frac{1}{K - \omega^2 M + i\omega C} \tag{A.21}$$

An example follows to demonstrate how straightforward the analysis using complex amplitudes is. The displacement in the time domain $u(t)$ due to a harmonic load $P(t)$ specified in the real magnitude-phase angle form, with specified magnitude $|P(\omega)|$ and phase angle $\varphi(\omega)$ (analogous to Eq. A.1)

$$P(t) = |P(\omega)|\cos(\omega t + \varphi(\omega)) \tag{A.22}$$

is to be calculated. The complex amplitude of the load $P(\omega) = \text{Re}\, P(\omega) + i\text{Im}\, P(\omega)$ is computed as (analogous to Eq. A.5)

$$P(\omega) = |P(\omega)|\cos\varphi(\omega) + i|P(\omega)|\sin\varphi(\omega) \tag{A.23}$$

150 Foundation Vibration Analysis: A Strength-of-Materials Approach

Substituting Eq. A.23 in Eq. A.20 with Eq. A.21 results in the complex amplitude $u(\omega) = \operatorname{Re} u(\omega) + i\operatorname{Im} u(\omega)$. Finally, the magnitude of the response $|u(\omega)|$ and phase angle $\varphi(\omega)$ follow from Eq. A.6, with $u(t)$ specified in Eq. A.1. Any analysis avoiding complex notation would be much more awkward.

Second, the two-degree-of-freedom system of Fig. A.1b is addressed as an example of a dynamic system with multiple degrees of freedom. The equations of motion in the time domain are

$$[M]\{\ddot{u}(t)\} + [C]\{\dot{u}(t)\} + [K]\{u(t)\} = \{P(t)\} \tag{A.24}$$

with the static-stiffness matrix $[K]$, damping matrix $[C]$, and mass matrix $[M]$

$$[K] = \begin{bmatrix} k_1 + k_2 & -k_2 \\ -k_2 & k_2 + k_3 \end{bmatrix}, \quad [C] = \begin{bmatrix} c_1 + c_2 & -c_2 \\ -c_2 & c_2 + c_3 \end{bmatrix}, \quad [M] = \begin{bmatrix} m_1 & 0 \\ 0 & m_2 \end{bmatrix} \tag{A.25a}$$

and the displacements $\{u(t)\}$ and loads $\{P(t)\}$

$$\{u(t)\} = \begin{Bmatrix} u_1(t) \\ u_2(t) \end{Bmatrix}, \quad \{P(t)\} = \begin{Bmatrix} P_1(t) \\ P_2(t) \end{Bmatrix} \tag{A.25b}$$

For harmonic excitation with amplitudes $\{P(\omega)\} = [P_1(\omega)\ P_2(\omega)]^T$ and $\{u(\omega)\} = [u_1(\omega)\ u_2(\omega)]^T$ as in Eq. A.7,

$$\{P(t)\} = \{P(\omega)\}e^{i\omega t} \tag{A.26a}$$

$$\{u(t)\} = \{u(\omega)\}e^{i\omega t} \tag{A.26b}$$

apply. Substituting Eqs A.26, A.8 and A.10 in Eq. A.24 yields the force-displacement relationship expressed in amplitudes

$$\{P(\omega)\} = [S(\omega)]\{u(\omega)\} \tag{A.27}$$

with the frequency dependent complex dynamic-stiffness matrix

$$[S(\omega)] = [K] - \omega^2[M] + i\omega[C] \tag{A.28}$$

$[S(\omega)]$ represents a complex frequency response function in matrix form, as does the dynamic-flexibility matrix $[G(\omega)]$ (Green's function) appearing in the inverse relationship of Eq. A.27

$$\{u(\omega)\} = [S(\omega)]^{-1}\{P(\omega)\} = [G(\omega)]\{P(\omega)\} \tag{A.29}$$

The extension from the scalar case to the matrix case is thus straightforward. In particular, a decomposition as in Eq. A.17 is possible for each coefficient of the dynamic-stiffness matrix $[S(\omega)]$.

In the time domain the equations of motion are described by a linear system of ordinary differential equations of second order in displacements, with time as the independent variable (Eq. A.24). For harmonic excitation, i.e. in the frequency domain, a linear system of equations in the displacement amplitudes with a complex coefficient matrix as a function of frequency arises (Eq. A.27). The solution of this system for each frequency is straightforward.

Eliminating the amplitude $u_2(\omega)$, for example, from Eq. A.27 yields a scalar relationship in $u_1(\omega)$. Its coefficient represents a dynamic-stiffness coefficient, which can be decomposed as in Eq. A.17. Both $k(a_0)$ and $c(a_0)$ will be rational functions of a_0.

A.3 Periodic excitation

A periodic excitation with period T repeats itself infinitely many times. As an example, a load $P(t)$ is plotted in Fig. A.2, represented by the solid and dashed lines. It can be expressed in terms of the sum of a series of harmonic components called the *Fourier series*. Each term j in the series represents a discrete harmonic excitation with a specific frequency ω_j, as addressed in Appendix A.1. This leads to

$$P(t) = a_0 + \sum_{j=1}^{\infty} a_j \cos \omega_j t + \sum_{j=1}^{\infty} b_j \sin \omega_j t \qquad (A.30)$$

where $\omega_j = j\omega_1$ and $\omega_1 = 2\pi/T$. Multiplying Eq. A.30 by $\cos \omega_k t$ ($k = 0, 1, \ldots, \infty$) and integrating over T, and with

$$\int_0^T \cos \omega_j t \cos \omega_k t \, dt = \begin{cases} 0 & k \neq j \\ \frac{T}{2} & k = j \end{cases} \qquad (A.31a)$$

$$\int_0^T \sin \omega_j t \cos \omega_k t \, dt = 0 \qquad (A.31b)$$

results in the coefficients

$$a_0 = \frac{1}{T} \int_0^T P(t) \, dt \quad \text{and} \quad a_j = \frac{2}{T} \int_0^T P(t) \cos \omega_j t \, dt \qquad (A.32a)$$

Proceeding analogously with $\sin \omega_k t$ yields

$$b_j = \frac{2}{T} \int_0^T P(t) \sin \omega_j t \, dt \qquad (A.32b)$$

In a practical analysis, only a few terms (not an infinite number) are processed.

The form with complex amplitudes and exponential functions, corresponding for a specific term to Eq. A.2 (Eq. A.12a), is derived by replacing the trigonometric functions in Eq. A.30 with

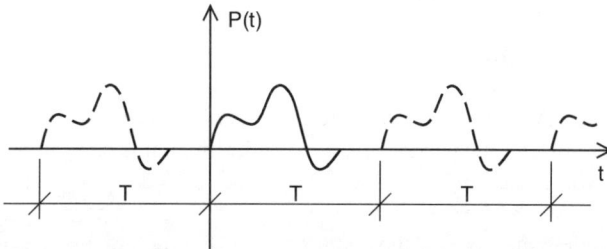

Figure A.2 Periodic excitation

exponential terms using the substitutions

$$\sin \omega_j t = -\frac{1}{2}i\left(e^{i\omega_j t} - e^{-i\omega_j t}\right) \tag{A.33a}$$

$$\cos \omega_j t = \frac{1}{2}\left(e^{i\omega_j t} + e^{-i\omega_j t}\right) \tag{A.33b}$$

This results in

$$P(t) = \sum_{j=-\infty}^{\infty} P(\omega_j) e^{i\omega_j t} \tag{A.34}$$

with (analogous to Eq. A.32)

$$P(\omega_j) = \frac{1}{T}\int_0^T P(t) e^{-i\omega_j t}\,dt = \frac{1}{T}\int_{-T/2}^{T/2} P(t) e^{-i\omega_j t}\,dt, \quad j = 0, \pm 1, \pm 2, \ldots \tag{A.35}$$

The limits of integration have been changed to $-T/2$ and $T/2$, which spans the total period. Note that $P(\omega_{-j})$ is equal to the complex conjugate of $P(\omega_j)$.

The procedure known as the *frequency-domain response analysis* for a linear dynamic system subjected to a periodic load is now discussed (as an example). It is assumed that the transient response caused by the intial displacement and velocity has decayed, which in a typical soil-structure-interaction problem will be the case, as sufficient radiation damping in the unbounded soil is present. Thus, only the steady-state response to the periodic load, which theoretically implies that the load has acted indefinitely, is of interest. Each term in the Fourier series represents a harmonic load with a discrete frequency, and the corresponding response can be determined using the complex frequency response function. The total response is obtained simply by summing the responses to the individual terms.

Turning to the example of Fig. A.1a, the load $P(t)$ is transformed from the time domain to the frequency domain yielding the complex amplitudes $P(\omega_j)$ (Eq. A.35). The response to the jth component in the frequency domain follows, based on the complex frequency response function for the displacement amplitude $u(\omega_j)$ using Eq. A.20 with the dynamic-flexibility coefficient $G(\omega_j)$ in Eq. A.21. The corresponding displacement to the jth component in the time domain is determined from Eq. A.2. Adding the responses to all harmonic loads results in the displacement $u(t)$ in the time domain, which corresponds to Eq. A.34 evaluated for a displacement and not a load.

A.4 Arbitrary excitation

The procedure of Appendix A.3 can be generalised to an excitation that is non-periodic and thus arbitrary. It is sufficient to extend the period T towards infinity. In this case the series expansion, modified appropriately, will represent (for instance) the load plotted as a solid line in Fig. A.2.

Taking this limit $T \to \infty$, the frequency increment $\omega_1 = 2\pi/T$ separating any two distinct frequencies becomes infinitesimal $d\omega$, and the distinct frequencies ω_j are replaced by a *continuous* function ω. The subscript j in Eq. A.34 is dropped, and the summation approaches an integration.

The definition

$$P(\omega) = \int_{-\infty}^{\infty} P(t) e^{-i\omega t} \, dt \tag{A.36}$$

is introduced. Substituting Eq. A.35 with $T = 2\pi/d\omega$ in Eq. A.34 and using Eq. A.36 yields

$$P(t) = \frac{1}{2\pi} \int_{-\infty}^{\infty} P(\omega) e^{i\omega t} \, d\omega \tag{A.37}$$

Thus, the Fourier series becomes a *Fourier integral*. The two integrals $P(\omega)$ (Eq. A.36) and $P(t)$ (Eq. A.37) are known as a *Fourier transform pair*, as the frequency function can be calculated from the time function and vice versa by analogous operations. $P(\omega)$ is the Fourier transform of $P(t)$, and $P(t)$ is the inverse Fourier transform of $P(\omega)$.

The frequency domain analysis of dynamic systems requires that transformations are performed. For instance, the Fourier transform of $P(t)$ must be calculated at the beginning of the analysis, and the inverse Fourier transform of $u(\omega)$ at the end. The integrals are, in general, evaluated numerically. The integrals over the infinite range are truncated, which means that an arbitrary excitation is processed as a periodic one. This leads to discrete Fourier transforms, which are evaluated efficiently as so-called fast Fourier transforms. Important differences exist between continuous and discrete Fourier transforms. Errors can be reduced using corrective solutions.

A discussion of these numerical topics lies outside the scope of this book. The reader is referred to books on structural dynamics, which present a detailed examination.

Appendix B

Dynamic soil-structure interaction

This appendix discusses the form in which the results of a foundation vibration analysis, the dynamic stiffness and, for seismic excitation, the effective foundation input motion, are used in a dynamic soil-structure-interaction analysis. In Appendix B.1 the basic equations of motion of the coupled structure-soil system are derived. The seismic excitation is converted to an equivalent load acting on the dynamic system, which is determined by the free-field motion. For easy reference, the latter is calculated for vertically propagating waves in Appendix B.2.

B.1 Equations of motion in total displacement

To analyse dynamic soil-structure interaction (Fig. B.1) two substructures are defined, the structure and the unbounded soil with the excavation, called system ground (Fig. B.2). The structure-soil interface is assumed to be rigid. Loads can either be applied to the structure, or seismic excitation in the form of vertically propagating waves can be introduced via the soil.

Subscripts are used to denote the nodes of the discretised system. The node at the centre of the base, being part of the structure-soil interface, is denoted as 0, and the remaining nodes of the structure by s. To differentiate between the different subsystems, superscripts are used when necessary. The structure is indicated by s (when used with a property matrix), and the soil with

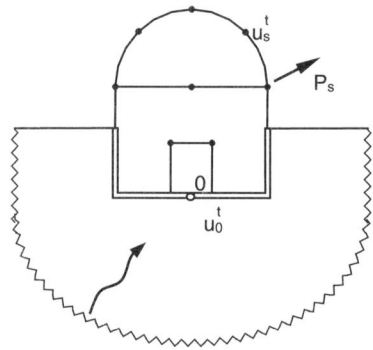

Figure B.1 Structure-soil system with rigid interface

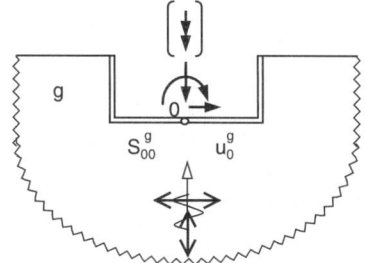

Figure B.2 Substructure soil (system ground) with degrees of freedom and effective foundation input motion (torsional motion vanishes for vertically propagating seismic excitation)

excavation by g (for ground). (Other reference soil systems are introduced later in this appendix and also in the main text of the book, as are other nodes.)

The derivation of the basic equations of motion of the structure-soil system is performed directly for harmonic excitation (in the frequency domain). The equations of motion of the structure are formulated in total displacement amplitudes $\{u^t(\omega)\}$ as

$$\begin{bmatrix} [S_{ss}(\omega)] & [S_{s0}(\omega)] \\ [S_{0s}(\omega)] & [S^s_{00}(\omega)] \end{bmatrix} \begin{Bmatrix} \{u^t_s(\omega)\} \\ \{u^t_0(\omega)\} \end{Bmatrix} = \begin{Bmatrix} \{P_s(\omega)\} \\ -\{P_0(\omega)\} \end{Bmatrix} \quad (B.1)$$

The word total (superscript t) is used to indicate that the motion is referred to a stationary origin. The dynamic-stiffness matrix $[S^s(\omega)]$ of the structure, which is a bounded system, is calculated as (Appendix A.2, Eqs A.28, A.19)

$$[S^s(\omega)] = [K](1 + 2i\zeta) - \omega^2[M] \quad (B.2)$$

where $[K]$ and $[M]$ are the static-stiffness and mass matrices, respectively, and ζ denotes the hysteretic structural damping ratio. The vector $\{u^t(\omega)\}$, of order equal to the number of dynamic degrees of freedom of the total discretised system, can be decomposed into the subvectors $\{u^t_s(\omega)\}$ and $\{u^t_0(\omega)\}$. The latter denotes the amplitudes of the rigid body motion (three translations and three rotations) of the structure-soil interface. $[S^s(\omega)]$ is decomposed accordingly. To avoid unnecessary symbols, the superscript s (for structure) is used only when confusion would otherwise arise. $\{P_s(\omega)\}$ denotes the amplitudes of the loads acting on the structure. Finally, $\{P_0(\omega)\}$ are the amplitudes of the interaction forces of the other substructure, the soil system ground. Note that in this formulation in total displacements, the nodes not in contact with the soil are not loaded by seismic excitation.

To express $\{P_0(\omega)\}$, the unbounded soil system ground with an excavation and a massless rigid structure-soil interface (Fig. B.2) is addressed. $[S^g_{00}(\omega)]$ denotes its dynamic-stiffness matrix, and $\{u^g_0(\omega)\}$ the displacement amplitudes of the soil system ground caused by the earthquake. These two quantities, which represent the results of the foundation vibration analysis, are calculated in Section 5.3 (Eqs 5.24 and 5.25 for a horizontal earthquake). As is evident from Eq. 5.25, $\{u^g_0(\omega)\}$ is a function of the amplitudes $\{u^f(\omega)\}$ of the free-field seismic motion at the locations where subdisks are introduced (Section 5.1, Fig. 5.1). $\{u^g_0(\omega)\}$ denotes the amplitudes of the effective foundation input motion.

For vertically propagating P-waves the free-field displacement is also vertical, and the effective foundation input motion consists of a vertical component, which represents some 'average' of the free-field displacement in the zone of embedment (see also Fig. 1.2b). For vertically propagating

S-waves the free-field displacement is horizontal, varying in the vertical direction, and the effective foundation input motion consists of a horizontal component, which again represents some average of the free-field displacement in the zone of embedment, and a rotational component (rocking) (see also Fig. 1.2a). For vertically propagating waves, no torsional component in the effective foundation input motion is present (Fig. B.2). In the case of a surface foundation, the effective foundation input motion is equal to the corresponding component of the free-field displacement at the free surface, and no rotation occurs.

Now that a qualitative description of $\{u_0^g(\omega)\}$ has been given, the derivation is continued. For the motion $\{u_0^g(\omega)\}$, the interaction forces acting at the node 0 vanish, because for this loading state the rigid structure-soil interface shown in Fig. B.2 is a free surface. The *interaction forces* of the soil will thus depend *on the motion relative to the effective foundation input motion* $\{u_0^g(\omega)\}$, and their amplitudes can be expressed as

$$\{P_0(\omega)\} = [S_{00}^g(\omega)](\{u_0^t(\omega)\} - \{u_0^g(\omega)\}) \tag{B.3}$$

Substituting Eq. B.3 into Eq. B.2 yields

$$\begin{bmatrix} [S_{ss}(\omega)] & [S_{s0}(\omega)] \\ [S_{0s}(\omega)] & [S_{00}^s(\omega)] + [S_{00}^g(\omega)] \end{bmatrix} \begin{Bmatrix} \{u_s^t(\omega)\} \\ \{u_0^t(\omega)\} \end{Bmatrix} = \begin{Bmatrix} \{P_s(\omega)\} \\ [S_{00}^g(\omega)]\{u_0^g(\omega)\} \end{Bmatrix} \tag{B.4}$$

which are the *basic equations of motion of the structure-soil system* with a rigid structure-soil interface expressed in total displacement amplitudes.

In this formulation the earthquake excitation is characterised by $\{u_0^g(\omega)\}$, the seismic rigid-body motion of the node 0 (Fig. B.2) of the reference soil system ground (taking the excavation into account). As $\{u_0^g(\omega)\}$ appears on the right-hand side of Eq. B.4 determining an *equivalent load*, it is called the effective foundation input motion. Note that $\{u_0^g(\omega)\}$ does not occur in the structure-soil system of Fig. B.1. By setting $\{u_0^t(\omega)\} = 0$ in Eq. B.4, it can be deduced that the right-hand side $\{P_0^g(\omega)\} = [S_{00}^g(\omega)]\{u_0^g(\omega)\}$ represents the amplitudes of the equivalent loads exerted on the rigid structure-soil interface at node 0 by the seismic motion when the base is kept fixed. They are also called driving loads (forces).

As discussed in Section 5.3, the equivalent load can be expressed directly as the product of the dynamic-stiffness matrix of the free field discretised at the location of the disks $[S^f(\omega)]$ and the free-field motion at the same points $\{u^f(\omega)\}$ (Eq. 5.25 for a horizontal earthquake), premultiplied by the transpose of the kinematic constraint matrix $[A]$, or

$$[S_{00}^g(\omega)]\{u_0^g(\omega)\} = [A]^T \begin{Bmatrix} [S^f(\omega)]\{u^f(\omega)\} \\ \{0\} \end{Bmatrix} \tag{B.5}$$

$[S^f(\omega)]$ is calculated as an intermediate result in Section 5.2 (Eq. 5.6). Thus there is no need to determine the effective foundation input motion $\{u_0^g(\omega)\}$. The free-field motion $\{u^f(\omega)\}$ is sufficient. Note that the dynamic-stiffness matrix of the embedded foundation $[S_{00}^g(\omega)]$ still has to be determined, as it appears in the coefficient matrix on the left-hand side of the system of equations (Eq. B.4).

The basic equations of motion in total displacements (Eq. B.4) for seismic excitation can be interpreted physically as illustrated in Fig. B.3. The discretised structure $[S^s(\omega)]$ is supported on a generalised spring (spring-dashpot system with frequency-dependent coefficients) characterised

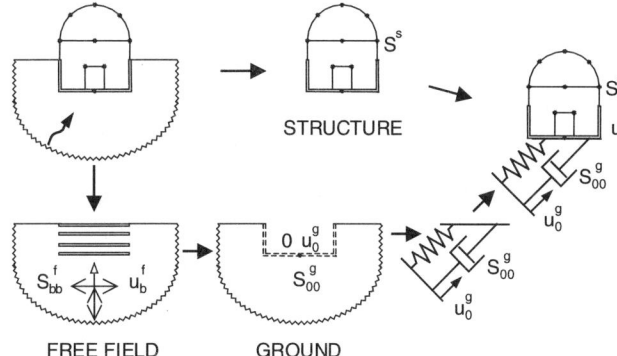

Figure B.3 Physical interpretation of basic equations of motion in total displacements with effective foundation input motion

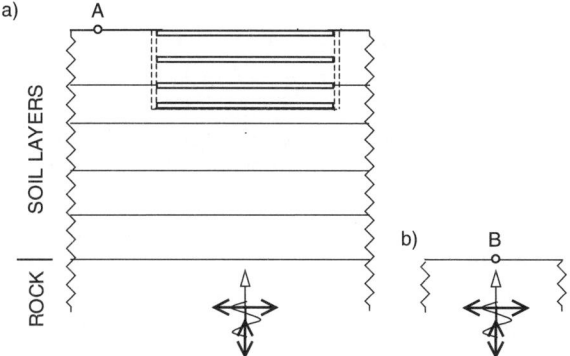

Figure B.4 Selection of control point for seismic input. a) On free surface of site. b) At fictitious outcrop of rock

by $[S_{00}^g(\omega)]$. The end of the generalised spring not connected to the structure is excited by the effective foundation input motion $\{u_0^g(\omega)\}$, which is calculated from the free-field information, $\{u^f(\omega)\}$ and $[S^f(\omega)]$.

B.2 Free-field response of site

A typical site consisting of soil layers overlying a flexible rock half-space is shown in Fig. B.4a. The free-field displacements are to be determined at the locations of the subdisks for vertically propagating S-waves (horizontal earthquake) and P-waves (vertical earthquake). The amplitude of the free-field displacement of the control motion $u_c(\omega)$ (subscript c for control motion) can be defined either at the free surface of the site (Point A), or at an assumed fictitious rock outcrop that is on the level of the rock under the assumption that there is no soil on top (Point B). These two control points are shown in Fig. B.4.

First, the horizontal earthquake is analysed, that is, S-waves with a horizontal particle motion propagating vertically with the shear-wave velocity c_s. This one-dimensional wave propagation can be modelled with a column of soil at the site (Fig. B.5), consisting of $n-1$ soil layers overlying flexible rock with n interfaces (nodes). In each layer j of depth d_j (and in the half-space n) the

158 Foundation Vibration Analysis: A Strength-of-Materials Approach

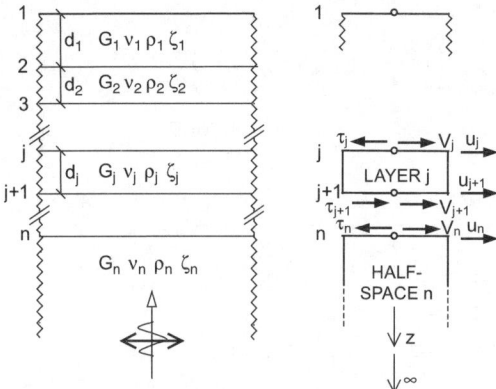

Figure B.5 Column of soil to represent free-field motion for vertically propagating S-waves and corresponding discrete model

shear modulus G_j, Poisson's ratio v_j (for the propagation of P-waves, addressed further on), mass density ρ_j and hysteretic damping ratio ζ_j are constant. The discrete model consists of n nodes with amplitudes $u_j(\omega)$ of the horizontal displacements ($j = 1, 2, \ldots, n$). The one-dimensional jth bar element between nodes j (negative face) and $j + 1$ (positive face) with a vertical axis coinciding with the z-axis (origin at node j) and the amplitudes of the shear stresses $\tau_j(\omega)$ and $\tau_{j+1}(\omega)$, and of the shear forces $V_j(\omega) = -\tau_j(\omega)$ and $V_{j+1}(\omega) = \tau_{j+1}(\omega)$, defined in the global coordinate system, is also shown in Fig. B.5. For the half-space n, the corresponding quantities with $V_n(\omega) = -\tau_n(\omega)$ are also indicated.

The dynamic-stiffness matrix of the jth layer, calculated *exactly* solving the wave equation, is now determined. In Appendix C, the wave equation of a prismatic bar in the time domain is derived as

$$u(z, t)_{,zz} - \frac{1}{c_{sj}^2} \ddot{u}(z, t) = 0 \tag{B.6}$$

using Eq. C.3 formulated for shear and not axial distortions, with shear stresses instead of normal stresses, and changing the wave velocity to c_{sj}. For harmonic excitation, substituting

$$u(z, t) = u(z, \omega) e^{i\omega t} \tag{B.7a}$$

$$\ddot{u}(z, t) = -\omega^2 u(z, \omega) e^{i\omega t} \tag{B.7b}$$

in Eq. B.6 yields

$$u(z, \omega)_{,zz} + \frac{\omega^2}{c_{sj}^2} u(z, \omega) = 0 \tag{B.8}$$

To solve this equation

$$u(z, \omega) = e^{i\gamma z} \tag{B.9}$$

is assumed. Substituting Eq. B.9 in Eq. B.8 leads to

$$\gamma = \pm \frac{\omega}{c_{sj}} \tag{B.10}$$

The solution is thus

$$u(z, \omega) = c_1 e^{i \frac{\omega}{c_{sj}} z} + c_2 e^{-i \frac{\omega}{c_{sj}} z} \tag{B.11}$$

To determine the integration constants c_1 and c_2 the boundary conditions

$$u(0, \omega) = u_j(\omega) \tag{B.12a}$$

$$u(d_j, \omega) = u_{j+1}(\omega) \tag{B.12b}$$

are enforced, resulting in

$$c_1 = \frac{-e^{-i(\omega/c_{sj})d_j} u_j + u_{j+1}}{e^{i(\omega/c_{sj})d_j} - e^{-i(\omega/c_{sj})d_j}} \tag{B.13a}$$

$$c_2 = \frac{e^{i(\omega/c_{sj})d_j} u_j - u_{j+1}}{e^{i(\omega/c_{sj})d_j} - e^{-i(\omega/c_{sj})d_j}} \tag{B.13b}$$

The amplitudes of the shear stresses are

$$\tau(z, \omega) = G_j u(z, \omega)_{,z} \tag{B.14}$$

$\tau_j(\omega)$ and $\tau_{j+1}(\omega)$ follow for $z = 0$ and $z = d_j$ respectively, which also determines the amplitudes of the shear forces

$$V_j(\omega) = -\tau_j(\omega) \tag{B.15a}$$

$$V_{j+1}(\omega) = \tau_{j+1}(\omega) \tag{B.15b}$$

Substituting Eq. B.11 with Eq. B.13 in Eq. B.14, then substituting in Eq. B.15 and expressing the exponential functions in trigonometric functions yields the force-displacement relationship

$$\begin{Bmatrix} V_j(\omega) \\ V_{j+1}(\omega) \end{Bmatrix} = [S_j(\omega)] \begin{Bmatrix} u_j(\omega) \\ u_{j+1}(\omega) \end{Bmatrix} \tag{B.16}$$

with the dynamic-stiffness matrix of the jth layer for vertically propagating S-waves

$$[S_j(\omega)] = \rho_j c_{sj} \frac{\omega}{\sin(\omega d_j/c_{sj})} \begin{bmatrix} \cos(\omega d_j/c_{sj}) & -1 \\ -1 & \cos(\omega d_j/c_{sj}) \end{bmatrix} \tag{B.17}$$

The dynamic-stiffness coefficient of the half-space, modelled as a one-dimensional semi-infinite column of rock, is now calculated. The solution of the wave equation, Eq. B.11, still applies. In the semi-infinite bar, only waves propagating in the positive z-direction (i.e. downwards towards infinity) can exist. As $u(z, \omega)$ is multiplied by $e^{i\omega t}$ (Eq. B.7a), the term with c_1 results in $c_1 e^{i\omega(t+z/c_{sn})}$, which corresponds to a wave propagating in the negative z-direction. This is verified as follows. If t is increased by any value $\bar{t}$ and simultaneously z is decreased by $c_{sn}\bar{t}$, the value $e^{i\omega(t+\bar{t}+z/c_{sn}-c_{sn}\bar{t}/c_{sn})} = e^{i\omega(t+z/c_{sn})}$ is not affected. Thus c_1 vanishes. Selecting the origin of the z-axis at node n and enforcing the boundary condition

$$u(0, \omega) = u_n(\omega) \quad (B.18)$$

the displacement amplitude is

$$u(z, \omega) = u_n(\omega) e^{-i\frac{\omega}{c_{sn}}z} \quad (B.19)$$

Eq. B.14 still applies, and $\tau_n(\omega)$ follows for $z = 0$. With

$$V_n(\omega) = -\tau_n(\omega) \quad (B.20)$$

and substituting accordingly leads to the force-displacement relationship

$$V_n(\omega) = S_n(\omega) u_n(\omega) \quad (B.21)$$

with the dynamic-stiffness coefficient of the half-space for vertically propagating S-waves

$$S_n(\omega) = iG_n \frac{\omega}{c_{sn}} = i\rho_n c_{sn} \omega \quad (B.22)$$

Equation B.21 can also be written as

$$V_n(\omega) = \rho_n c_{sn} \dot{u}_n(\omega) \quad (B.23)$$

which is the same result (for harmonic excitation) as derived in Appendix C (Eq. C.17 with Eq. C.18 for the prismatic bar of unit area, but subjected to shear deformation in the time domain). The one-dimensional wave propagation in the rock half-space leads to a dashpot with the coefficient $\rho_n c_{sn}$.

For hysteretic damping, G and c_s are replaced by G^* and c_s^* (the correspondence principle, Eqs 3.32b, 3.33a) in Eqs B.17 and B.22.

The equations of motion for harmonic excitation of the site can be established assembling the dynamic-stiffness matrices of the layers and the half-space as

$$[S(\omega)]\{u(\omega)\} = \{Q(\omega)\} \quad (B.24)$$

$\{u(\omega)\}$ denotes the displacement amplitudes $u_1(\omega), u_2(\omega) \ldots u_n(\omega)$, $[S(\omega)]$ the assembled dynamic-stiffness matrix and $\{Q(\omega)\}$ the load amplitudes, which depend on the location of the control point. $\{u(\omega)\}$ at those points corresponding to disks are equal to $\{u^f(\omega)\}$ in Eq. B.5.

If the control motion is specified at the free surface (Point A in Fig. B.4a), $u_1(\omega) = u_c(\omega)$. The calculation proceeds from this point downwards with a zero right-hand side $\{Q(\omega)\} = 0$. The equation of motion (Eq. B.24) at node j then leads to the $u_{j+1}(\omega)$ ($j = 1, 2, \ldots, n-1$). The free-field motion of node j is thus independent of the properties of the site below this node.

For a prescribed control motion at the rock outcrop (Point B in Fig. B.4b), the amplitudes of the loads $\{Q(\omega)\}$ can be calculated analogously as the load vector in the basic equations of motion of soil-structure interaction, Eq. B.4. For the site, the soil layers and the rock half-space represent the two substructures (Fig. B.4), the first corresponding to the structure and the second to the soil in the structure-soil system (Fig. B.1). This determines the coefficient matrix on the left-hand side of the equations of motion (Eq. B.24). As mentioned, the rock (Fig. B.4b) represents the system ground. The equivalent load is equal to the load exerted on node n by the seismic motion with amplitude $u_c(\omega)$ acting in the rock half-space when the node n is fixed (Fig. B.2). The amplitude of the equivalent load at node n (driving load) thus equals

$$Q_n(\omega) = S_n(\omega) u_c(\omega) \tag{B.25}$$

All other elements of $\{Q(\omega)\}$ vanish. The free-field motion again follows from Eq. B.24. It is a function of the properties of the total system (site).

Second, the vertical earthquake is addressed, that is P-waves with a vertical particle motion propagating vertically with the dilatational-wave velocity c_p. Again, this one-dimensional wave propagation is modelled with a column of soil at the site. The analysis in analogous to that discussed above for the horizontal earthquake, but with vertical displacements, axial strains, normal stresses and normal forces present. Replacing c_s by c_p, the dynamic stiffnesses (Eqs B.17 and B.22) are

$$[S_j(\omega)] = \rho_j c_{pj} \frac{\omega}{\sin(\omega d_j / c_{pj})} \begin{bmatrix} \cos(\omega d_j / c_{pj}) & -1 \\ -1 & \cos(\omega d_j / c_{pj}) \end{bmatrix} \tag{B.26}$$

and

$$S_n(\omega) = \mathrm{i} \rho_n c_{pn} \omega \tag{B.27}$$

The assemblage process of the site and introduction of the control motion for the two choices of the control point are, again, analogous.

Appendix C

Wave propagation in a semi-infinite prismatic bar

To illustrate the fundamentals of one-dimensional wave propagation, a semi-infinite prismatic bar is examined. After characterising wave motion, the wave equation is derived and its solution is discussed. Then wave propagation towards infinity is examined, and the energy balance is addressed. Finally, the wave reflection and refraction that occur at a material discontinuity are analysed. All investigations are performed working in the time domain.

At the origin O an axial displacement as a prescribed function of time $u_0(t)$ is applied to the bar with area A, modulus of elasticity E and mass density ρ (Fig. C.1a). This enforced displacement starts at $t = 0$ and exhibits any variation up to $t = t_0$. $u_0(t)$ thus vanishes for a negative argument and for an argument larger than t_0. The disturbance propagates along the bar causing a displacement $u(z, t)$, which is called a wave. It takes time for a disturbance to propagate from its source to other positions. As the disturbance, or wave, is transmitted from one cross-section to the next, internal elastic forces and inertial forces are activated. This leads to strain and kinetic energies, which are transmitted by the wave. This transmission of energy occurs because the motion is passed on from one cross-section to the next around an equilibrium position, and not by some global motion of the bar.

To derive the equation of motion, equilibrium of an infinitesimal element (Fig. C.1b) is formulated. With the inertial load mass times acceleration and the normal force $N(z, t)$,

$$-N(z, t) + N(z, t) + N(z, t),_z \, dz - \rho \, A \, dz \, \ddot{u}(z, t) = 0 \tag{C.1}$$

applies. Substituting the force-displacement relationship

$$N(z, t) = E \, A \, u(z, t),_z \tag{C.2}$$

yields the partial differential equation of motion, also called the (one-dimensional) wave equation

$$u(z, t),_{zz} - \frac{1}{c^2} \ddot{u}(z, t) = 0 \tag{C.3}$$

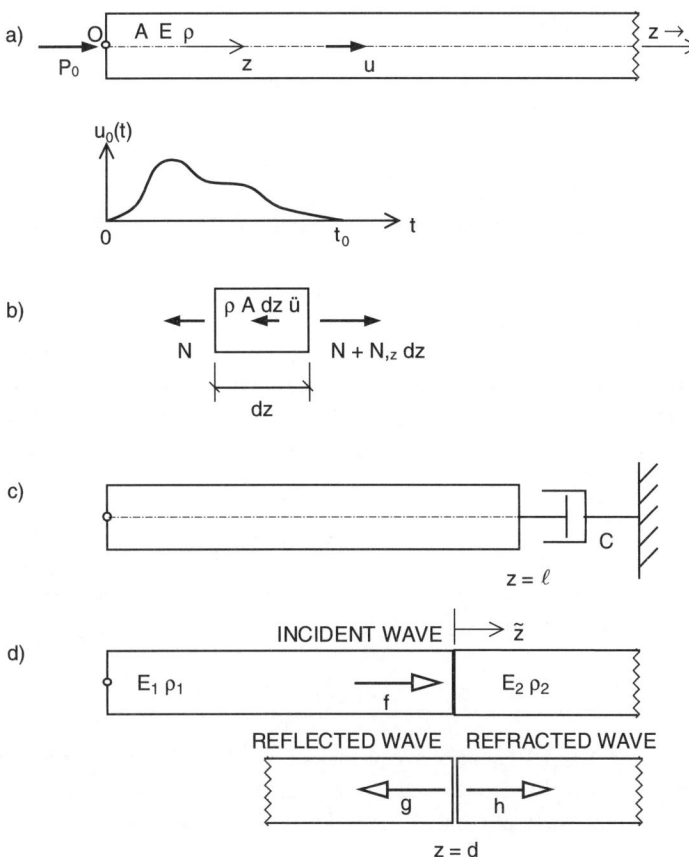

Figure C.1 Wave propagation in semi-infinite prismatic bar. a) Bar with outward wave propagation. b) Equilibrium of infinitesimal element. c) Viscous dashpot modelling outward bar up to infinity. d) Waves at material discontinuity

where c denotes the wave velocity in the bar

$$c = \sqrt{\frac{E}{\rho}} \tag{C.4}$$

The wave will propagate along the axis of the bar, and the motion of the cross-section (particle motion) is in the same direction.

To solve the wave equation (Eq. C.3) directly in the time domain, the variables z and t are changed to

$$\xi = t - \frac{z}{c} \tag{C.5a}$$

$$\eta = t + \frac{z}{c} \tag{C.5b}$$

This transforms Eq. C.3 to

$$u(\xi, \eta),_{\xi\eta} = 0 \tag{C.6}$$

which can be integrated to give

$$u(\xi, \eta) = f(\xi) + g(\eta) \tag{C.7}$$

or

$$u(z, t) = f\left(t - \frac{z}{c}\right) + g\left(t + \frac{z}{c}\right) \tag{C.8}$$

where f and g are two arbitrary functions of their arguments. Equation C.8 is the general solution of the one-dimensional wave equation.

The functions $f(t - z/c)$ and $g(t + z/c)$ can easily be interpreted physically. If t is increased by any value $\bar{t}$ and simultaneously z is increased by $c\bar{t}$, the value $f(t + \bar{t} - (z + c\bar{t})/c)$ is not altered ($= f(t - z/c)$). The function $f(t - z/c)$ thus represents a wave propagating in the positive z-direction with the constant velocity c, without changing its shape. Analogously, $g(t + z/c)$ is a wave propagating in the negative z-direction with the velocity c, without changing its shape.

The actual functions $f(t - z/c)$ and $g(t + z/c)$ are determined by the boundary condition. For the prescribed displacement $u_0(t)$ at O (Fig. C.1a)

$$u(z = 0, t) = u_0(t) \tag{C.9}$$

and postulating an *outward propagating wave away from the source* of the disturbance O (in the positive z-direction, thus involving the function $f(t - z/c)$ and not $g(t + z/c)$) yields

$$u(z, t) = u_0\left(t - \frac{z}{c}\right) \tag{C.10}$$

Thus, it follows from Eq. C.10 that the wave propagation along the semi-infinite bar as a function of time and axial coordinate is determined by the same function u_0 describing the enforced displacement at O but with the argument $t - z/c$ (Fig. C.2a). Note that $u_0(t - z/c)$ vanishes for a negative argument and for an argument larger than t_0. At the origin O at $z = 0$, the displacement $u(t) = u_0(t)$ as a function of time is plotted in Fig. C.1a. At the specific axial coordinate $z = z_1$ (Fig. C.2a), $u(t - z/c)$ is presented as a function of t in Fig. C.2b. It is the same function, just displaced by the propagation time z_1/c of the wave up to z_1. The displacement as a function of the axial coordinate is presented for two specific times ($t_2 < t_0$ and $t_3 > t_0$) in Fig. C.2c. Due to the negative sign of z in the argument $t - z/c$, the function is reversed. For instance, for the time t_3, the wave propagates up to the axial coordinate $z = ct_3$. The displacement vanishes for $z < c(t_3 - t_0)$, as the argument $t_3 - z/c > t_3 - t_3 + t_0 (= t_0)$ is larger than t_0, and u_0 is thus equal to zero (Fig. C.1a).

The reaction force $P_0(t)$ at the origin O (Fig. C.2a), which is applied to generate the displacement $u_0(t)$ is determined as follows. $P_0(t)$ is equal to the negative normal force $N(z = 0, t)$

$$P_0(t) = -N(z = 0, t) \tag{C.11}$$

with $N(z, t)$ specified in Eq. C.2. Substituting Eq. C.10 yields

$$u_0(t - z/c), z\bigg|_{z=0} = \frac{du_0(t - z/c)}{d(t - z/c)} \frac{\partial (t - z/c)}{\partial z}\bigg|_{z=0} = \dot{u}_0(t)\left(-\frac{1}{c}\right) \tag{C.12}$$

Substituting Eqs C.12 and C.2 in Eq. C.11 yields with Eq. C.4

$$P_0(t) = \rho c A \dot{u}_0(t) \tag{C.13}$$

This relationship represents a viscous dashpot with the coefficient $\rho c A$.

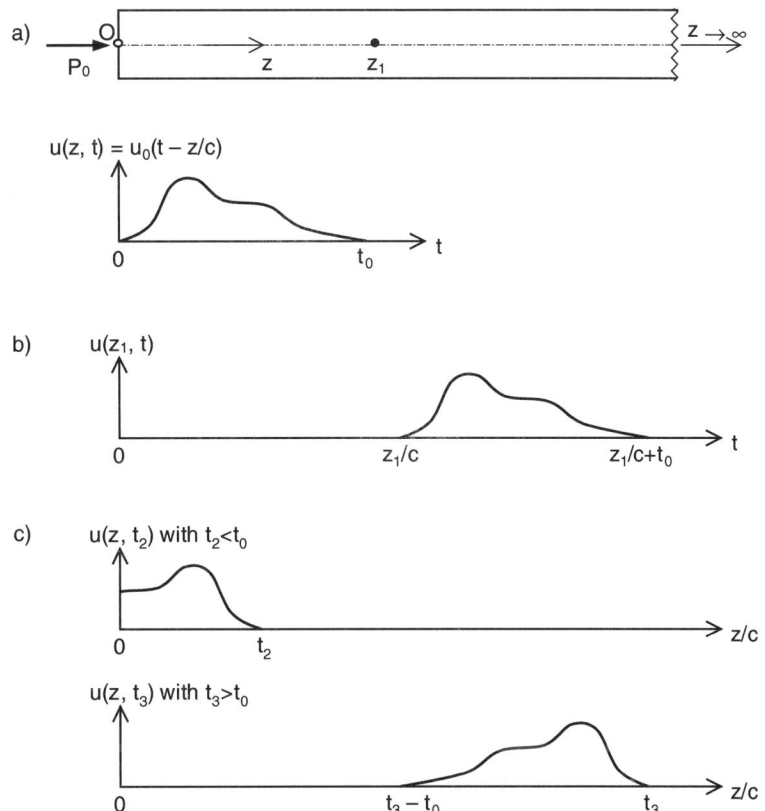

Figure C.2 Wave propagation in semi-infinite prismatic bar as function of time and of axial coordinate. a) Outward wave propagation. b) Wave at specific axial coordinate as function of time. c) Wave at two specific times as function of axial coordinate

It is of interest to determine the differential equation an outward propagating wave satisfies. This can then be used as the boundary condition to be enforced at some coordinate $z = \ell$ (Fig. C.1c), where an artificial boundary can be introduced. The bar outside this artificial boundary up to infinity is thus modelled rigorously. When the incident (outward propagating) wave f encounters the artificial boundary as an incident wave, this wave must pass through this boundary without any modification, so that it can continue propagating towards $z \to \infty$. No reflected wave g, which would propagate in the negative z-direction, may arise. The wave f satisfies the boundary condition, as already mentioned, but g does not. Because the functions $f(t - z/c)$ and $g(t + z/c)$ differ by the signs of the z/c terms in the arguments, a differentiation with respect to z is appropriate. This involves a differentiation of f and g with respect to the argument – for instance $f'(t - z/c) = \mathrm{d}f(t - z/c)/\mathrm{d}(t - z/c)$, which is equal to $\dot{f}(t - z/c)$. The differential equations

$$f(t - z/c),_z + \frac{\dot{f}(t - z/c)}{c} = 0 \tag{C.14a}$$

$$g(t + z/c),_z - \frac{\dot{g}(t + z/c)}{c} = 0 \tag{C.14b}$$

apply. Selecting Eq. C.14a, which is identically satisfied for f, as the boundary condition for the displacement u at $z = \ell$

$$u(z = \ell, \ t),_z + \frac{1}{c}\dot{u}(z = \ell, \ t) = 0 \tag{C.15}$$

results in $g = 0$. This is easily verified by substituting Eq. C.8 in Eq. C.15, which leads to $g'(t + z/c) = \mathrm{d}g(t + z/c)/\mathrm{d}(t + z/c) = \dot{g}(t + z/c) = 0$ at $z = \ell$. Equation C.15 is also called the *radiation condition*.

The physical interpretation of the boundary condition at $z = \ell$ becomes apparent when Eq. C.15 is multiplied by EA, leading to

$$EA\, u(z = \ell, \ t),_z + \frac{EA}{c}\dot{u}(z = \ell, \ t) = 0 \tag{C.16}$$

After substituting Eqs C.2 and C.4,

$$N(z = \ell, \ t) + C\dot{u}(z = \ell, \ t) = 0 \tag{C.17}$$

results, with

$$C = \rho c A \tag{C.18}$$

Equation C.17 expresses equilibrium at the artificial boundary, involving the normal force and the force of a viscous dashpot with a coefficient C (Fig. C.1c). This is also expressed in Eq. C.13. ρc is called the *impedance*. The dashpot thus replaces the part of the bar past the boundary up to infinity.

The energy of the incident wave is continuously being totally dissipated in the dashpot. This is verified as follows. In the infinitesimal element the wave travels a distance $\mathrm{d}z = c\,\mathrm{d}t$ in time $\mathrm{d}t$. The strain energy is equal to half of the product of the normal force and the strain $u(z,t),_z$ multiplied by $\mathrm{d}z$

$$\mathrm{d}E_s = \frac{1}{2}N(z,t)\,u(z,t),_z\,\mathrm{d}z = \frac{1}{2}\rho A\,\dot{u}^2(z,t)\,\mathrm{d}z \tag{C.19}$$

where Eqs C.2, C.4 and the radiation condition Eq. C.15 are substituted. The kinetic energy is equal to half the mass of the infinitesimal element multipled by the square of the velocity,

$$\mathrm{d}E_k = \frac{1}{2}\rho A\,\mathrm{d}z\,\dot{u}^2(z,t) = \frac{1}{2}\rho A\,\dot{u}^2(z,t)\mathrm{d}z \tag{C.20}$$

The energy dissipated in the dashpot is calculated as the product of the force $\rho c A\,\dot{u}(z, \ t)$ and the displacement $\dot{u}(z, \ t)\,\mathrm{d}t$

$$\mathrm{d}E_d = \rho c A\,\dot{u}(z,t)\,\dot{u}(z,t)\,\mathrm{d}t = \rho A\,\dot{u}^2(z,t)\,\mathrm{d}z \tag{C.21}$$

using $\mathrm{d}t = \mathrm{d}z/c$. The term $\mathrm{d}E_d$ is equal to the total energy of the wave $\mathrm{d}E_s + \mathrm{d}E_k$.

It follows that in the unbounded (semi-infinite) bar, although elastic, the *strain and kinetic energies* of the wave *are dissipated through wave propagation towards infinity*, which is called *radiation damping*.

The effect of a discontinuity in the material properties of the bar is now studied. At the interface at $z = d$ with material properties E_1 and $\rho_1 (c_1 = \sqrt{E_1/\rho_1})$ on the left and E_2 and $\rho_2 (c_2 = \sqrt{E_2/\rho_2})$ on the right (Fig. C.1d), displacement compatibility and equilibrium must be satisfied. This requires that the incident wave propagating in the positive z-direction in bar 1 generates two additional waves, a *reflected* wave propagating in the negative z-direction in bar 1, and a *refracted* wave propagating in the positive z-direction in bar 2. Note that both additional waves *propagate outwards away from the discontinuity, which acts as the source of the disturbance*. The incident wave of the displacement follows from Eq. C.10 with the wave velocity c_1 as $f(t - z/c_1) = u_0(t - z/c_1)$. To simplify the nomenclature, a coordinate transformation is introduced, with $z = \tilde{z} + d$ and $t = \tilde{t} + d/c_1$, i.e. the origin of the $\tilde{z}$- and $\tilde{t}$-coordinates lies at the interface at the time when the incident wave $f(\tilde{t} - \tilde{z}/c_1) = u_0(\tilde{t} - \tilde{z}/c_1)$ starts to arrive. The reflected wave, also propagating with velocity c_1, is $g(\tilde{t} + \tilde{z}/c_1)$, and the refracted wave propagating with velocity c_2 is $h(\tilde{t} - \tilde{z}/c_2)$. The stresses follow as $Eu_{,z} = \rho c^2 u_{,\tilde{z}}$ (Eq. C.4), yielding for the three stress waves $\sigma_f(\tilde{t} - \tilde{z}/c_1) = -\rho_1 c_1 \dot{f}(\tilde{t} - \tilde{z}/c_1)$, $\sigma_g(\tilde{t} + \tilde{z}/c_1) = \rho_1 c_1 \dot{g}(\tilde{t} + \tilde{z}/c_1)$ and $\sigma_h(\tilde{t} - \tilde{z}/c_2) = -\rho_2 c_2 \dot{h}(\tilde{t} - \tilde{z}/c_2)$. Note that the derivative of the displacement with respect to the argument is the same as that with respect to time t or $\tilde{t}$. Formulating compatibility of the displacements at the interface $\tilde{z} = 0$ results in

$$f(\tilde{t}) + g(\tilde{t}) = h(\tilde{t}) \tag{C.22}$$

and equilibrium of the stresses at $\tilde{z} = 0$ leads to

$$\sigma_f(\tilde{t}) + \sigma_g(\tilde{t}) = \sigma_h(\tilde{t}) \tag{C.23a}$$

or

$$-\rho_1 c_1 \dot{f}(\tilde{t}) + \rho_1 c_1 \dot{g}(\tilde{t}) + \rho_2 c_2 \dot{h}(\tilde{t}) = 0 \tag{C.23b}$$

Integrating Eq. C.23b with respect to time leads, together with Eq. C.22, to two equations that permit the reflected and refracted waves to be expressed as a function of the incident wave. This yields

$$g(\tilde{t}) = -\alpha\, f(\tilde{t}) \tag{C.24a}$$
$$h(\tilde{t}) = (1 - \alpha)\, f(\tilde{t}) \tag{C.24b}$$

with the reflection coefficient

$$-\alpha = \frac{\rho_1 c_1 - \rho_2 c_2}{\rho_1 c_1 + \rho_2 c_2} \tag{C.25}$$

The reflection coefficient equals the difference of the impedances of the two materials divided by the sum. For the case $\rho_2 c_2 > \rho_1 c_1$, $-\alpha < 0$ results, yielding a reflected wave with the opposite sign of the initial wave.

Fixed and free boundaries can be considered as special cases of the material discontinuity. For a fixed boundary $\rho_2 c_2 \to \infty$, resulting in $-\alpha = -1$ and thus $g(\tilde{t}) = -f(\tilde{t})$, i.e. the incident displacement wave is completely reflected with a change in sign. For a free boundary $\rho_2 c_2 = 0$, leading to $-\alpha = 1$ and thus $g(\tilde{t}) = f(\tilde{t})$, i.e. the incident displacement wave is again completely reflected, but with the same sign.

Appendix D

Historical note

It is appropriate to review the pioneering research that formed the basis of the computational procedure to analyse vibrations of a foundation embedded in a layered half-space using cone models. Only key papers are mentioned which, in retrospect, influenced decisively the development.

The development can be divided into three phases. Phase I consists of the pioneering work up to 1974, and covers the analysis of a surface foundation on a homogeneous half-space with outward propagating waves in a truncated semi-infinite cone. Phase II, with significant breakthroughs from 1990 to 1994 yielding a dependable practical method (although limited in scope), addresses reflected and refracted waves in a cone occurring at a material discontinuity, and introduces modelling of an embedded foundation with a stack of disks modelled as double cones. These concepts permit the analysis of a foundation on or embedded in a layer overlying a flexible half-space (where the layer can also be fixed at its base) and of a foundation embedded in a homogeneous half-space. The accuracy for these cases is sufficient, comparable to or even better than that achieved for a surface foundation on a homogeneous half-space. A multiple-layered half-space is also examined using so-called cone frustums, which, however, requires further assumptions, some of which are difficult to justify. The analysis can be less accurate, with some physically impossible results such as negative radiation damping in the low frequency range. Phase III sees significant streamlining in 2001 and 2002, and generalises the concepts mentioned above for a layer overlying a half-space to a multiple-layered half-space without introducing any additional assumptions. A thorough evaluation demonstrates sufficient accuracy for a large range of practical cases.

The pioneering paper (Ref. [8]) dating back to 1942, which was long before its time, addresses the translational truncated semi-infinite cone to model a foundation on the surface of a homogeneous half-space for vertical and horizontal motions. The rocking motion of a surface foundation on a homogeneous half-space is examined more than 30 years later (in 1974) using shear distortions in a cone (Ref. [17]). A spring-dashpot-mass model with frequency-independent coefficients and one additional degree of freedom is also developed in this paper, which represents the rotational cone exactly, establishing the basis of the lumped-parameter models. The torsional motion of a surface foundation on a homogeneous half-space is analysed using a cone model in Refs [32] and [33]. In all these cases of a homogeneous half-space, the key aspect of wave motion consisting of the outward propagation of waves away from the source of disturbance at the surface foundation with spreading (increase of the tributary area) in the direction of propagation is modelled with a truncated semi-infinite cone.

Again, during a long period of time no research in cones is pursued. It takes more than 15 years until significant progress is reported in three areas. First, the formulation of the semi-infinite cone is made consistent, the application is expanded and the cone's features are better understood. The formulation for the homogeneous half-space is systematically addressed for all degrees of freedom in Ref. [19]. For the rocking motion, the truncated semi-infinite cone exhibits axial distortions, which is the logical choice. The cone model for nearly incompressible soil discussed in Ref. [21] limits the dilatational-wave velocity and introduces a trapped mass. Reservations against using cone models are proved to be unfounded in Ref. [22]. A powerful well-founded formulation to analyse surface foundations on a homogeneous half-space thus exists, but is only of academic interest as layering and embedment are not covered. However, a sound basis exists for further developments.

Second, wave reflection and refraction at a material discontinuity corresponding to a layer interface in a cone is developed. A surface foundation on a layer fixed at its base is analysed with cone segments in Refs [18] and [20]. Waves reflected at the fixed boundary and the free surface propagate in their own cones with the sectional property of the cone model increasing in the direction of wave propagation. The generalisation to a surface foundation on a layer overlying a flexible half-space is discussed for translational and rotational motions in Refs [38] and [39]. Besides reflected waves, refracted waves propagating in their own cones are present.

Third, the extension to a foundation embedded in a homogeneous half-space using a stack of embedded disks modelled with double cones is performed in Ref. [23]. In addition, the effective foundation input motion for seismic excitation is determined. A single pile in a homogeneous half-space is analysed with cones in Ref. [41]. A foundation embedded in a dynamic system consisting of a layer with half-spaces on top and bottom is discussed in Ref. [24].

The research activities in these three areas permit, for a limited but important class of sites (homogeneous half-space, layer overlying half-space), surface and embedded foundations to be analysed with sufficient engineering accuracy.

The concept of using cones is generalised to a surface foundation on and a foundation embedded in a multiple-layered half-space in Ref. [40]. A so-called backbone cone is generated yielding cone frustums for which the dynamic-stiffness matrices can be determined making stringent assumptions. After assemblage, the dynamic-stiffness matrix of the multiple-layered half-space is obtained, which can be used to calculate the dynamic-stiffness matrix of the foundation. Although this procedure, called cone frustums, is generally applicable, it is based on assumptions that are, to a certain extent, difficult to justify. Computational experience is limited, but, for the surface foundation addressed, a significant decrease in accuracy is observed. Radiation damping can even become negative, which is physically impossible.

For a complete description of cone models covering the development up to 1994, the textbook [37] can be consulted, where further references in neighbouring areas are also listed.

Only very recently, from 2001 onwards, the physically appealing concept of one-dimensional wave propagation in cone segments with reflections and refractions occurring at material interfaces as mentioned above is generalised to a multiple-layered half-space without introducing additional assumptions. An explicit formulation is not possible, as infinite sums of infinite sums occur. It is, however, possible to track the reflection and refraction of each incident wave sequentially and determine the resulting wave pattern up to a certain stage by superposition. This important concept is derived in Ref. [42] where the accuracy is also evaluated in a parametric study, and is shown to be more than sufficient. The procedure based on cone frustums of Ref. [40] can thus be regarded as outdated. An efficient computer implementation with recursive evaluation of the reflected and refracted waves is discussed in Ref. [7].

Appendix E

Program CONAN (CONe ANalysis) – user's guide

The methods described in this book are implemented in an executable program named 'CONAN' (CONe ANalysis), available from the book website, http://www.civil.uwa.edu.au/~deeks/conan/. The program executes the analysis procedures much faster than the MATLAB environment, but the logic of the procedures is identical to those described in Appendix F. This appendix is a self-contained user's guide for the program CONAN. Familiarity with dynamic-stiffness coefficients and effective foundation input motion is necessary, but no detailed understanding of the strength-of-materials approach to foundation vibration analysis is required to use the program. The appendix commences with an overview of the program. This is followed by a detailed description of the preparation of input data for the program, including guidance for the construction of suitable layered models. Execution of the program is then described. The final section provides suggestions regarding the use of the results of the program.

E.1 Program overview

CONAN is an executable program that can be used to compute the dynamic-stiffness coefficients and/or the effective foundation input motion for surface and embedded foundations. The site may be horizontally layered, and the foundation may be partially or fully embedded. The wall and horizontal base of the foundation are assumed to be rigid (providing accurate results when the foundation is very stiff in comparison to the site, which is usually the case). The foundation is also assumed to be axi-symmetric. If this is not the case, an equivalent radius must be used. The radius of an embedded foundation may change with the depth.

An input text file is first prepared describing the site and the foundation. The format of this text file is defined in Section E.2. The text file can be prepared using the Microsoft Windows' text editor 'Notepad', which can be accessed through the 'Accessories' submenu of the 'Programs' submenu of the Windows' 'Start' menu, or any other software package capable of creating and modifying text files (which are normally named with a file extension 'txt').

The CONAN program is then executed, and the input text file describing the problem is processed. A simple menu is used to direct the program, indicating which coefficients to calculate, which range of frequencies to process, and whether to normalise the results or not.

The program stores the results into an output text file, and also displays them on the screen as they are calculated. The output text file can also be opened with Notepad, and the results can be printed or transferred to another software package (such as MATLAB or EXCEL) for further processing.

E.2 Problem description

Every site is taken to consist of an upper homogeneous half-space (which may have zero shear modulus and density), a number of homogeneous layers of finite thickness (this number may be zero), and a lower homogeneous half-space (which may have infinite shear modulus and density), as illustrated in Fig. E.1. For convenience the upper and lower half-spaces are referred to as 'layers', even though they have infinite thickness. It is possible to specify an upper half-space with infinite shear modulus and density and/or a lower half-space with zero shear modulus and density, although such situations are unusual.

The radius of the foundation is specified at each layer interface. To simplify the problem description, the foundation radius at the interface immediately below a layer is included in the description of that layer. This radius is specified as zero where the interface is below or above the foundation. If two adjacent interfaces have non-zero radii specified, the radius of the foundation is assumed to vary linearly between them. However, if the radius is non-zero at one interface and zero at the next, the foundation is assumed to extend only to the interface with the non-zero radius.

For each layer of the site (including any homogeneous half-spaces) the material properties of the soil must also be specified.

Each line of the input text file describes one layer of the site, and commences with a character or word indicating the type of layer present. Three characters ('F', 'R' and 'H') or the corresponding words ('FREE', 'RIGID' and 'HALFSPACE') are used to specify the properties of the upper and lower half-spaces, and are only used on the first and last lines of the input file.

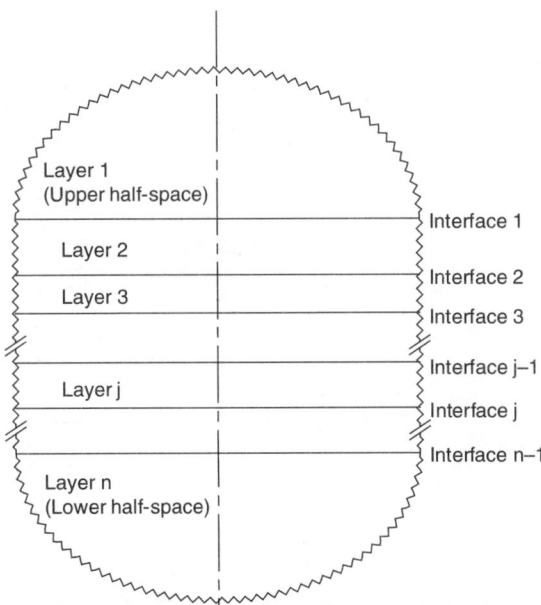

Figure E.1 General description of layered full-space

'F' or 'FREE' designates a free surface, or a half-space with zero shear modulus and zero density. It is usually used to describe the upper half-space on the first line of the input file. The radius of the foundation on the interface (which in the case of the upper half-space is the free surface of the site) is specified as the second item.

'R' or 'RIGID' designates a rigid half-space with infinite shear modulus and infinite density. It is usually used to describe an underlying rock layer on the final line of the input file. The foundation radius in this case is specified as zero, and is ignored in the computations.

'H' or 'HALFSPACE' designates a homogeneous half-space. It is usually used on the last line of the file to describe the underlying half-space, but may also be used on the first line of the file for the analysis of foundations embedded in a full-space. When describing an underlying half-space, the foundation radius is specified as zero and has no effect. However, when used to describe an upper half-space, the foundation radius is significant. After the foundation radius, the input file line specifies the shear modulus (G), Poisson's ratio (ν), density (ρ) and hysteretic damping ratio (ζ) of the soil.

'L' or 'LAYER' designates a layer of finite thickness, and cannot be used on the first or last line of the input file. The letter or word is followed by the foundation radius at the interface below the layer, the shear modulus (G), Poisson's ratio (ν), density (ρ) and hysteretic damping ratio (ζ) of the soil, and the thickness of the layer (d).

For each type of layer, different items on the same line are separated by spaces or tabs. It is necessary to use a consistent set of units for r, G, ρ and d. Newtons, metres and kilograms are usually most convenient, although kilonewtons, metres and tonnes are useful alternatives. The dynamic-stiffness coefficients will be computed in the set of units used for the input data.

Two examples are now presented illustrating how the input text files are prepared. The first example consists of a surface foundation on a layered site, as illustrated in Fig. E.2. The corresponding input file is presented in Listing E.1. The second example consists of an embedded

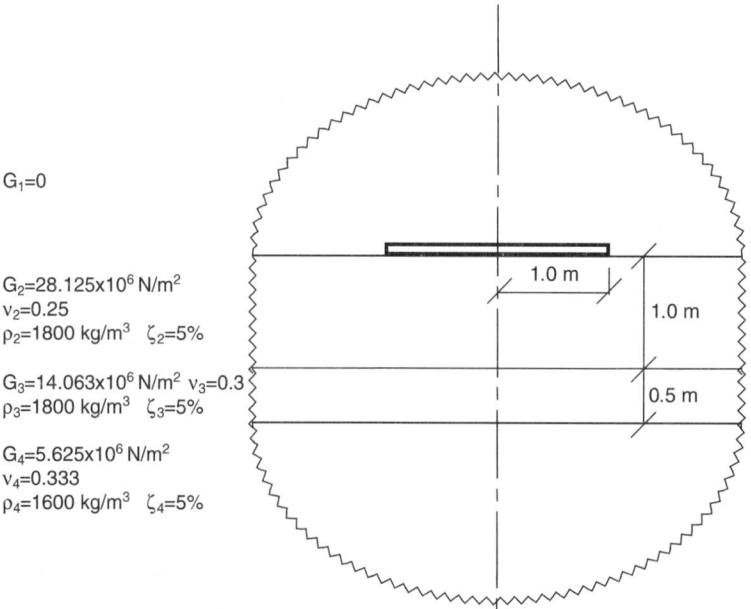

Figure E.2 Example 1 – Surface foundation on layered site overlying homogeneous half-space

```
F     1.0
L     0.0      28.125e6      0.25      1800      0.05      1.0
L     0.0      14.063e6      0.30      1800      0.05      0.5
H     0.0       5.625e6      0.333     1600      0.05
```

Listing E.1 Text input file for Example 1

```
F     1.0
L     1.0      28.125e6      0.25      1800      0.05      0.1
L     1.0      28.125e6      0.25      1800      0.05      0.1
L     1.0      28.125e6      0.25      1800      0.05      0.1
L     1.0      28.125e6      0.25      1800      0.05      0.1
L     1.0      28.125e6      0.25      1800      0.05      0.1
L     1.0      28.125e6      0.25      1800      0.05      0.1
L     1.0      28.125e6      0.25      1800      0.05      0.1
L     1.0      28.125e6      0.25      1800      0.05      0.1
L     1.0      28.125e6      0.25      1800      0.05      0.1
L     1.0      28.125e6      0.25      1800      0.05      0.1
L     1.0      14.063e6      0.30      1800      0.05      0.0625
L     1.0      14.063e6      0.30      1800      0.05      0.0625
L     1.0      14.063e6      0.30      1800      0.05      0.0625
L     1.0      14.063e6      0.30      1800      0.05      0.0625
L     0.0      14.063e6      0.30      1800      0.05      0.25
L     0.0       5.625e6      0.333     1600      0.05      1.0
R     0.0
```

Listing E.2 Text input file for Example 2

foundation in a similar site, but with an underlying rock layer. The situation is illustrated in Fig. E.3. In this case the embedded foundation is discretised in the vertical direction, with the actual layers subdivided into fictitious layers such that the thickness d of each layer satisfies (Eq. 5.1a)

$$d \leq \frac{\pi c}{5\omega} \tag{E.1}$$

ω represents the highest frequency the dynamic model must accurately be able to represent, and c designates the appropriate wave velocity (in general the shear-wave velocity $c_s = \sqrt{G/\rho}$). This procedure is described in Section 5.1.

E.3 Using CONAN

The program is executed in the Windows environment by double-clicking on the CONAN icon. The user is first prompted to enter the name of the input text file containing the problem description.

174 Foundation Vibration Analysis: A Strength-of-Materials Approach

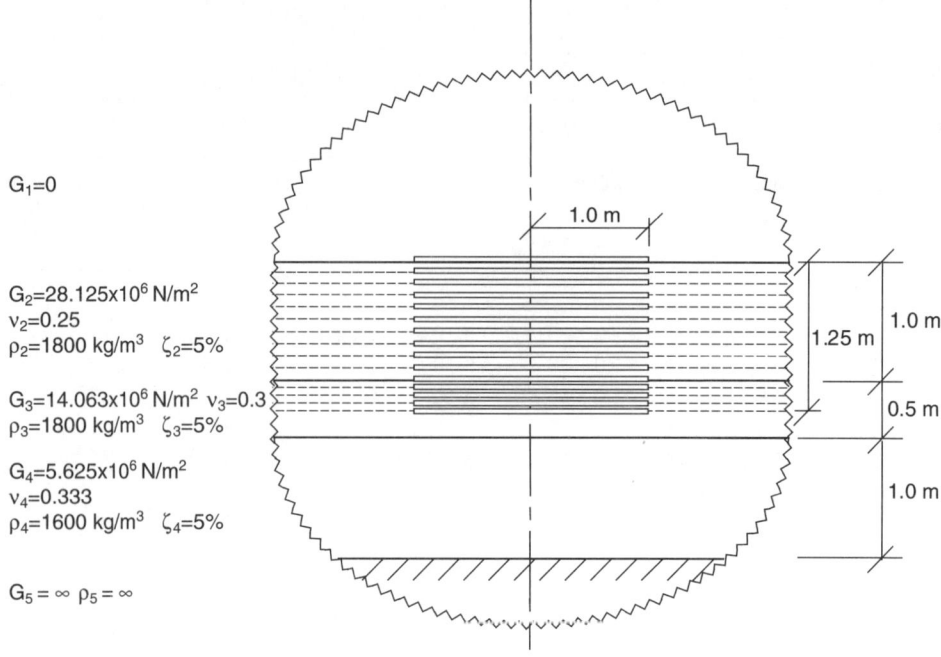

Figure E.3 Example 2 – Foundation embedded in layered site overlying rigid base

The full name of the file should be entered, including the 'txt' extension, if present. The contents of the text file are echoed to the program window for checking.

A menu controlling the execution of the program is then presented. Each item in the menu is addressed separately in the following paragraphs.

1 *Dynamic-stiffness coefficient.* This option computes the dynamic-stiffness coefficient at one or more excitation frequencies for a chosen degree of freedom. When this option is chosen, the user is prompted to enter the degree of freedom. This is specified using a single character. 'H' indicates the horizontal degree of freedom, in which case the rocking degree of freedom (for an embedded foundation) is constrained to be zero. 'V' indicates the vertical degree of freedom, while 'T' indicates the torsional degree of freedom. 'R' indicates the rocking degree of freedom, but also computes the dynamic-stiffness coefficients for the horizontal degree of freedom and for the cross-coupling between the horizontal and rocking degrees of freedom. The user is then prompted to enter the first excitation frequency (in rad/s), followed by the last excitation frequency. If these are not identical, the frequency interval at which computation is required (the frequency step) is requested. The user is then asked whether or not normalisation of the results is required. If 'N' is entered, for 'No', the results are output in dimensional form, tabulated against excitation frequency ω specified in rad/s. If 'Y' is entered, for 'Yes', the static-stiffness coefficient K for the foundation is computed first, and used to decompose the dynamic-stiffness coefficient into the form $S(a_0) = K[k(a_0) + ia_0\, c(a_0)]$. The static-stiffness coefficient is output, while $k(a_0)$ and $c(a_0)$ are tabulated against the real part of dimensionless frequency, $a_0 = \omega r_0/c_s$. If for an embedded foundation the radius varies, the largest radius is used for r_0. The material properties of the layer below the upper half-space are used to determine c_s, ignoring hysteretic damping. Finally the user is requested to enter the name of

the output text file. As the results are computed, they are written both on the screen and to this output text file.

2 *Effective foundation input motion.* When this option is chosen, the user is requested to select between the horizontal 'H' and vertical 'V' degrees of freedom by entering the appropriate single character code. This degree of freedom indicates the direction of the ground motion generated by the earthquake. For a horizontal earthquake the computed effective foundation input motion consists of both horizontal and rocking motions, while for a vertical earthquake it consists of vertical motion only. The user is then requested to enter the control point location. This is the position where the control motion (free field) is to be specified, and can either be on the surface of the site (taken to be at the interface beneath the upper half-space) or on the free surface of a fictitious rock outcrop (taken to be at the interface above the lower half-space). The first, last and step frequencies (in rad/s) are then specified as described for the dynamic-stiffness coefficient. The user is asked whether or not normalisation of the results is required. The normalisation just causes the excitation frequency for the output to be selected as a_0 in dimensionless terms in the final tabulation. The dimensionless frequency a_0 is determined as described for the dynamic-stiffness coefficient. Whether or not normalisation is specified, the effective foundation input motion is normalised by the control motion amplitude. Finally, the user is requested to enter the name of the output text file. As the results are computed they are written both on the screen and to this text file.

3 *Both.* This option computes both the dynamic-stiffness coefficient and the effective foundation input motion for a range of frequencies. The degree of freedom is limited to horizontal 'H' and vertical 'V' motion. When the character for horizontal motion is entered, horizontal, rocking and coupled dynamic-stiffness coefficients are computed, along with horizontal and rocking effective foundation input motions. In the case that the character for vertical motion is entered, only the dynamic-stiffness coefficient and effective foundation input motion for vertical motion will be computed. The control point location, the frequency range for computation and whether or not normalisation of the results is required must also be specified, as described above.

4 *Read new model description from file.* This option discards the current site description and reads a new problem description from another input text file specified by the user. The option allows analysis of multiple models without having to execute the program separately for each one.

5 *Change the calibration method (currently <method>).* This option indicates the calibration method for the cone aspect ratios currently being used in the analysis. By default <method> is 'half-space', but this can be changed to 'full-space' by selecting this option. For details regarding the two calibration methods the reader should refer to Section 3.6. Selecting the option again will change the calibration method back to 'half-space'. The option thus 'toggles' the calibration method.

6 *Change reference point (currently <location>).* This option changes the point on the embedded foundation where the horizontal and rocking degrees of freedom are defined when computing the horizontal, rocking and coupling dynamic-stiffness coefficients. This is the point at which the structural model will be connected when a dynamic soil-structure interaction analysis is undertaken. By default <location> is 'bottom', but this can be changed to 'top' by selecting this option. The option toggles between 'bottom' and 'top'.

7 *Change internal soil treatment (currently <treatment>).* This option changes how the soil inside the foundation is treated during the analysis. By default <treatment> is 'excavate', indicating that the soil is analytically excavated, as described in Section 5.3. However, in some situations, such as the suction caisson addressed in Section 7.3, the soil is not excavated. This option toggles the treatment between 'excavate' and 'none'.

8 *Quit.* This option exits the program.

E.4 Further processing of results

The results from CONAN are saved in an output text file, the name of which is specified by the user. Columns are separated by tab characters, so the file can easily be transferred into a spreadsheet program, such as Excel. Real and imaginary components are tabulated in consecutive columns. The data can be transferred to the MATLAB environment by copying from the output file, then performing a 'Paste special…' in MATLAB, choosing the option to paste the columns as real vectors. Columns containing real and imaginary components must be combined together to form complex vectors before use. This procedure is discussed further in Appendix F.8.

Alternatively, further programs can be written that access the output file to obtain the dynamic stiffness of the foundation and then provide a higher level of structural modelling than that discussed in this book.

For simpler problems analysis can be performed in the Excel environment, although the description of vibration with real quantities must be used (Appendix A.1), as complex numbers cannot be handled readily.

Appendix F

MATLAB® Procedures for cone analysis

This appendix describes and lists functions allowing foundation vibration analysis to be carried out in the MATLAB environment using the cone approach described in this book. In addition to permitting the reader to carry out all the examples presented in the book, the listings of the functions illustrate how easily the approach may be implemented. The reader is also able to combine and expand the functions to handle problems beyond those discussed, or to implement them in other computer languages. Updates to these procedures will be posted on the book website (http://www.civil.uwa.edu.au/~deeks/conan/), and implementations in other languages (such as Fortran, C, C++ and Pascal) are envisaged. The reader should check the website from time to time.

F.1 MATLAB overview

MATLAB provides an interactive technical computation environment in which a user can manipulate mathematical entities, such as matrix and scalar quantities, functions and equations. Results of calculations can be presented numerically or graphically. The user can define custom functions using a simple procedural programming language in *M-files*. MATLAB is a commercial product of The MathWorks, Inc., and is widely used in industry and education. Further information can be obtained from the website http://www.mathworks.com.

Since the user works interactively with the data, a single 'program' is not presented in this appendix. Instead, a series of functions (M-files) is detailed, which the user can employ in the interactive environment to perform the required analysis. In particular, calculations are performed using dimensional quantities, and non-dimensionalisation of the results must be done by the user in the MATLAB environment, as required. A significant advantage of this approach is that the user can employ MATLAB capabilities, such as forward and inverse Fourier transforms, to perform the entire analysis for a dynamic structure-soil system using the techniques described in Appendix A. However, a detailed understanding of the material presented in this book is necessary to take full advantage of the MATLAB approach. The casual user should employ the executable program described in Appendix E.

F.2 Problem description

The layered site and the dimensions of the foundation are described in a single table, as discussed in Section 4.5 and Appendix E.2. Each line in the table represents a layer in the site and the interface below it (Fig. E.1). For readability of the program code each table line is represented by a data structure with fields corresponding to the columns of the table. In MATLAB the problem geometry and material properties are then contained in a single array of these data structures. Any layer can be referred to in its entirety as a member of this array. For example, if the name of the array is *layers*, the following interpretations are made.

`layers`	: array containing all data needed to describe the model
`layers(j)`	: variable containing all data describing the jth layer and underlying interface
`layers(j).type`	: 'H' = half-space, 'L' = finite layer, 'F' = free surface, 'R' = rigid half-space
`layers(j).disk`	: radius of disk at jth interface, zero if no disk
`layers(j).G`	: shear modulus of jth layer if type is 'H' or 'L', otherwise ignored
`layers(j).nu`	: Poisson's ratio of jth layer if type is 'H' or 'L', otherwise ignored
`layers(j).rho`	: material density of jth layer if type is 'H' or 'L', otherwise ignored
`layers(j).damping`	: hysteretic damping ratio of jth layer if type is 'H' or 'L', otherwise ignored
`layers(j).thickness`	: thickness of jth layer if type is 'L', otherwise ignored

The array containing the model data can be introduced into the MATLAB environment member by member from the command line. For the example illustrated in Fig. E.2, the data contained in Listing E.1 can be brought into the MATLAB environment in the following way.

```
>> layers(1).type='F'; layers(1).disk=1;
>> layers(2).type='L'; layers(2).G=28.125e6; layers(2).nu=0.25;
>> layers(2).rho=1800; layers(2).damping=0.05; layers(2).thickness=1;
>> layers(3).type='L'; layers(3).G=14.063e6; layers(3).nu=0.3;
>> layers(3).rho=1800; layers(3).damping=0.05; layers(3).thickness=0.5;
>> layers(4).type='H'; layers(4).G=5.625e6; layers(4).nu=1/3;
>> layers(4).rho=1600; layers(4).damping=0.05;
```

However, this method of data entry becomes tedious for models with many layers. Often the construction of a data file describing the model is a more convenient (and permanent) approach. To permit a text data file of the form described in Appendix E to be read into the MATLAB environment, a function *ReadLayers()* is provided in Listing F.1. This allows the same data files to be used with the executable CONAN program and in the interactive MATLAB environment.

On entry *ReadLayers()* requires the name of the text file containing the data as a string, which is stored in the variable *fileName*. The function reads through each line of the file. If the line begins with 'F', 'L', 'H' or 'R', the data for a layer and interface corresponding to a row in the data table are read. All other lines are ignored, and can be used for comments. (Use of a consistent symbol at the beginning of comment lines is recommended, such as '*'. This avoids inadvertent use of

```
function layer=ReadLayers(fileName)
    infile = fopen(fileName,'r');
    n = 1;
    while ~feof(infile)
        c = fscanf(infile,'%s',[1,1]);
        if isequal(c,'F') | isequal(c,'L') | isequal(c,'H') | isequal(c,'R')
            layer(n).type = c;
            layer(n).disk = fscanf(infile,'%f',[1,1]);
            if isequal(layer(n).type,'L') | isequal(layer(n).type,'H')
                layer(n).G = fscanf(infile,'%f',[1,1]);
                layer(n).nu = fscanf(infile,'%f',[1,1]);
                layer(n).rho = fscanf(infile,'%f',[1,1]);
                layer(n).damping = fscanf(infile,'%f',[1,1]);
                if isequal(layer(n).type,'L')
                    layer(n).thickness = fscanf(infile,'%f',[1,1]);
                end
            end
            n = n + 1;
        end
    end
    fclose(infile);
```

Listing F.1 Function to read model data from text file

one of the reserved characters at the beginning of a comment line.) On exit the function returns the array containing the data structures to the MATLAB environment. This should be assigned an array name in the following way.

```
>> layers = ReadLayers('ExampleE1.txt');
```

If the file 'ExampleE1.txt' contains Listing E.1, after this command is executed the array *layers* contains the same information as if the information had been entered interactively at the command line, as indicated above.

F.3 General functions

This section lists functions which can be used independently to return certain information about wave propagation within the cones, but which are also used by the higher level functions (presented later), which compute the dynamic stiffness of the foundation and effective foundation input motion for a soil-structure interaction analysis. The functions calculate the wave velocity within a layer, the aspect ratio of a particular cone, the dynamic stiffness of a truncated semi-infinite cone, and the trapped mass term required for nearly-incompressible and incompressible materials. In each case the function must be told which degree of freedom is to be processed, and for this purpose the characters 'H' (horizontal), 'V' (vertical), 'T' (torsional) and 'R' (rocking) are employed.

The function *WaveVelocity()* calculates the wave velocity within a single layer. It is called with the data structure for a single layer in which the velocity is required (*layer*) and a character indicating the degree of freedom for the wave (*dof*). On completion it returns the wave velocity, which is complex if the hysteretic damping ratio of the material is non-zero. The equations employed are referenced in Listing F.2. As an example of use of the procedure, assuming *layers*

180   Foundation Vibration Analysis: A Strength-of-Materials Approach

```
function c = WaveVelocity(layer, dof)
    if isequal(layer.type,'L') | isequal(layer.type,'H')
        G = layer.G*(1.0+i*2.0*layer.damping); % correspondence principle, Sec 3.1
        nu = layer.nu;
        rho = layer.rho;
        if isequal(dof,'H') | isequal(dof,'T')
            c = sqrt(G/rho);                    % Eq. 2.3
        elseif isequal(dof,'V') | isequal(dof,'R')
            if nu>1/3 % nearly incompressible or incompressible
                c = 2.0*sqrt(G)/rho;            % limited to 2cs, Sec 3.4
            else % compressible
                c = sqrt(2.0*G*(1.0-nu)/(rho*(1.0-2.0*nu))); % Eq. 2.1
            end
        end
    else
        c = 0.0;
    end
```

Listing F.2   Function to determine wave velocity within layer

contains the data for Example 1 in Appendix E.2 (Fig. E.2).

```
>> c1 = WaveVelocity(layers(2),'V')
```

stores the wave velocity for the vertical wave in the layer beneath the disk into the variable $c1$.

In a similar manner, the function *AspectRatio()* calculates the aspect ratio of the cone corresponding to a single layer (*layer*) for a particular degree of freedom (*dof*). As discussed in Section 3.6, the cone's static stiffness may be calibrated against a half-space or against a full-space. The *AspectRatio()* function provides both options. By default it will return the aspect ratio based on a half-space calibration. However, if the MATLAB environment contains a global variable *calibration*, and the value of this variable is 'full-space', the function will return the aspect ratio based on a full-space calibration. The equations used to determine the aspect ratio in each case are indicated in Listing F.3. The function is used as follows. To determine the aspect ratio for the vertical degree of freedom of the fourth layer (for example, the half-space in Example 1 in Appendix E.2 (Fig. E.2)),

```
>> arhs = AspectRatio(layers(4),'V')
```

Alternatively, to obtain the full-space calibration,

```
>> global calibration, calibration = 'full-space'
>> arfs = AspectRatio(layers(4),'V')
```

The full-space calibration remains in force until the global calibration variable is changed. Consequently it also affects any higher order functions which use the *AspectRatio()* function.

The *ConeStiffness()* function returns the frequency dependent dynamic-stiffness coefficient of a truncated semi-infinite cone with the properties of a layer (*layer*) and a specified initial radius (r). On entry to the procedure the angular frequency (w) and degree of freedom (*dof*) are also required. Calling the function with an angular frequency of zero yields the static-stiffness coefficient of the cone (the real part of the result must be taken if the hysteretic damping is not zero). Since the dynamic-stiffness coefficient of the truncated cone is the same as that of a disk of radius r on a homogeneous half-space with the properties of the layer (provided the half-space calibration is used for the aspect ratio), the function can also be used to determine this coefficient. The static-stiffness coefficient of the top layer may, in some cases, be used to non-dimensionalise the dynamic stiffness of a foundation on a layered site. However, this function

```matlab
function ar = AspectRatio(layer, dof)
    global calibration;
    nu = layer.nu;
    if(~strcmp(calibration,'full-space')) % i.e. standard half-space calibration
        switch dof
            case 'H'
                ar = 0.125*pi*(2-nu);  % Eq. 3.31
            case 'V'
                    if nu>1/3
                        ar = pi*(1-nu);  % Eq. 3.28 with c = 2*cs
                    else
                        ar = 0.5*pi*(1-nu)^2/(1-2*nu);  % Eq. 2.16
                    end
            case 'R'
                    if nu>1/3
                        ar = 1.125*pi*(1-nu);  % Eq. 3.62 with c = 2*cs
                    else
                        ar = 0.5625*pi*(1-nu)^2/(1-2*nu);  % Eq. 3.62 with c = cp
                    end
            case 'T'
                ar = 0.28125*pi;  % Eq. 3.55
        end
    else % alternate full-space calibration
        switch dof
            case 'H'
                ar = pi/32*(7-8*nu)/(1-nu);  % Eq. 3.106
            case 'V'
                    if nu>1/3
                        ar = pi/4*(3-4*nu)/(1-nu);  % Eq. 3.104 with c = 2*cs
                    else
                        ar = pi/8*(3-4*nu)/(1-2*nu);  % Eq. 3.104 with c = cp
                    end
            case 'R'
                    if nu>1/3
                        ar = 9*pi/32*(3-4*nu)/(1-nu);  % Eq. 3.102 with c = 2*cs
                    else
                        ar = 9*pi/64*(3-4*nu)/(1-2*nu);  % Eq. 3.102 with c = cp
                    end
            case 'T'
                ar = 0.28125*pi;  % Eq. 3.55
        end
    end
```

Listing F.3 Function to determine cone aspect ratio for layer

is mainly used to determine the initial force applied to a disk in the computational procedure for determining the dynamic-stiffness coefficients of the foundation. The equations applied are indicated in Listing F.4.

For nearly-incompressible and incompressible materials ($1/3 < \nu \leq 1/2$) with vertical or rocking motion, a trapped mass term or mass moment of inertia term, depending on the degree of freedom, appears. The value of this term is computed in the function *TrappedMass()*, presented in Listing F.5. The value of the trapped mass term depends on the properties of the layer (*layer*) and the radius of the disk (r).

F.4 Heart of the procedure

The functions described in this section implement the two building blocks of the computational procedure described in this book. The heart of the computational procedure is the recursive

182 Foundation Vibration Analysis: A Strength-of-Materials Approach

```
function S=ConeStiffness(layer, r, w, dof)
   if isequal(layer.type,'F')
      S = 0.0;
   elseif isequal(layer.type,'R')
      S = Inf;
   else % isequal(layer.type,'L') | isequal(layer.type,'H')
      c = WaveVelocity(layer,dof);
      z0 = AspectRatio(layer,dof)*r;
      if isequal(dof,'H') | isequal(dof,'V')
         A0 = pi*r^2;
         S = layer.rho*c^2*A0/z0 + i*w*layer.rho*c*A0;   % Eq. 3.16
      else % isequal(dof,'R') | isequal(dof,'T')
         b0 = w*z0/c;
         if isequal(dof,'T')
            I0 = 0.5*pi*r^4;
         else % isequal(dof,'R')
            I0 = 0.25*pi*r^4;
         end
         S = 3.0*layer.rho*c^2*I0/z0*(1.0-(b0^2-i*b0^3)/(3.0+3.0*b0^2));
            % Eqs 3.45-3.47
      end
   end
```

Listing F.4 Function to determine dynamic stiffness of truncated semi-infinite cone

```
function M=TrappedMass(layer, r, dof)
   M = 0;
   if isequal(layer.type,'H') | isequal(layer.type,'L')
     nu = layer.nu;
     if nu>1/3
       if isequal(dof,'V')
           M = 2.4*(nu-1/3)*layer.rho*pi*r^3;       % Eq. 3.84
       elseif isequal(dof,'R')
           M = 1.2*(nu-1/3)*layer.rho*0.25*pi*r^5;  % Eq. 3.85
       end
     end
   end
```

Listing F.5 Function to determine trapped mass or mass moment of inertia

function *Transmit()* (Listing F.6). This function tracks an incident wave through a layer to an interface, computes the reflected and refracted waves propagating in their own cones, then calls itself to process these waves as new incident waves. If one of these new waves impinges on an interface, further reflected and refracted waves are generated, which the function again processes by calling itself. The recursion continues until either the magnitude of the wave is less than 0.0001 of the initial wave, or the depth in the wave tree is greater than $20 + 2\times$ the number of layers.

Due to the importance and sophistication of this procedure, it will now be described in more detail. Before the procedure is called, a global array containing a copy of the array describing the layered model is created. The array is made global for reasons of efficiency and memory usage. If the entire array were passed to the *Transmit()* function as a parameter, each instance of the function would make its own copy of the data, which would be time-consuming and wasteful of stack space. Also, the displacement amplitudes at the interfaces are cumulated in fields of the array, and MATLAB does not allow function parameters to be altered by a function.

```
function Transmit(n, dirn, r0, u0, w, dof)
   global laYers
   global coneCounter
   global depthCounter
   coneCounter = coneCounter + 1;
   depthCounter = depthCounter + 1;
   maxDepth = 20 + 2*(size(laYers,2)-2);
   if dirn==1 % DOWN
      layerA = laYers(n);
      layerB = laYers(n+1);
   else % dirn==-1 %UP
      layerA = laYers(n+1);
      layerB = laYers(n);
   end
   r = r0 + layerA.thickness/AspectRatio(layerA,dof);   % Eq 4.4
   f = Attenuation(layerA,r0,r,w,dof)*u0;               % Eq 4.5 or 4.26b
   if norm(f)>0.0001
      if depthCounter > maxDepth
         f = f*(1.0 - 0.1*(depthCounter-maxDepth));   % kill off wave within 10 steps
      end
      g = -Alpha(layerA,layerB,r,w,dof)*f;             % Eq 4.19
      h = f + g;                                        % Eq 4.10
      laYers(n).amplitude = laYers(n).amplitude + h;
      % reflected wave
      Transmit(n-dirn,-dirn,r,g,w,dof);
      % refracted wave
      if isequal(layerB.type,'L')
         Transmit(n+dirn,dirn,r,h,w,dof);
      end
   end
   depthCounter = depthCounter - 1;
```

Listing F.6 Function to compute all reflections and refractions generated by incident wave

The reason for using a global *copy* rather than making the original data global will become clearer when the higher level functions are considered in the next section. In essence the higher level functions create the global copy, and the user does not need to be concerned with it. Hiding the global copy allows the user to retain several data arrays containing data for various models in the MATLAB environment, then call the higher level function passing the name of the model data of immediate interest. The higher level function makes a global copy of this data named *laYers*, and calls *Transmit()* in due course. *Transmit()* operates on the data in the global *laYers* array.

Two other global variables are also declared in the procedure. *coneCounter* is used to count the total number of reflections and refractions processed in the course of an analysis. This is for information only. The calling function initialises *coneCounter* to zero before calling *Transmit()*. Each time *Transmit()* is called (to process a wave propagating in a cone), *coneCounter* is incremented. The user can inspect the value of *coneCounter* after the analysis is completed. The second global variable, *depthCounter*, is used in the termination procedure. It is also initialised to zero by the calling function, and is incremented each time the *Transmit()* function is entered. However, unlike *coneCounter*, it is decremented when the procedure is exited. The consequence is that the value of the variable at any stage in the analysis indicates the depth of the recursion. This information is used in the termination criterion.

The *Transmit()* function receives, as parameters, the number of the interface towards which the wave is propagating (n), the direction the wave is propagating (conveyed in the integer *dirn*, which is either 1 for down or -1 for up), the initial radius of the cone segment in which the wave

184 Foundation Vibration Analysis: A Strength-of-Materials Approach

is propagating (*r0*) and the initial amplitude of this wave (*u0*). The angular frequency (*w*) and degree of freedom (*dof*) are also provided. The function uses the number of the interface and the direction of the wave to determine which layer the wave is currently propagating in (*layerA*), and which layer is on the other side of the interface (*layerB*). The radius of the cone at the interface is computed from the initial radius, the thickness of *layerA* and the cone aspect ratio for *layerA*. The *Attenuation()* function described below is then used to compute the amplitude of the incident wave at the interface (*f*) from its initial amplitude.

The magnitude of the incident wave is compared with 0.0001 to determine whether further reflections and refractions should be ignored (implicitly assuming that the first initial wave in the sequence has unit amplitude). The depth counter is checked, and if it is greater than the maximum specified depth (20 + 2× number of finite thickness layers), the amplitude is progressively attenuated (effectively performing a weighted average of the last ten reflections and refractions of the wave). The minimum magnitude and maximum depth together specify the termination criterion.

The amplitudes of the reflected wave (*g*) and the refracted wave (*h*) are computed using the reflection coefficient (calculated by the function *Alpha()* described below). The contribution of this wave to the amplitude of the interface vibration is equal to the amplitude of the refracted wave, which is cumulated into the amplitude field for the *n*th interface.

The *Transmit()* function is only called for a wave propagating in a layer of finite thickness, as a wave propagating into a half-space leads to no further reflections and refractions. Consequently the layer denoted by *layerA* in the function (the layer in which the wave is propagating when *Transmit()* is called) always has finite thickness, and the reflected wave *g* propagates back across this finite layer. Further reflections and refractions due to the reflected wave are processed by calling *Transmit()*, this time with the next interface in the opposite direction to the direction of incident wave propagation (*n − dirn*), with the wave propagating in the opposite direction to the initial wave (*−dirn*), with an initial radius equal to the radius of the cone at the current interface (*r*) and with an initial amplitude of *g*.

The layer into which the refracted wave propagates may be a half-space. Further reflections and refractions are only processed if the type of *layerB* is 'L', indicating a layer of finite thickness. In this case *Transmit()* is called with the next interface in the direction of incident wave propagation (*n + dirn*), with the wave propagating in the same direction (*dirn*), with an initial radius equal to the radius of the cone at the current interface (*r*) and with an initial amplitude of *h*.

All further reflections and refractions are processed recursively automatically until the magnitudes of the amplitudes are less than 0.0001. On exit the cumulated vibration amplitudes at each interface are available in the *amplitude* fields of the global *laYers* array.

The *Attenuation()* function (Listing F.7) computes the attenuation and phase shift occurring as the initial wave propagates through a cone segment from the properties of the layer (*layer*) and

```
function a = Attenuation(layer, r0, r, w, dof)
   c = WaveVelocity(layer,dof);
   ra = r0/r;

   a = exp(-i*w*layer.thickness/c);
   if isequal(dof,'H') | isequal(dof,'V')
      a = ra*a;                                                         % Eq. 4.5
   else % isequal(dof=='R') | isequal(dof=='T')
      a = ra*ra*(1.0+(ra-1.0)/(1.0+i*w*r0*AspectRatio(layer,dof)/c))*a; % Eq. 4.26b
   end
```

Listing F.7 Function to determine attenuation through layer

```
function a = Alpha(layerA, layerB, r, w, dof)
   if isequal(layerB.type,'R')
      a = 1.0;
      return
   end
   if isequal(layerB.type,'F')
      a = -1.0;
      return
   end
   cA = WaveVelocity(layerA,dof);
   cB = WaveVelocity(layerB,dof);
   zA = r*AspectRatio(layerA,dof);  % distance from cone apex to interface layer A
   zB = r*AspectRatio(layerB,dof);  % distance from cone apex to interface layer B
   if isequal(dof,'H') | isequal(dof,'V')
      betaA = layerA.rho*cA^2*(1/zA + i*w/cA);  % Eq 4.16
      betaB = layerB.rho*cB^2*(1/zB + i*w/cB);  % Eq 4.17
   else % isequal(dof,'R') | isequal(dof,'T')
      betaA = layerA.rho*cA^2*(3/zA+3*i*w/cA+zA*(i*w/cA)^2)/(1+i*w*zA/cA);  % Eq 4.35
      betaB = layerB.rho*cB^2*(3/zB+3*i*w/cB+zB*(i*w/cB)^2)/(1+i*w*zB/cB);  % Eq 4.36
   end
   a = (betaB-betaA)/(betaB+betaA);  % Eq 4.18 and Eq. 4.38
```

Listing F.8 Function to determine reflection coefficient for interface

the initial radius ($r0$) and final radius (r) of the cone segment, which are dependent on the angular frequency (w) and the degree of freedom (*dof*).

The *Alpha()* function (Listing F.8) generates the reflection coefficient for a wave impinging on the interface between two layers. The wave is assumed to be propagating in the layer denoted by the data structure *layerA* and impinging on an interface with the layer denoted by *layerB*. The radius of the cone at the interface is r, and the reflection coefficient is dependent on both the angular frequency (w) and the degree of freedom excited (*dof*).

F.5 Dynamic stiffness of the free field

This section provides two functions that allow the dynamic-stiffness matrix of the discretised free field to be calculated, as detailed in Section 5.2.

The *Transmit()* function described in Appendix F.4 must be called from another function which initialises the wave propagation, then computes a final result at the end of the process. The *DiskGreens()* function presented in Listing F.9 does just that, and is thus a driver function for the recursive *Transmit()* function. The method of wave reflection and refraction in cones is used to calculate the value of the Green's function for a particular disk (at the interface n) at all the disks present in the model described by the data array *layers*. The function is evaluated for a particular angular frequency (w) and degree of freedom (*dof*).

The *DiskGreens()* function commences by copying the data array *layers* into the global array *laYers* and declaring and initialising a global *depthCounter*, as discussed in Appendix F.4. The radius of the disk for which the Green's function is required becomes the initial radius of the initial cone, and the amplitude of displacement at this interface is initialised to unit value. The amplitudes of displacements at all other interfaces are initialised to zero. If the layer above the disk has finite thickness, the *Transmit()* procedure is called to process the wave propagating upwards from the disk. Similarly, if the layer below the disk has finite thickness, the *Transmit()* procedure is called to process the wave propagating downwards from the disk. After these two procedures

```
function Gn = DiskGreens(layers, n, w, dof)
   global laYers      % a globally visible and changeable copy of the layer data
   laYers = layers;
   global depthCounter % global depth counter to allow Transmit to terminate by depth
   depthCounter = 0;

   % initial radius of cone
   r0 = laYers(n).disk;
   % initial amplitudes of displacement
   u0 = 1.0;
   for k=1:size(laYers,2), laYers(k).amplitude = 0.0; end
   laYers(n).amplitude = u0;

   if laYers(n).type=='L' % transmit wave up across the layer above the disk
      Transmit(n-1,-1,r0,u0,w,dof);
   end

   if laYers(n+1).type=='L' % transmit wave down across the layer below the disk
      Transmit(n+1,1,r0,u0,w,dof);
   end

   % determine amplitude of force applied to disk, considering both initial cones
   P = ConeStiffness(laYers(n),r,w,dof) + ConeStiffness(laYers(n+1),r,w,dof);

   % determine value of Green's function at each disk
   numInterfaces = size(laYers,2) - 1; % number of interfaces between layers
   k1 = 1;
   for k=1:numInterfaces
      if laYers(k).disk>0.0 % only require value at disks
         Gn(k1) = laYers(k).amplitude/P; % displacement due to force of unit amplitude
         k1 = k1 + 1;
      end
   end
end
```

Listing F.9 Function for determining Green's function of single disk

have recursively computed all generated reflections and refractions up to the stage indicated by the termination criterion, the amplitude of displacement at each interface for the force amplitude $P(\omega)$ applied at the disk is known. The amplitude of this force is equal to the sum of the dynamic-stiffness coefficients of the initial cones above and below the disk, as the initial displacement amplitude is unity. Since the Green's function for the disk represents the displacement amplitudes due to a force of unit amplitude at the disk, the values of the Green's function are determined at each disk by dividing the calculated amplitude of the displacement at the disk by the amplitude of the exciting force. The function returns a one-dimensional array of size equal to the number of disks in the model.

A function computing the dynamic-stiffness matrix of the free field, discretised by a number of disks, as specified in a data array *layers*, is presented in Listing F.10. This function *FreeField()* simply assembles the values of the Green's function for each disk k (computed by the function *DiskGreens()*) into the kth column of the dynamic-flexibility matrix. The dynamic-flexibility matrix is then inverted to provide the dynamic-stiffness matrix for the free field. If just a single disk is present in the model description, this function returns the dynamic-stiffness coefficient of the single disk (with no trapped mass terms for incompressible or nearly-incompressible materials).

F.6 Dynamic stiffness of the foundation

The dynamic stiffness of a single disk is often of interest. When a cone model is used, there is no interaction between any of the four degrees of freedom, so a single function can be used to compute

```
function Sf = FreeField(layers, w, dof)
  % assemble dynamic-flexibility matrix column by column
  numInterfaces = size(layers,2) - 1;
  k = 1;
  for n=1:numInterfaces
    if layers(n).disk>0.0
       G(:,k) = DiskGreens(layers,n,w,dof);
       k = k + 1;
    end
  end
  % invert dynamic-flexibility to obtain dynamic-stiffness matrix of free field
  Sf = inv(G); % Eq. 5.6
```

Listing F.10   Function for determining dynamic-stiffness matrix of free field

the horizontal, vertical, torsional or rocking dynamic stiffness, as illustrated by the *FreeField()* function described above. In the main text a function *DiskStiffness()* (Listing 4.2) is introduced to compute the dynamic-stiffness coefficient of a single disk. This is necessary to introduce the concepts step by step. Here, however, one function is used to handle both the single disk and the embedded foundation (multiple-disk) cases. If multiple disks are present, the function cannot determine the dynamic-stiffness coefficient for rocking. An additional function is introduced later in this appendix to handle the rocking case.

For an embedded rigid foundation, the rigid-body constraint resulting from the foundation must be introduced, and the soil mass in the volume occupied by the foundation analytically excavated. As discussed in Section 5.3, the kinematic-constraint vector $\{A\}$ is a column of ones for rigid-body translation of the foundation in the horizontal and vertical directions, and for a rigid-body torsional rotation of the foundation about the axis of symmetry. In each of these cases the triple product $\{A\}^T [S^f(\omega)]\{A\}$ is just equal to the scalar sum of the elements of $[S^f(\omega)]$. The MATLAB function *sum()* produces a vector of column sums when performed on a matrix, and a scalar when performed on a vector. Thus *sum(sum(Sf))* produces the scalar sum of all elements of *Sf*. In the case of a single disk $[S^f(\omega)]$ is 1×1, and so $sum(sum(Sf)) = Sf$.

The mass (or polar mass moment of inertia in the case of the torsional degree of freedom) of the soil in the volume occupied by the foundation is computed separately in the function *Mass()* (included in Listing F.11). The mass of soil between each pair of disks is calculated (using the exact formula for the volume of a cone frustum), and the total mass summed. The soil is excavated analytically by adding the square of the angular frequency times this mass to the dynamic stiffness (Listing F.11). In the case of a single disk the excavated mass is zero.

The effect of any trapped mass terms due to the layers above or below the foundation being nearly-incompressible or incompressible on the dynamic-stiffness coefficient is included by subtracting the trapped mass from the excavated mass. This is performed in the *Mass()* function. The *TrappedMass()* function (Listing F.5) described in Appendix F.3 is used to compute the trapped mass terms. Should the relevant layer not be nearly-incompressible, the *TrappedMass()* function returns zero.

For an embedded foundation (i.e. a model with more than one disk) the horizontal and rocking degrees of freedom are not independent. Although the function of Listing F.11 permits computation of the horizontal dynamic stiffness for an embedded foundation, this is done on the assumption that rocking motion of the foundation is prevented. The moment required to prevent this rocking motion is not computed. The dynamic stiffness for rocking cannot be computed in this way, since simultaneous rocking of each disk without horizontal motion causes shearing of the foundation, which has been specified to be rigid. Instead, the 2×2 dynamic-stiffness matrix describing the

188   Foundation Vibration Analysis: A Strength-of-Materials Approach

```
function Sg = DynamicStiffness(layers, w, dof)
   % dynamic-stiffness coefficient from Eq. 5.14 or Eq. 5.26
   Sg = sum(sum(FreeField(layers,w,dof))) + w^2*Mass(layers,dof);
function M = Mass(layers, dof)
   % compute mass of soil in excavation
   M = 0.0;
   numInterfaces = size(layers,2) - 1;
   nTopDisk = 64; % just a large number
   nBottomDisk = 0;
   for n=1:numInterfaces
      if layers(n).disk>0.0
         if n < nTopDisk, nTopDisk = n; end
         if n > nBottomDisk, nBottomDisk = n; end
         if layers(n+1).disk > 0.0
            r = layers(n+1).disk;
            rr = layers(n).disk/r; % ratio of first disk radius to second
            dm = pi*layers(n+1).thickness*layers(n+1).rho*r^2;
            if isequal(dof,'T')
               M = M + dm*r^2*(1+rr+rr^2+rr^3+rr^4)/10;
            else
               M = M + dm*(1+rr+rr^2)/3;
            end
         end
      end
   end
   % subtract any trapped mass due to soil incompressibility
   M = M - TrappedMass(layers(nTopDisk),layers(nTopDisk).disk,dof);
   M = M - TrappedMass(layers(nBottomDisk+1),layers(nBottomDisk).disk,dof);
```

Listing F.11 Functions for determining dynamic stiffness of disk or embedded foundation (except for rocking degree of freedom)

coupling of the horizontal and rocking degrees of freedom must be calculated. The process is described in detail in Section 5.3.

The implementation provided in Listing F.12 differs in minor detail from the equations provided in the text (Eqs 5.19 to 5.24) in two respects. First, provision is made to be able to reference the rocking motion to any point on the foundation (specified as a depth from the first interface, *zRef*). Second, the kinematic-constraint matrix $[A]$ is split into an upper portion $[AH]$ (with the rows relating to the horizontal degrees of freedom of the disks) and a lower portion $[AR]$ (with the rows relating to the rocking degrees of freedom of the disks). Since there is no interaction between the horizontal and rocking degrees of freedom in the free-field dynamic stiffness matrix, Eq. 5.24 can be written as

$$[S_{00}^g(\omega)] = [AH]^{\mathrm{T}}[S^f(\omega)][AH] + [AR]^{\mathrm{T}}[S_\vartheta^f(\omega)][AR] + \omega^2[M] \qquad (F.1)$$

where $[M]$ is the rigid-body mass matrix such as described in Eq. 5.22 for the specific case of Fig. 5.4 and computed in the sub-function *HorzAndRockMass()*.

Should either (or both) of the layers above and below the foundation be nearly-incompressible or incompressible, a trapped mass moment of inertia term results for the rocking degree of freedom. This term is included in the dynamic-stiffness matrix by subtracting the trapped mass moment of inertia from the diagonal element of $[M]$ corresponding to the rocking degree of freedom.

```
function Sg = HorzAndRockStiffness(layers, w, zRef)
  SfH = FreeField(layers,w,'H'); % free field dynamic-stiffness matrix horz dof
  SfR = FreeField(layers,w,'R'); % free field dynamic-stiffness matrix rocking dof

  numDisks = size(SfH,1);
  AH = zeros(numDisks,2); %upper portion of constraint matrix relating to horz dof
  AR = zeros(numDisks,2); %lower portion of constraint matrix relating to rocking dof

  % set up constraint matrix - first column horizontal translation
  %                         - second column rotation around zRef
  numInterfaces = size(layers,2) - 1;
  z = 0.0; % depth of interface
  k = 1;
  for n=1:numInterfaces
    if layers(n).disk>0.0
        AH(k,1) = 1.0;
        AH(k,2) = zRef - z;
        AR(k,2) = 1.0;
        k = k + 1;
    end
    z = z + layers(n+1).thickness;
  end

  Sg = AH'*SfH*AH + AR'*SfR*AR + w^2*HorzAndRockMass(layers, zRef); %Eq. 5.24 or 5.27

function M = HorzAndRockMass(layers, zRef)
  M = zeros(2,2);
  z = 0.0;
  numInterfaces = size(layers,2) - 1;
  nTopDisk = 64; % just a large number
  nBottomDisk = 0;
  for n=1:numInterfaces
     if layers(n).disk>0.0
        if n<nTopDisk, nTopDisk = n; end
        if n>nBottomDisk, nBottomDisk = n; end
        if layers(n+1).disk>0.0
           r = layers(n+1).disk;
           rr = layers(n).disk/r; % ratio of first disk radius to second
           t = layers(n+1).thickness;
           h = zRef - z - t;
           dm = pi*layers(n+1).rho*r^2*t;
           a = (1+rr+rr^2)/3;
           b = (1+2*rr+3*rr^2)/6;
           c = r^2*(1+rr+rr^2+rr^3+rr^4)/20 + t^2*(1+3*rr+6*rr^2)/30;
           M(1,1) = M(1,1) + dm*a;
           M(1,2) = M(1,2) + dm*(a*h + b*t/2);
           M(2,2) = M(2,2) + dm*(c + a*h^2 + b*h*t);
        end
     end
     z = z + layers(n+1).thickness;
  end
  M(2,1) = M(1,2);
  % subtract any trapped mass moment of inertia due to soil incompressibility
  M(2,2) = M(2,2) - TrappedMass(layers(nTopDisk),layers(nTopDisk).disk,'R');
  M(2,2) = M(2,2) - TrappedMass(layers(nBottomDisk+1),layers(nBottomDisk).disk,'R');
```

Listing F.12 Functions for determining coupled horizontal and rocking dynamic stiffness of embedded foundation

F.7 Effective foundation input motion

This section presents functions for determining the effective foundation input motion for vertically propagating *S*-waves (horizontal earthquake) and for vertically propagating *P*-waves (vertical earthquake). First the free-field motion is determined from the prescribed ground motion at the control point. The effective foundation input motion is then computed from this free-field motion.

The variation of the free-field motion with depth can be computed from the displacement amplitude at a control point defined either at the free surface of the site (Point A in Fig. B.4), or at an assumed fictitious rock outcrop on the level of the underlying half-space (Point B). This procedure is detailed in Appendix B.2. An implementation of the procedure is presented in Listing F.13. The description of the layered site is provided in the *layers* data array in the usual way, along with the angular frequency (w) and the degree of freedom (dof). In addition, the amplitude of the control displacement (uc) and the position of the control point (*controlPoint*)

```
function uf = FreeFieldMotion(uc,controlPoint,layers,w,dof)
    n = size(layers,2) - 1; % number of interfaces
    S = zeros(n,n); Q = zeros(n,1);
    for i=1:n-1 % layers of finite thickness
        j = i + 1; % layer number
        c = WaveVelocity(layers(j),dof);
        d = layers(j).thickness;
        rho = layers(j).rho;
        Sj(1,1) = cos(w*d/c); Sj(1,2) = -1;
        Sj(2,1) = -1; Sj(2,2) = Sj(1,1);
        S(i:i+1,i:i+1) = (rho*c*w/sin(w*d/c))*Sj; % Eq B.17, B.26
    end

    if isequal(layers(n+1).type,'H') % lower half-space
        Sn = i*layers(n+1).rho*WaveVelocity(layers(n+1),dof)*w; % Eq B.22, B.27
        S(n,n) = S(n,n) + Sn;
        if isequal(controlPoint,'B') % control motion on fictitious rock outcrop
            Q(n) = Sn*uc; % Eq B.25
            u = inv(S)*Q;
        end
    elseif isequal(layers(n+1).type,'R') & isequal(controlPoint,'B')
        % control motion on rigid base
        Q(n-1) = -S(n-1,n)*uc;
        u(1:n-1) = inv(S(1:n-1,1:n-1))*Q(1:n-1);
        u(n) = uc;
    end

    if isequal(controlPoint,'A') % control motion on first interface (free surface)
        Q(2) = -S(2,1)*uc;
        u(1) = uc;
        u(2:n) = inv(S(2:n,2:n))*Q(2:n);
    end

    % extract motions at disks
    k = 1;
    for i=1:n
        if layers(i).disk>0
            uf(k) = u(i);
            k = k+1;
        end
    end
end
```

Listing F.13 Function to determine free-field motion for layered site from control motion

```
function ug = VerticalInputMotion(layers, w, uf)
    Sf = FreeField(layers,w,'V');  % find free field dynamic-stiffness matrix
    Sg = sum(sum(Sf)) + w^2*Mass(layers,'V');  % Eq. 5.14
    ug = 1/Sg * sum(Sf*uf);                    % Eq. 5.17
```

Listing F.14 Function for determining effective foundation input motion for vertical earthquake

are also specified. The *controlPoint* is assigned the character 'A' or 'B', indicating the appropriate position.

The function *FreeFieldMotion()* of Listing F.13 implements the theory described in Appendix B.2 straightforwardly. However, if the underlying half-space is rigid and the control motion is specified at 'B', the amplitude at this interface is treated as prescribed, since $S_n(\omega)$ in Eq. B.25 becomes infinite. The function returns a vector of free-field displacement amplitudes at the disks specified in the *layers* data array.

In the functions determining the effective foundation input motion described below, a vector of free-field displacement amplitudes at the disks is required. This vector is determined from a control motion using the *FreeFieldMotion()* function described in Listing F.13.

The effective foundation input motion for a vertical earthquake is obtained by implementing Eq. 5.17. Since $\{A\}$ is a vector of ones, pre-multiplication of the vector $[S^f(\omega)]\{u^f(\omega)\}$ by $\{A\}^T$ is equivalent to summing the elements of the vector. Consequently, the implementation is very simple, as demonstrated by the function *VerticalInputMotion()*, presented in Listing F.14. On entry *uf* is a vector containing the vertical free-field motion amplitudes at the levels of the disks. For efficiency, the dynamic-stiffness coefficient of the embedded foundation is calculated directly from Eq. 5.11. Any trapped mass terms due to nearly-incompressible or incompressible soil layers are subtracted from the excavated mass in the function *Mass()*.

The effective foundation input motion for a horizontal earthquake can be found in a similar way, implementing Eq. 5.25. A function *HorizontalInputMotion()* is presented in Listing F.15 that accomplishes this efficiently. On entry to the procedure the vector *uf* contains the horizontal free-field motion amplitudes at the levels of the disks. The function returns the horizontal input motion and the rocking input motion for the embedded foundation. Any trapped mass moment of inertia terms due to nearly-incompressible or incompressible soil layers are subtracted from the excavated mass moment of inertia in the function *HorzAndRockMass()*. The dynamic-stiffness coefficients of the foundation are also returned. The cross-coupling terms are averaged, since the method using cones may induce a slight asymmetry into these terms.

F.8 Worked example: seismic response

To illustrate how the procedures can be used together in MATLAB to perform a complete analysis, the seismic response example addressed in Section 7.2 is considered. The response of the reactor building illustrated in Fig. 7.7a to the 1989 Newcastle earthquake (Fig. 7.9a) is calculated. The reactor building is situated on a site similar to Fig. 6.1c, with the depth of the first layer being 5 m, and is modelled as a single-degree-of-freedom structure (Fig. 7.8).

A text file containing the acceleration time history at 0.02 s intervals (Fig. 7.9a) is assumed to be available. This data is transferred to MATLAB via the clipboard by copying the data from the file, while open in a text editor or a spreadsheet such as Excel, then selecting 'Paste Special...' to pass it into MATLAB. The data is pasted as a vector named a. A discrete Fourier transformation is then

192 Foundation Vibration Analysis: A Strength-of-Materials Approach

```
function [ug0,vg0,Sh,Sr,Shr] = HorizontalInputMotion(layers, w, zRef, uf)
    SfH = FreeField(layers,w,'H'); % free field dynamic-stiffness matrix horz dof
    SfR = FreeField(layers,w,'R'); % free field dynamic-stiffness matrix rocking dof

    numDisks = size(SfH,1);
    AH = zeros(numDisks,2); %upper portion of constraint matrix relating to horz dof
    AR = zeros(numDisks,2); %lower portion of constraint matrix relating to rocking dof

    % set up constraint matrix - first column horizontal translation
    %                          - second column rotation around z0
    numInterfaces = size(layers,2) - 1;
    z = 0.0; % depth of interface
    k = 1;
    for n=1:numInterfaces
        if layers(n).disk>0.0
            AH(k,1) = 1.0;
            AH(k,2) = zRef - z;
            AR(k,2) = 1.0;
            k = k + 1;
        end
        z = z + layers(n+1).thickness;
    end

    Sg = AH'*SfH*AH + AR'*SfR*AR;
    Sg = Sg + w^2*HorzAndRockMass(layers, zRef); % Eq. 5.24
    ug = inv(Sg)*(AH')*SfH*uf; % Eq. 5.25

    ug0 = ug(1); vg0 = ug(2);
    Sh = Sg(1,1); Sr = Sg(2,2); Shr = (Sg(1,2)+Sg(2,1))/2;
```

Listing F.15 Function for determining effective foundation input motion for horizontal earthquake

```
% first copy acceleration time history from text file
% 1024 points at 0.02 sec intervals
% 'Paste Special' into MATLAB creating a vector a(1..1024)
dt=0.02;
N=1024;
dw=2*pi/(N*dt);
w=dw:dw:410*dw; % just use first 410 harmonics -> up to 125 rad/s
X=fft(a); % convert to frequency domain
for j=7:410, uc(j)=2*X(j+1)/N/(-w(j)^2); end % divide by -wj^2 to
                                             % convert to control disp
% X(1) assumed to be zero .. no constant term, lower frequency terms discarded

k=46e9; m=73e6; h=20; dr=0.025; % structure properties (reactor)
```

Listing F.16 MATLAB commands to perform discrete Fourier transformation

performed by entering the commands presented in Listing F.16 to obtain $u_c(\omega_j)$ ($j = 1, \ldots, 410$) with $\omega_j = j \Delta \omega$ and $\Delta \omega = 2\pi/(1024*0.02)$ rad/s. Acceleration amplitudes are converted to displacement amplitudes in the frequency domain by dividing by $-\omega_j^2$. The properties of the structure are also entered.

The foundation of the reactor building is embedded to a depth of 6.25 m. As described in Section 7.2, 14 layers and 15 disks are used to model the foundation, with $\Delta e = 0.5$ m for the first layer and $\Delta e = 0.3125$ m for the second layer. The properties of the first layer are the shear modulus $G_0 = 1124 \times 10^6$ N/m^2 and the mass density $\rho_0 = 1800$ kg/m^3, as well as the Poisson's ratios and the hysteretic damping ratios specified in Fig. 6.1c. The equivalent foundation radius is 30 m. A text file describing this situation is constructed in accordance with Appendix E, and is detailed in Listing F.17.

For each ω_j the dynamic-stiffness matrix and the effective foundation input motion can be found and Eq. 7.22 solved to obtain the amplitude of the structural distortion for that particular harmonic. This is effectively achieved within a loop, which can also be entered at the MATLAB command line. However, calculation of the dynamic-stiffness coefficients is a time-consuming process that has to be repeated 410 times. Since the compiled program CONAN executes this analysis far faster than MATLAB, two alternative approaches are detailed here.

```
F   30.0
L   30.0    1124e6      0.25        1800    0.05    0.5
L   30.0    1124e6      0.25        1800    0.05    0.5
L   30.0    1124e6      0.25        1800    0.05    0.5
L   30.0    1124e6      0.25        1800    0.05    0.5
L   30.0    1124e6      0.25        1800    0.05    0.5
L   30.0    1124e6      0.25        1800    0.05    0.5
L   30.0    1124e6      0.25        1800    0.05    0.5
L   30.0    1124e6      0.25        1800    0.05    0.5
L   30.0    1124e6      0.25        1800    0.05    0.5
L   30.0    1124e6      0.25        1800    0.05    0.5
L   30.0    562e6       0.30        1800    0.05    0.3125
L   30.0    562e6       0.30        1800    0.05    0.3125
L   30.0    562e6       0.30        1800    0.05    0.3125
L   30.0    562e6       0.30        1800    0.05    0.3125
L   0.0     562e6       0.30        1800    0.05    1.25
H   0.0     224.8e6     0.333333    1600    0.05
```

Listing F.17 Data file for site of Fig. 6.1c with $d = 5\,\text{m}$, $r_0 = 30\,\text{m}$, $G_0 = 1124 \times 10^6\,\text{N/m}^2$ and $\rho_0 = 1800\,\text{kg/m}^3$

```
% First option: all in MATLAB
layers=ReadLayers('Eg7-2a.txt'); % foundation properties
% calculate response u(wj) at each frequency
for j=1:410, ...
    uf = FreeFieldMotion(uc(j),'A',layers,w(j),'H'); ...
    w(j), ... % print frequencies as they are calculated
    [ug0 vg0 Sh Sr Shr] = HorizontalInputMotion(layers,w(j),0.0,uf), ...
    w2m = w(j)^2*m; ...
    A = [k/w2m*(1+2*dr*i)-1,-1,-1;-1,Sh/w2m-1,Shr/w2m/h-1; ...
         -1,Shr/w2m/h-1,Sr/w2m/h^2-1]; ...
    B(1,1) = ug0 + h*vg0; ...
    B(2,1) = B(1,1); ...
    B(3,1) = B(1,1); ...
    C = A\B; ...
    u(j) = C(1,1); ...
end
```

Listing F.18 MATLAB commands to compute structural distortion amplitudes

On a very fast computer the analysis can be performed completely in MATLAB using the procedures detailed in this appendix. The commands required at the command line are listed in F.18.

However, as a much faster alternative, the executable CONAN program can be used. This program is executed to obtain the dynamic-stiffness coefficients for horizontal, rocking and coupled motion and to obtain the effective foundation input motion. The real and imaginary parts of these are stored in the columns of a text file by CONAN. The content of this output text file is copied onto the clipboard and then transferred to the MATLAB environment by performing a 'Paste Special...', creating a named vector for each column. It is then necessary to combine the real and imaginary components within MATLAB to obtain the dynamic-stiffness coefficients and effective foundation input motion as complex numbers. The structural distortion amplitudes can finally be calculated using Eq. 7.22. The complete set of commands is specified in Listing F.19.

Once the structural distortion amplitudes have been found at each frequency by one of the techniques described above, the response at any time can be found by combining Eqs 7.23 and 7.24. This can be done efficiently for a sequence of times by storing the times in a vector, then using matrix operations to perform the summation of Eq. 7.24 implicitly. Listing F.20

```
% Second option: use CONAN.exe to perform the time consuming calculations
%
% Use CONAN to calculate Sh, Sr, Shr, ug0 & vg0 at each wj
% Copy from CONAN output file including headings and 'Paste Special' into MATLAB
% as vectors, importing real and imaginary parts individually then combining.

% The following are all column vectors of length 410.
Sh = Sh_real+i*Sh_imag;
Sr = Sr_real+i*Sr_imag;
Shr = Shr_real+i*Shr_imag;
ug0 = ugh_real+i*ugh_imag; % computed for unit control displacement amplitude
vg0 = ugr_real+i*ugr_imag; % computed for unit control displacement amplitude

% calculate response u(wj) at each frequency
for j=1:410, ...
    w2m = w(j)^2*m; ...
    A = [k/w2m*(1+2*dr*i)-1,-1,-1;-1,Sh(j)/w2m-1,Shr(j)/w2m/h-1; ...
         -1,Shr(j)/w2m/h-1,Sr(j)/w2m/h^2-1]; ...
    B(1,1) = uc(j)*(ug0(j) + h*vg0(j)); ... % scale for control displacement amplitude
    B(2,1) = B(1,1); ...
    B(3,1) = B(1,1); ...
    C = A\B; ...
    u(j) = C(1,1); ...
end
```

Listing F.19 Computing structural distortion amplitudes utilising CONAN output

```
% Finish is common
t=0:dt:(N-1)*dt;       % times at which response is calculated
ut = u*exp(i*w'*t);    % u, w and t are row vectors, this sums automatically for each tj
max(abs(real(ut)))     % print the max structural displacement
plot(t,ut)             % plot structural distortion time history
```

Listing F.20 Computing maximum amplitude and time history of structural distortion

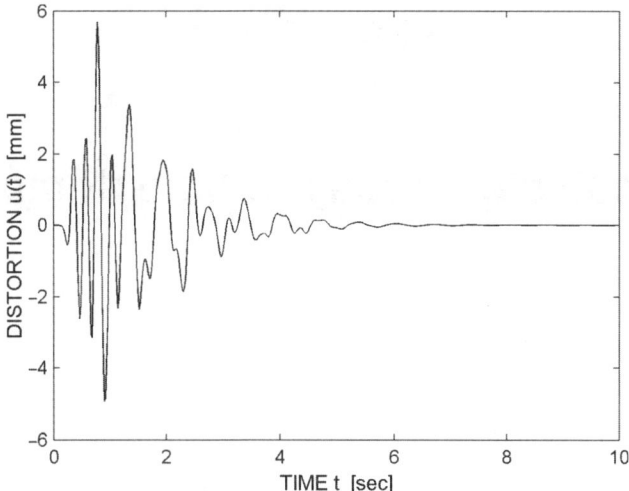

Figure F.1 Structural distortion time history

shows the commands that are used to accomplish this. After these commands the row vector *ut* contains the structural distortion at each time in the row vector *t*. The resulting structural distortion time history plot is presented in Fig. F.1. Note that the maximum value coincides with that specified in Table 7.2.

Appendix G

Analysis directly in time domain

To determine the response for an arbitrary excitation specified in the time domain, a Fourier transformation of the load to the frequency domain is performed at the beginning of the analysis. This allows a frequency-domain response analysis to be conducted, and an inverse Fourier transformation of the result, the displacement, to the time domain is performed at the end. In this indirect procedure, which only applies for linear systems, the soil is rigorously modelled using cones. No assumptions apply other than those of the one-dimensional strength-of-materials theory of cones. In particular, the reflection and refraction of the waves at material discontinuities leading to frequency-dependent reflection coefficients for cones, being tapered bars and beams, are considered exactly. An example consisting of an analysis of seismic soil-structure interaction is discussed in Section 7.2.

However, the one-dimensional wave equation for cones can be solved directly in the time domain. This is presented for translational motion in Section 2.2 (Eq. 2.17b) (and for rotational motion in Appendix A2 of Ref. [37]). This outward wave propagation is described by a function which depends on the coordinate z with the argument $t - z/c +$ constant (for propagation in the positive z-direction). The necessity to work in the frequency domain in a rigorous consistent analysis stems from the wave mechanism at a material discontinuity, as the reflection coefficient is a function of ω (with the exception of a fixed or free boundary). The reflection coefficient $-\alpha(\omega)$ is specified in Eq. 4.18 with β and β' being functions of ω (Eqs 4.16 and 4.17 for the translational cone and Eqs 4.35 and 4.36 for the rotational cone). The low frequency limit $-\alpha(0)$ (Eq. 4.23) and the high frequency limit $-\alpha(\infty)$ (Eq. 4.24) are, however, independent of ω, being a function of the material properties of the two layers at the interface. Thus, if one of these limits is used as an approximation, a frequency-independent formulation of the wave mechanisms at a material discontinuity follows, which permits the analysis to be performed directly in the time domain without any transformation to and from the frequency domain. This procedure is illustrated for a special case, a disk on the surface of a single layer fixed at its base, in Section 2.3 (Fig. 2.5) with the key equation specified in Eq. 2.34 with Eq. 2.27. The method can be extended straightforwardly to a multi-layered site by tracking the reflections and refractions of each incident wave sequentially and superposing in the time domain, analogous to the procedure in the frequency domain.

The formulation directly in the time domain not only covers the analysis of a layered half-space for specified loads acting on the foundation, but can be extended to the analysis of soil-structure interaction. To be able to formulate the equations of motion of the structure-soil system in the time domain (analogous to Eq. B.4), the interaction forces of the soil (analogous to Eq. B.3)

Analysis directly in time domain   197

must also be expressed in the time domain. As can be demonstrated, this involves the so-called unit-impulse response function, which is defined as the interaction forces as a function of time caused by unit displacement impulses (Dirac function). It is equal to the inverse Fourier transform of the dynamic stiffness. Luckily, this property does not have to be used to calculate the unit-impulse response function, as it would require the determination of the dynamic stiffness for many frequencies. A direct determination based on the definition of the interaction force-displacement relationship is possible. For the translational degree of freedom of a surface foundation modelled with cone theory, the interaction forces of the soil can be expressed as the superposition of the spring and the dashpot forces at the current time and at past times determined by propagation times of the generated wave pattern, multiplied by factors. Each term of the sum without the factor is analogous to the interaction force-displacement relationship (Eq. 2.24) involving the displacement and velocity of the homogeneous half-space with the properties of the first layer. This interaction force-displacement relationship of the soil can also be used when the structure exhibits non-linear behaviour.

As the wave pattern in the time domain for a rotational cone is not addressed in this book, the detailed discussion is limited to the translational cone. A translational degree of freedom of a surface foundation on a layered half-space is examined. However, extension to a rotational degree of freedom using the equations in Appendices A2 and A3 of Ref. [37] is possible. Generalisation to an embedded foundation based on the concepts of Chapter 5 as described in Section 4.3 of Ref. [37] also seems feasible.

In Section G.1 the displacements of a layered half-space for prescribed loads acting on the surface foundation are calculated directly in the time domain. In Section G.2 the interaction force-displacement relationship in the time domain of a surface foundation is determined. This permits the analysis of soil-structure interaction directly in the time domain, whereby the structure can behave non-linearly. In Section G.3, as an example, a rigid block on the surface of a layered half-space for vertical seismic excitation is analysed in the time domain. In Section G.4, the time domain analysis for the rotational degree of freedom of a surface foundation, as mentioned above, is also sketched.

G.1   Flexibility analysis for translation

To develop the computational procedure, a disk on a site with two layers overlying a flexible half-space for the vertical degree of freedom is addressed (Fig. G.1). This is the same example as discussed in Section 4.3 (Fig. 4.11). The displacement of the disk $u_0(t)$ (but also any other displacement under the disk within the site) is to be calculated as a function of the applied load $P_0(t)$.

The computational procedure is an extension of that described in Section 2.3.

The general wave pattern with the corresponding wave tree shown in Fig. 4.11b is also plotted in Fig. G.1. Outward wave propagation across cone segments and the reflections and refractions at interfaces, both leading to a decrease in the wave magnitudes, occur repeatedly. The wave pattern in the layered half-space consists of the superposition of the contributions of the various cone segments.

The load $P_0(t)$ yields waves propagating downward from the disk in an initial cone modelling a disk on a homogeneous half-space with the material properties of the first layer (Fig. G.2). The construction of the cone with opening angle and the wave velocity c follows from Table 3.1. The interaction force displacement relationship (Eq. 2.24) with the generating function $\bar{u}_0(t)$

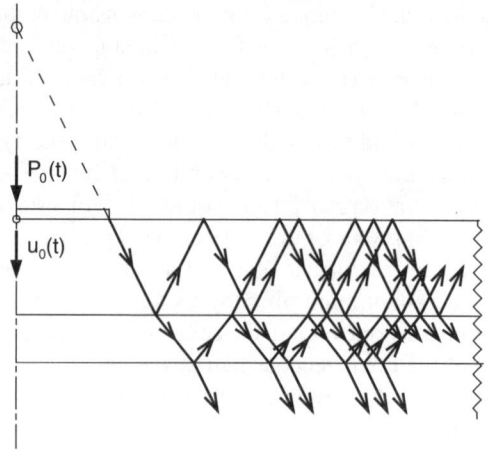

Figure G.1 Wave pattern for disk on layered half-space

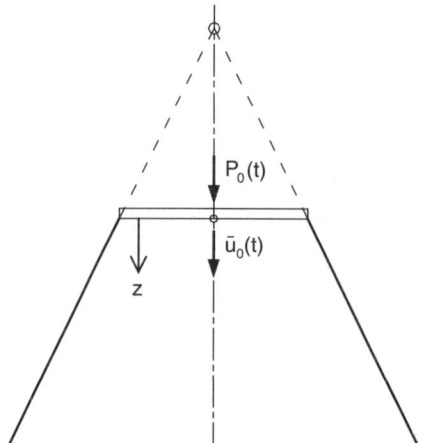

Figure G.2 Initial cone modelling disk on homogeneous half-space with material properties of first layer with generating wave pattern

replacing $u_0(t)$ is

$$C\dot{\bar{u}}_0(t) + K\bar{u}_0(t) = P_0(t) \tag{G.1}$$

which, for a general variation of $P_0(t)$ can be solved numerically for $\bar{u}_0(t)$. The spring and dashpot coefficients K and C are specified in Eq. 2.25. At depth z the displacement is (Section 2.2)

$$u(z) = \frac{z_0}{z_0 + z}\bar{u}_0\left(t - \frac{z}{c}\right) \tag{G.2}$$

Wave propagation across a cone segment and the wave mechanism leading to reflected and refracted waves in the time domain are now addressed. The procedure is analogous to that in the frequency domain (Section 4.1). Figure G.3 presents a layer with depth d and Poisson's ratio v, wave velocity c and mass density ρ. At the origin of the (local) coordinate axis z where a (fictitious) disk with radius r_0 is present, the displacement, of which the propagation is to be

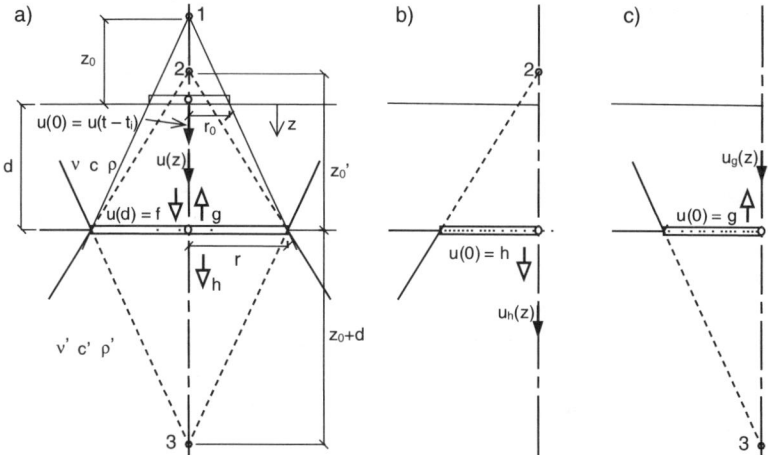

Figure G.3 Wave propagation across cone segment and at material discontinuity interface. a) Incident wave impinging on interface generating reflected and refracted waves. b) Refracted wave as new incident wave. c) Reflected wave as new incident wave

studied, is denoted as $u(0) = u(t - t_i)$ with t_i equal to the propagation time across all previous layers (depth of wave tree). $u(t - t_i)$ is determined as follows. If the disk is on the free surface and is loaded by $P_0(t)$, $u(t) = \bar{u}_0(t)$ follows from Eq. G.1 ($t_i = 0$) or it is equal to the fictitious disk's displacement of the reflected or refracted wave, as discussed below. The interface with the other layer with material properties v', c', and ρ' is located at distance d.

The outward wave propagation in the cone with apex 1 leads to the displacement at coordinate z

$$u(z) = \frac{z_0}{z_0 + z} u\left(t - t_i - \frac{z}{c}\right) \tag{G.3}$$

In the denominator the distance from the apex is present. The displacement of the incident wave f at the interface follows from Eq. G.3 with $z = d$ as

$$u(d) = \frac{z_0}{z_0 + d} u\left(t - t_i - \frac{d}{c}\right) = f \tag{G.4}$$

The reflected wave g and refracted wave h are specified using a frequency-independent reflection coefficient $-\alpha$ (Eqs 4.19 and 4.20)

$$g = -\alpha f \tag{G.5}$$
$$h = (1 - \alpha) f \tag{G.6}$$

with $-\alpha$ either determined for the low frequency limit $-\alpha(0)$ (Eq. 4.23)

$$-\alpha(0) = \frac{\dfrac{\rho c^2}{z_0 + d} - \dfrac{\rho' c'^2}{z'_0}}{\dfrac{\rho c^2}{z_0 + d} + \dfrac{\rho' c'^2}{z'_0}} \tag{G.7a}$$

which for $\nu = \nu'$ simplifies to

$$-\alpha(0) = \frac{\rho c^2 - \rho' c'^2}{\rho c^2 + \rho' c'^2} \tag{G.7b}$$

or for the high frequency limit $-\alpha(\infty)$ (Eq. 4.24)

$$-\alpha(\infty) = \frac{\rho c - \rho' c'}{\rho c + \rho' c'} \tag{G.8}$$

z'_0 denotes the apex height of the cone with apex 2 for the propagation of the refracted wave. As $\alpha(\infty)$ is the same as the reflection coefficient (for all frequencies) of the prismatic bar (Eq. C.25), for the wave mechanisms determining the reflected and refracted waves, the cones are replaced at the interface locally by prismatic bars when $\alpha(\infty)$ is used.

g and f become new initial waves propagating in their own cones. For the reflected wave (Eqs G.5 and G.4)

$$g = -\frac{z_0}{z_0 + d} \alpha\, u\left(t - t_i - \frac{d}{c}\right) \tag{G.9}$$

applies. The decay in magnitude of the wave is expressed by the product of two factors, the first describing the propagation across the cone segment and the second arising from the wave mechanism at the interface. Note that the time $t_{i+1} = t_i + d/c$ is the propagation time to the fictitious disk governing the propagation in the next cone segment with apex 3. g in Eq. G.9 thus corresponds to $u(t - t_{i+1})$, that is to $u(0)$ for the next cone segment. The same is also true for the refracted wave h.

For each path (Fig. G.1) yielding a contribution to $u_0(t)$, propagation across several cone segments and reflection/refraction at several interfaces occur. Equations G.4, G.5 and G.6 are thus applied repeatedly sequentially. It is obvious that for this time domain analysis the same logic of the computational algorithm as described in the computer implementation in Section 4.5 for the frequency domain analysis applies. In addition, the displacement $u(0) = u(t - t_i)$ of Fig. G.3 can be expressed as the product of $\bar{u}_0(t - t_i)$ and a factor describing the decay in magnitude of previous propagation and reflection/refraction.

Superposing the displacements at the free surface of all the wave paths yields for the displacement of the disk on the layered half-space

$$u_0(t) = \bar{u}_0(t) + \sum_{j=1}^{n} e_j^F \bar{u}_0(t - t_j) \tag{G.10}$$

The first term on the right-hand side represents the (first) incident wave. A term of the sum corresponds to the contribution of the jth wave path that impinges on the free surface. n is the number of these wave paths. The constant e_j^F is equal to the factor addressed above for the decay of magnitude arising from propagation across cone segments and reflection/refraction at interfaces. e_j^F can be called a flexibility *echo constant*, as the term

$$\sum_{j=1}^{n} e_j^F \bar{u}_0(t - t_j)$$

expresses that, as in acoustics, an explosive-type wave returns at distinct intervals of time with decreasing magnitude. The superscript F indicates that a flexibility formulation is processed.

Equation G.10, applicable to a multi-layered site, exhibits the same structure as Eq. 2.34 for a single layer.

Thus, *for a specified load $P_0(t)$ acting on a surface foundation, the displacement $u_0(t)$ of the surface foundation on a layered half-space is calculated directly in the time domain*. As an intermediate step, the generating function $\bar{u}_0(t)$ is determined, i.e. the displacement of the foundation on a homogeneous half-space with the material properties of the first layer with the same $P_0(t)$ applied. This efficient procedure, *avoiding transformations to the frequency domain and back to the time domain*, is based on the assumption that the *frequency-independent reflection coefficients* $-\alpha$ are used. The effect of this assumption remains small, as is systematically substantiated in Section 3.6.4 of Ref. [37].

G.2 Interaction force-displacement relationship in the time domain for translation

To be able to formulate the equations of motion for dynamic soil-structure interaction, the interaction force-displacement relationship of the soil in the time domain must be established. For the surface foundation in translational motion, P_0 must be expressed as a function of u_0. In Eq. G.1, P_0 is expressed as a function of $\bar{u}_0$, and in Eq. G.10 u_0 is expressed as a function of $\bar{u}_0$. The latter relationship can be 'inverted' leading to $\bar{u}_0$ as a function of u_0, which can then be substituted into the former relationship, yielding the desired equation P_0 as a function of u_0.

In detail, the derivation proceeds as follows. Equation G.10 is reformulated as

$$u_0(t) = \sum_{j=0}^{n} e_j^F \bar{u}_0(t - t_j) \tag{G.11}$$

with

$$e_0^F = 1 \tag{G.12a}$$
$$t_0 = 0 \tag{G.12b}$$

To be able to perform the inversion mentioned above in a concise manner, in the general case fictitious interfaces are introduced, so that after subdivision of the layers the propagation time across each layer of the site is the same. The propagation time across each original layer (thickness divided by wave velocity) is calculated and the largest common denominator determines the additional layering. This permits Eq. G.11 to be formulated as

$$u_0(t) = \sum_{j=0}^{m} e_j^F \bar{u}_0(t - jT) \tag{G.13}$$

with $0.5T$ denoting the common propagation time across each layer (after subdivision) and m the corresponding number of wave paths. Since when a wave impinges on a fictitious interface no reflected wave is generated, some e_j^Fs will vanish.

As an example, for the site consisting of two layers on a half-space with the geometry and material properties shown in Fig. G.4, the first and second layers are subdivided into two and three sublayers, respectively. $0.5T$ for a sublayer in the first layer equals $(1/2) r_0/c$ which is the

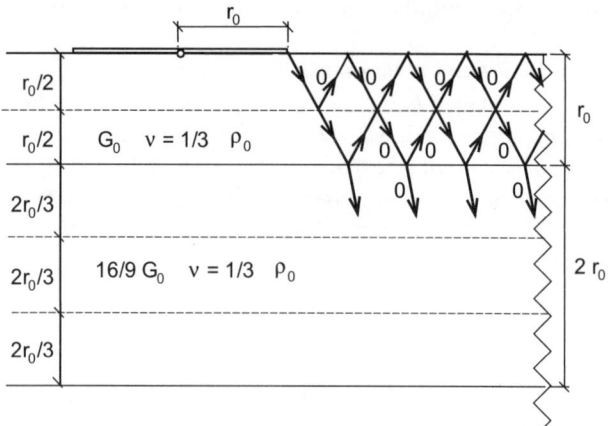

Figure G.4 Disk on layered half-space with fictitous interfaces shown as dashed lines

same as for a sublayer in the second layer $(2/3)r_0/(4/3\,c) = (1/2)r_0/c$. The waves with vanishing values are indicated in the figure.

As will be demonstrated (see from Eq. G.21 onwards), the generating function $\bar{u}_0(t)$ can be expressed as a function of displacements $u_0(t - jT)$

$$\bar{u}_0(t) = \sum_{j=0}^{m} e_j^K\, u_0(t - jT) \tag{G.14}$$

with a new set of echo constants e_j^K. The superscript K denotes the stiffness formulation. It will become apparent that e_j^K may be determined step by step from the known e_j^F. The same equation also applies to convert the velocities

$$\dot{\bar{u}}_0(t) = \sum_{j=0}^{m} e_j^K\, \dot{u}_0(t - jT) \tag{G.15}$$

Substituting Eqs G.14 and G.15 into Eq. G.1 results in the *interaction force-displacement relation of the disk on the surface of a layered half-space in translational motion*

$$P_0(t) = K \sum_{j=0}^{m} e_j^K\, u_0(t - jT) + C \sum_{j=0}^{m} e_j^K\, \dot{u}_0(t - jT) \tag{G.16}$$

It describes the dynamic behaviour of the substructure soil (corresponding to Eq. B.3 in the frequency domain with seismic excitation) to be used when formulating the equations of motion of soil-structure interaction in the time domain (corresponding to Eq. B.4). The equations of motion of the structure in the time domain are established straightforwardly.

For instance, for seismic excitation with the free-field motion $u_0^g(t) = u_0^f(t)$ in the vertical direction, the vertical interaction force-displacement relationship is

$$P_0(t) = K \sum_{j=0}^{m} e_j^K (u_0^t(t-jT) - u_0^f(t-jT)) + C \sum_{j=0}^{m} e_j^K (\dot{u}_0^t(t-jT) - \dot{u}_0^f(t-jT)) \quad \text{(G.17)}$$

with the superscript t indicating the total motion. If the structure consists of a rigid block with mass m the equation of motion of the coupled system in the vertical direction is formulated as

$$m \ddot{u}_0^t(t) + P_0(t) = 0 \quad \text{(G.18)}$$

It will be demonstrated that

$$e_0^K = 1 \quad \text{(G.19)}$$

applies (Eq. G.21a). Substituting Eq. G.17 in Eq. G.18 yields

$$m \ddot{u}_0^t(t) + C \dot{u}_0^t(t) + K u_0^t(t) = C \dot{u}_0^f(t) + K u_0^f(t) - C \sum_{j=1}^{m} e_j^K (\dot{u}_0^t(t-jT) - \dot{u}_0^f(t-jT))$$

$$- K \sum_{j=1}^{m} e_j^K (u_0^t(t-jT) - u_0^f(t-jT)) \quad \text{(G.20)}$$

Note that the right-hand side of Eq. G.20 is known at time t in a time integration scheme. The desired e_j^K may be determined step by step from the known e_j^F as follows

$$e_0^K = 1 \quad \text{(G.21a)}$$

and for $j \geq 1$

$$e_j^K = -\sum_{l=0}^{j-1} e_l^K e_{j-l}^F \quad \text{(G.21b)}$$

Equation G.21 is verified as follows. Formulating the flexibility echo-constants formula (Eq. G.13) for $u_0(t)$ at times $t, t-T, t-2T, \ldots, t-jT$ including terms up to $\bar{u}_0(t-jT)$ yields

$$u_0(t) = e_0^F \bar{u}_0(t) + e_1^F \bar{u}_0(t-T) + e_2^F \bar{u}_0(t-2T) + \cdots$$
$$+ e_{j-1}^F \bar{u}_0(t-(j-1)T) + e_j^F \bar{u}_0(t-jT) \quad \text{(G.22a)}$$

$$u_0(t-T) = e_0^F \bar{u}_0(t-T) + e_1^F \bar{u}_0(t-2T) + e_2^F \bar{u}_0(t-3T) + \cdots$$
$$+ e_{j-2}^F \bar{u}_0(t-(j-1)T) + e_{j-1}^F \bar{u}_0(t-jT) \quad \text{(G.22b)}$$

$$u_0(t-2T) = e_0^F \bar{u}_0(t-2T) + e_1^F \bar{u}_0(t-3T) + e_2^F \bar{u}_0(t-4T) + \cdots$$
$$+ e_{j-3}^F \bar{u}_0(t-(j-1)T) + e_{j-2}^F \bar{u}_0(t-jT) \quad \text{(G.22c)}$$

$$\cdots$$

$$u_0(t-(j-1)T) = e_0^F \bar{u}_0(t-(j-1)T) + e_1^F \bar{u}_0(t-jT) \quad \text{(G.22d)}$$

$$u_0(t-jT) = e_0^F \bar{u}_0(t-jT) \quad \text{(G.22e)}$$

The stiffness echo-constant formulation (Eq. G.14), also with terms up to $u_0(t - jT)$, yields

$$\bar{u}_0(t) = e_0^K u_0(t) + e_1^K u_0(t - T) + e_2^K u_0(t - 2T) + \cdots + e_{j-1}^K u_0(t - (j-1)T)$$
$$+ e_j^K u_0(t - jT) \tag{G.23}$$

Substituting Eq. G.22 in Eq. G.23 and equating the coefficients of $\bar{u}_0(t)$, $\bar{u}_0(t - T)$, $\bar{u}_0(t - 2T), \cdots, \bar{u}_0(t - jT)$ results in

$$e_0^K e_0^F = 1 \tag{G.24a}$$
$$e_0^K e_1^F + e_1^K e_0^F = 0 \tag{G.24b}$$
$$e_0^K e_2^F + e_1^K e_1^F + e_2^K e_0^F = 0 \tag{G.24c}$$
$$\cdots$$
$$e_0^K e_j^F + e_1^K e_{j-1}^F + e_2^K e_{j-2}^F + \cdots + e_{j-1}^K e_1^F + e_j^K e_0^F = 0 \tag{G.24d}$$

With $e_0^F = 1$ (Eq. G.12a), it follows from this sequence

$$e_0^K = 1 \tag{G.25a}$$
$$e_1^K = -e_0^K e_1^F \tag{G.25b}$$
$$e_2^K = -e_0^K e_2^F - e_1^K e_1^F \tag{G.25c}$$
$$\cdots$$
$$e_j^K = -e_0^K e_j^F - e_1^K e_{j-1}^F - e_2^K e_{j-2}^F - \cdots - e_{j-1}^K e_1^F$$
$$= -\sum_{l=0}^{j-1} e_l^K e_{j-l}^F \tag{G.25d}$$

This is the same result as specified in Eq. G.21.

G.3 Seismic analysis of a rigid block on the surface of a layered half-space

To demonstrate the accuracy of this approach, the seismic analysis of a rigid block on the surface of a layered half-space is undertaken. The layered half-space corresponds to that calculated in Section 6.1 (Fig. 6.1a), but without hysteretic damping. The block and half-space are illustrated in Fig. G.5. The mass of the block is 1.2×10^6 kg, and the soil properties of the upper layer are shear modulus $G_0 = 28.1 \times 10^6$ N/m² and density $\rho_0 = 1800$ kg/m³. The radius of the block r_0 is 5 m.

The block on the layered half-space is subjected to an earthquake equivalent to the 1989 Newcastle earthquake (Fig. 7.9a), but acting in the vertical direction. This defines $u_0^f(t)$ in Eq. G.20. To permit unbiased comparison with the frequency domain solution, the acceleration time history (at 0.02 s intervals) is transformed to the frequency domain using a fast Fourier transformation. Only the first 410 terms are used, and the free-field displacement and velocity amplitudes are obtained in the frequency domain using Eqs A.10 and A.9. An inverse transformation is then performed to obtain $\dot{u}_0^f(t)$ and $u_0^f(t)$, which are used to drive the time integration of Eq. G.20.

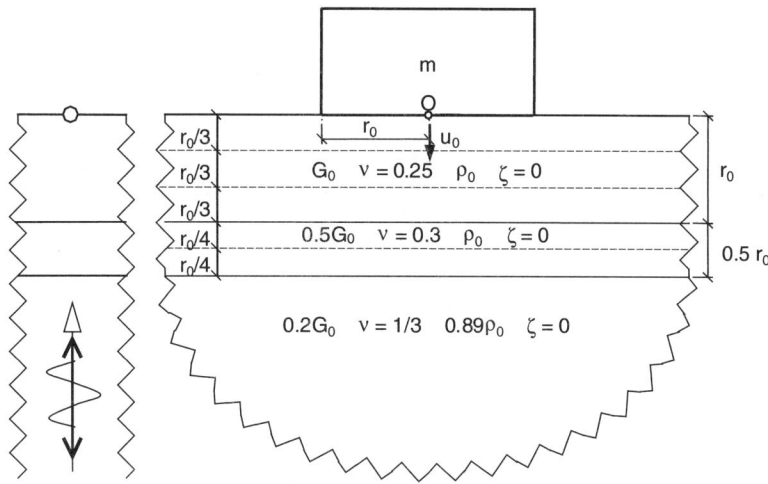

Figure G.5 Seismic analysis of block foundation on multi-layered site

The wave velocity in the first layer follows as $c_p = 216.51$ m/s, resulting in the propagation time across the first layer of 0.0231 s. The wave velocity in the second layer, $c_p = 165.36$ m/s, leads to a propagation time of 0.0151 s. The total propagation time across the layers is 0.0382 s, and the fraction of this time spent in the first layer is $0.0231/0.0382 \approx 0.6 = 3/5$, while the fraction of time spent in the second layer is $0.0151/0.0382 \approx 0.4 = 2/5$. The greatest common denominator of the two propagation times is thus 0.0382/5, yielding $T = 2*0.0382/5 = 0.0153$ s. The upper layer is subdivided into three equal layers, while the second layer is subdivided into two equal layers. The fictitious interfaces creating these layers are shown by dashed lines in Fig. G.5. The propagation time across each of these layers is approximately $0.5\,T$.

Time integration of Eq. G.20 is performed using the constant-average-acceleration method (otherwise known as the Newmark $\beta = 1/2$ method) with a time step of 0.02 s. Two analyses are performed. In the first the stiffness echo constants are computed from the flexibility echo constants using reflection coefficients formulated with the low frequency limit (Eq. G.7a), while in the second the stiffness echo constants are computed from the flexibility echo constants using reflection coefficients formulated with the high frequency limit (Eq. G.8).

To provide a comparison, the response of the block is also determined in the frequency domain using the first 410 terms of the transformed ground motion time history, as described in Section 7.2. In this calculation frequency-dependent reflection coefficients are used.

Figure G.6 shows the vertical displacement $u_0(t)$ of the block (where $u_0(t) = u_0^t(t) - u_0^f(t)$) calculated in the three ways described. Close agreement is evident, indicating that the frequency dependence of the reflection coefficients has little effect on the overall response in this case. The good agreement between the time domain analyses and the frequency domain analysis demonstrates the validity of the technique described in this appendix.

G.4 Rotation analysis

For the sake of completeness it is also appropriate to address the rotational degree of freedom of a surface foundation on a layered half-space. In contrast to the general approach governing the derivations in this book providing all details, only key expressions are examined. The reader

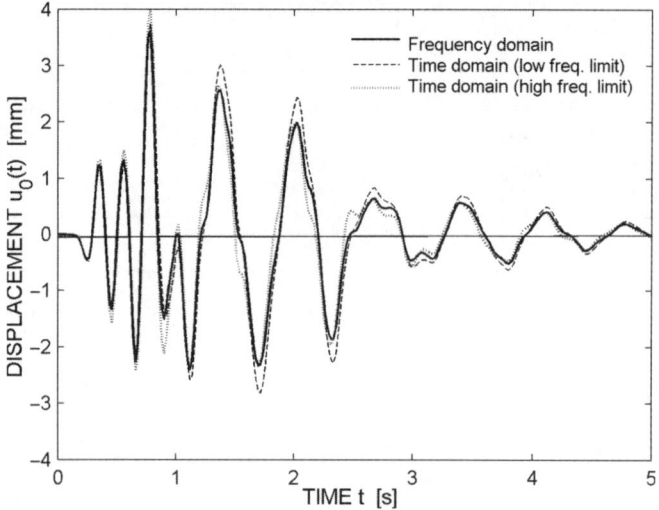

Figure G.6 Displacement of block foundation on multi-layered site calculated in three different ways

is referred to the corresponding parts of Ref. [37] for further explanations of the time domain analyses (Sections A.2, A.3, 2.3, 3.1.2, 3.3.1, 3.3.2). Also, it should be remembered that a variable or equation in the frequency domain and the corresponding relationship in the time domain form a Fourier transform pair.

The equation of motion of the rotational truncated semi-infinite cone is addressed. Substituting into the equilibrium equation formulated in the time domain (Fig. 3.3)

$$T(z, t),_z \, dz + \rho \, I(z) \, dz \, \ddot{\vartheta}(z, t) = 0 \qquad (G.26)$$

the moment-rotation relationship

$$T(z, t) = \rho c^2 \, I(z) \, \vartheta(z, t),_z \qquad (G.27)$$

yields (Eq. 3.39)

$$\vartheta(z, t),_{zz} + \frac{4}{z} \vartheta(z, t),_z - \frac{1}{c^2} \ddot{\vartheta}(z, t) = 0 \qquad (G.28)$$

Its solution is

$$\vartheta(z, t) = \frac{z_0^3}{z^3} \vartheta_1 \left(t - \frac{z - z_0}{c} \right) + \frac{z_0}{c} \frac{z_0^2}{z^2} \vartheta_1' \left(t - \frac{z - z_0}{c} \right) \qquad (G.29)$$

where $\vartheta_1(t - (z - z_0)/c)$ represents an arbitrary function of the argument for outward propagating waves (in the positive z-direction) and $\vartheta_1'(t - (z - z_0)/c)$ denotes differentiation with respect to the argument $t - (z - z_0)/c$.

Enforcing the boundary condition $\vartheta(z = z_0, \, t) = \vartheta_0(t)$ in Eq. G.29 leads to

$$\vartheta_1(t) + \frac{z_0}{c}\dot{\vartheta}_1(t) = \vartheta_0(t) \tag{G.30}$$

which can be solved, yielding

$$\vartheta_1(t) = \frac{c}{z_0}\int_0^t e^{-\frac{c}{z_0}(t-\tau)}\vartheta_0(\tau)\,d\tau \tag{G.31}$$

Equation G.29 is transformed to

$$\vartheta(z,\,t) = \frac{z_0^2}{z^2}\vartheta_0\left(t - \frac{z-z_0}{c}\right) + \left(\frac{z_0^3}{z^3} - \frac{z_0^2}{z^2}\right)\vartheta_1\left(t - \frac{z-z_0}{c}\right) \tag{G.32}$$

Within the half-space ($z > z_0$), part of the rotation is due to the convolution ϑ_1 (second term on the right-hand side of Eq. G.32). In passing it is worth mentioning that the convolution integral does not have to be evaluated from $t = 0$ onwards if a so-called recursive evaluation is performed (Section 2.3.2 and Appendix C of Ref. [37]).

The moment $M_0(t)$ acting on the surface disk yields waves propagating downwards from the disk in an initial cone modelling a disk on a homogeneous half-space with the material properties of the first layer (analogous to Fig. G.2). The interaction moment-rotation relationship of the rotational cone is conveniently formulated rigorously in the time domain using the spring-dashpot-mass model of Fig. 3.12 (Eq. 3.93) with a bar denoting rotation of the generating function as

$$C_\vartheta \dot{\overline{\vartheta}}_0(t) - C_\vartheta \dot{\overline{\vartheta}}_1(t) + K_\vartheta \overline{\vartheta}_0(t) = M_0(t) \tag{G.33a}$$

$$M_\vartheta \ddot{\overline{\vartheta}}_1(t) - C_\vartheta \dot{\overline{\vartheta}}_0(t) + C_\vartheta \dot{\overline{\vartheta}}_1(t) = 0 \tag{G.33b}$$

M_ϑ, C_ϑ and K_ϑ are specified in Eqs 3.92, 3.59 and 3.58. Applying numerical time integration yields $\overline{\vartheta}_0(t)$ and $\overline{\vartheta}_1(t)$ for any variation of $M_0(t)$. $\overline{\vartheta}_1(t)$ is thus calculated without actually evaluating the convolution integral. At depth z measured from the loaded disk, the (transformed) Eq. G.32 leads to

$$\vartheta(z,\,t) = \frac{z_0^2}{(z_0+z)^2}\overline{\vartheta}_0\left(t - \frac{z}{c}\right) + \left(\frac{z_0^3}{(z_0+z)^3} - \frac{z_0^2}{(z_0+z)^2}\right)\overline{\vartheta}_1\left(t - \frac{z}{c}\right) \tag{G.34}$$

Note that in the denominators the distance from the apex, squared and cubed, is present.

The outward wave propagation in the cone with apex 1 (analogous to Fig. G.3) leads at depth d to the rotation of the incident wave f at the interface, yielding a reflected wave g (cone with apex 3) and a refracted wave h (cone with apex 2), calculated with the same frequency-independent reflection coefficient as for the translational cone. These three rotations have contributions of $\overline{\vartheta}_0(t - d/c)$ and $\overline{\vartheta}_1(t - d/c)$, with both terms multiplied by corresponding coefficients. The reflected wave will propagate upwards, leading to a new incident wave f. This process will continue. Outward wave propagation across cone segments and the reflections and refractions at interfaces, both leading to a decrease in the wave magnitudes, will occur repeatedly. Superposing the displacements at the free surface of all wave paths yields for the rotation of the disk on the surface of the layered half-space

$$\vartheta_0(t) = \sum_{j=0}^{n} e_{0j}^F \overline{\vartheta}_0(t - t_j) + \sum_{j=0}^{n} e_{1j}^F \overline{\vartheta}_1(t - t_j) \tag{G.35}$$

with the flexibility echo constants e_{0j}^F and e_{1j}^F of the layered half-space modelled by rotational cones. These follow from the same logic of the computational procedure as in the frequency domain analysis. t_i is the propagation time over all the previous cone segments (depth of wave tree). Obviously, Eq. G.35 for the rotational cone is in the same form as Eq. G.11 (Eq. G.10) for the translational cone. Thus, a surface foundation loaded by a moment on a layered half-space can be analysed just as efficiently in the time domain as for a force.

It is also possible to determine the interaction moment-rotation relationship in the time domain directly without calculating the dynamic stiffnesses followed by an inverse Fourier transformation. Again, Eq. G.35 is 'inverted' after introducing fictitious interfaces. However, as $\overline{\vartheta}_1$ is a function of $\overline{\vartheta}_0$, so-called pseudo flexibility echo constants are first calculated permitting ϑ_0 to be expressed as a function of $\overline{\vartheta}_0$ only. This then permits $\overline{\vartheta}_0$ to be formulated as a function of ϑ_0. Insertion of $\overline{\vartheta}_0$ in the interaction moment-rotation relationship of the disk on a homogeneous half-space yields the desired moment. It turns out that this process just increases the number of terms in the equations (increase in the number of t_j and thus n). The reader is referred to Sections 3.3.1 and 3.3.2 of Ref. [37].

References

1. Aspel RJ, Luco JE. Torsional response of rigid embedded foundation. *Journal of the Engineering Mechanics Division*, ASCE 1976; **102**(6):957–970.
2. Aspel RJ, Luco JE. Impedance functions for foundations embedded in a layered medium: an integral equation approach. *Earthquake Engineering and Structural Dynamics* 1987; **15**:213–231.
3. Byrne BW, Houlsby GT. Investigating novel foundations for offshore windpower generation. *Proceedings of the 21st International Conference on Offshore Mechanics and Arctic Engineering*, OMAE, Oslo 2002; vol. 4, 567–576.
4. Chadwick P, Trowbridge EA. Oscillations of a rigid sphere embedded in an infinite elastic solid. I. Torsional oscillations. *Proceedings of the Cambridge Philosophical Society* 1967; **63**:1189–1205.
5. Chadwick P, Trowbridge EA. Oscillations of a rigid sphere embedded in an infinite elastic solid. II. Rectilinear oscillations. *Proceedings of the Cambridge Philosophical Society* 1967; **63**:1207–1227.
6. Cheng PW, Bierbooms WAAM. Distribution of extreme gust loads of wind turbines. *Journal of Wind Engineering and Industrial Aerodynamics* 2001; **89**:309–324.
7. Deeks AJ, Wolf JP. Recursive procedure for Greens function of disk embedded in layered half-space using strength-of-materials approach. (Submitted to *Earthquake Engineering and Structural Dynamics* for review and possible publication).
8. Ehlers G. The effect of soil flexibility on vibrating systems. *Beton und Eisen* 1942; **41**(21/22):197–203 (in German).
9. Emperador JM, Dominguez J. Dynamic response of axisymmetric embedded foundations. *Earthquake Engineering and Structural Dynamics* 1989; **18**:1105–1117.
10. Ferguson MC (Ed.), Kühn M, Bierbooms WAAM, Cockerill TT, Göransson B, Harland LA, van Bussel GJW, Vugts JH, Hes R. *Opti-OWECS Final Report Vol. 4: A Typical Design Solution for an Offshore Wind Energy Conversion System*, Institute for Wind Energy, Delft University of Technology, Delft, 1998.
11. Kausel E. Personal communication 1990.
12. Kausel E. Personal communication 2002.
13. Lam NTK. Personal communication 2003.

14. Lam NTK, Wilson JL, Hutchinson G. Generation of synthetic earthquake accelerograms using seismological modelling: a review. *Journal of Earthquake Engineering* 2000; **4**(3):321–354.
15. Luco JE, Mita A. Response of a circular footing on a uniform half-space to elastic waves. *Earthquake Engineering and Structural Dynamics* 1987; **15**:105–118.
16. Luco JE, Wong HL. Seismic response of foundations embedded in a layered half-space. *Earthquake Engineering and Structural Dynamics* 1987; **15**:233–247.
17. Meek JW, Veletsos AS. Simple models for foundations in lateral and rocking motion. *Proceedings of the 5th World Conference on Earthquake Engineering*, IAEE, Rome 1974; vol. 2, 2610–2613.
18. Meek JW, Wolf JP. Insights on cutoff frequency for foundation on soil layer. *Earthquake Engineering and Structural Dynamics* 1991; **20**:651–665, also in *Proceedings of the 9th European Conference on Earthquake Engineering*, EAEE, Moscow 1990; vol. 4-A, 34–43.
19. Meek JW, Wolf JP. Cone models for homogeneous soil. I. *Journal of the Geotechnical Engineering Division*, ASCE 1992; **118**(5):667–685.
20. Meek JW, Wolf JP. Cone models for soil layer on rigid rock. II. *Journal of the Geotechnical Engineering Division*, ASCE 1992; **118**(5):686–703.
21. Meek JW, Wolf JP. Cone models for nearly incompressible soil. *Earthquake Engineering and Structural Dynamics* 1993; **22**:649–663.
22. Meek JW, Wolf JP. Why cone models can represent the elastic half-space. *Earthquake Engineering and Structural Dynamics* 1993; **22**:759–771.
23. Meek JW, Wolf JP. Cone models for an embedded foundation. *Journal of the Geotechnical Engineering Division*, ASCE 1994; **120**(1):60–80.
24. Meek JW. Continuum model for built-in piles in layered soil. *Bautechnik* 1995; **72**(2):116–123 (in German).
25. Pais A, Kausel E. Approximate formulas for dynamic stiffnesses of rigid foundations. *Soil Dynamics and Earthquake Engineering* 1988; **7**:213–227.
26. Pak RYS, Gobert AT. Forced vertical vibration of rigid discs with arbitrary embedment. *Journal of Engineering Mechanics*, ASCE 1991; **117**:2527–2548.
27. Pak RYS, Abedzadeh F. Forced torsional oscillation from the interior of a half-space. *Journal of Sound and Vibration* 1993; **160**(3):401–415.
28. Sedgewick R. *Algorithms*. 2nd Edition, Addison-Wesley 1988.
29. Selvadurai APS. The dynamic response of a rigid circular foundation embedded in an isotropic medium of infinite extent. *International Symposium on Soils under Cyclic and Transient Loading*, Swansea 1980; 597–608.
30. Selvadurai APS. Rotary oscillations of a rigid disc inclusion embedded in an isotropic elastic infinite space. *International Journal of Solids and Structures* 1981; **17**:493–498.
31. Tassoulas JL. Personal communication 2002.
32. Veletsos AS, Nair VD. Torsional vibrations of viscoelastic foundations. *Journal of the Geotechnical Engineering Division*, ASCE 1974; **100**(3):225–246.
33. Veletsos AS, Nair VD. Response of torsionally excited foundations. *Journal of the Geotechnical Engineering Division*, ASCE 1974; **100**(3):476–482.
34. Veletsos AS, Verbic B. Basic response functions for elastic foundations. *Journal of the Soil Mechanics and Foundation Division*, ASCE 1974; **100**:189–202.
35. Veletsos AS, Wei YT. Lateral and rocking vibration of footings. *Journal of the Soil Mechanics and Foundation Division*, ASCE 1971; **97**:1227–1248.

36. Waas G, Hartmann HG, Werkle H. Damping and stiffness of foundations on inhomogeneous media. *Proceedings of the 9th World Conference on Earthquake Engineering*, IAEE, Tokyo and Kyoto 1988; vol. 3, 343–348.
37. Wolf JP. *Foundation Vibration Analysis Using Simple Physical Models*. Prentice-Hall: Englewood Cliffs, NJ, 1994.
38. Wolf JP, Meek JW. Cone models for a soil layer on a flexible rock half-space. *Earthquake Engineering and Structural Dynamics* 1993; **22**:185–193.
39. Wolf JP, Meek JW. Rotational cone models for a soil layer on flexible rock half-space. *Earthquake Engineering and Structural Dynamics* 1994; **23**:909–925.
40. Wolf JP, Meek JW. Dynamic stiffness of foundation on layered soil half-space using cone frustums. *Earthquake Engineering and Structural Dynamics* 1994; **23**:1079–1095.
41. Wolf JP, Meek JW, Song Ch. Cone models for a pile foundation. *Piles Under Dynamic Loads*, edited by S. Prakash. Geotechnical Special Publication No. 34, ASCE 1992, 94–113.
42. Wolf JP, Preisig M. Dynamic stiffness of foundation embedded in layered half-space based on wave propagation in cones. *Earthquake Engineering and Structural Dynamics* 2003; **32**:1075–1098.

Dictionary

For the reader who has not studied civil engineering at an Anglo-Saxon university, some key expressions are translated into French and German. Only words that are specific to the topic discussed in this book are listed. For the others, the dictionaries in soil mechanics and foundation engineering can be consulted. Expressions which are the same or very similar in the three languages, such as *foundation* and *vibration* are not addressed. In general, only one form, that which is used most in the book, is listed. For instance, only the adjective *embedded* is translated and not the noun *embedment* or the verb *to embed*. Coined expressions are specified as used in the text, for instance *trapped mass* and not *mass, trapped*.

ENGLISH	FRENCH	GERMAN
amplification	amplification (f)	Aufschaukelung (f)
apex	sommet (m)	Spitze (f)
bar	barre (f)	Stab (m)
beam	poutre (f)	Balken (m)
binary tree	arborescence binaire (f)	Binaerbaum (m)
blade	pale (f)	Fluegel (m)
bounded	fini	endlich
boundary-element method	méthode (f) des éléments aux frontières	Randelement-Methode (f)
caisson	caisson (m)	Senkkasten (m)
characteristic length	longueur (f) caractéristique	charakteristische Laenge (f)
chimney stack	cheminée (f)	Kaminschlott (m)
compatibility	compatibilité (f)	Vertraeglichkeit (f)
complex frequency response function	fonction (f) complexe de réponse en fréquence	komplexe Frequenz-Antwort Funktion (f)
cone	cone (m)	Kegel (m)
constrained modulus	module (m) de compressibilité	Kompressionsmodul (m)
crank	manivelle (f)	Kurbel (f)
cutoff frequency	fréquence (f) de limite	Grenzfrequenz (f)

ENGLISH	FRENCH	GERMAN
damping	amortissement (m)	Daempfung (f)
dashpot	amortisseur (m)	Daempfer (m)
degree of freedom	degré (m) de liberté	Freiheitsgrad (m)
dilatational wave	onde (f) de dilatation	Dilatationswelle (f)
disk	disque (m)	Scheibe (f)
dynamic flexibility	flexibilité (f) dynamique	dynamische Flexibilitaet (f)
dynamic stiffness	rigidité (f) dynamique	dynamische Steifigkeit (f)
effective foundation input motion	mouvement (m) effectif imposé sur la fondation	effektive Fundation-Eingabe Bewegung (f)
elastic modulus	module (m) élastique	elastischer Modul (m)
elastodynamics	elastodynamique (f)	Elastodynamik (f)
embedded	enterré	eingebettet
equilibrium	équilibre (m)	Gleichgewicht (n)
excavation	excavation (f)	Aushub (m)
excitation	excitation (f)	Erregung (f)
frame	cadre (m)	Rahmen (m)
free field	champ (m) libre	freies Feld (n)
free surface	surface (f) libre	freie Oberflaeche (f)
frequency response function	fonction (f) de réponse en fréquence	Frequenz-Antwort Funktion (f)
full space	espace (m) infini	Vollraum (m)
half-space	demi-espace (m)	Halbraum (m)
harmonic	harmonique	harmonisch
hemi-ellipsoid	hemi-ellipsoide (m)	Halb-Ellipsoid (n)
hysteretic damping	amortissement (m) hystérétique	Hysteresis Daempfung (f)
incident wave	onde (f) incidente	ankommende Welle (f)
inclusion	inclusion (f)	Einschluss (m)
incompressible	incompressible	unzusammendrueckbar
inertial load	charge (f) d'inertie	Traegheitslast (f)
interaction force	force (f) d'interaction	Interaktionskraft (f)
impedance	impedance (f)	Impedanz (f)
layer	couche (f)	Schicht (f)
lift	portance (f)	Auftrieb (m)
lumped-parameter model	modèle (m) avec paramètres concentrés	Model (n) mit konzentrierten Parametern
mass density	masse (f) volumique	Massendichte (f)
mass moment of inertia	moment (m) d'inertie massique	Massentraegheitsmoment (n)
moment of inertia	moment (m) d'inertie	Traegheitsmoment (n)
particle motion	mouvement (m) des particules	Teilchenbewegung (f)
piston	piston (m)	Kolben (m)
polar moment of inertia	moment (m) d'inertie polaire	polares Massentraegheitsmoment (n)

ENGLISH	FRENCH	GERMAN
primary system	système (m) de base	Grundsystem (n)
propagation	propagation (f)	Ausbreitung (f)
radiation	radiation (f), rayonnement (m)	Abstrahlung (f)
reactor building	bâtiment (m) de reacteur	Reaktorgebaeude (n)
reciprocating engine	moteur (m) alternatif	Kolbenmotor (m)
redundant	hyperstatique (f)	ueberzaehlige Groesse (f)
reflection	reflection (f)	Reflektion (f)
refraction	refraction (f)	Refraktion (f)
rock outcrop	affleurement (m) rocheux	zu tage tretender Fels (m)
response spectrum	spectre (m) de réponse	Antwortspektrum (n)
rocking	rotation (f)	Rotation (f)
rotary inertia	inertie (f) de rotation	Rotationstraegheit (f)
scaled boundary finite-element method	méthode (f) des éléments aux frontières similaires	skalierte Rand Finite Element Methode (f)
seismic	sismique	seismisch
shear distortion	distorsion (f) de cisaillement	Schubverzerrung (f)
shear modulus	module (m) de cisaillement	Schubmodul (m)
shear wave	onde (f) de cisaillement	Schubwelle (f)
site	site (m)	Standort (m)
soil-structure interaction	interaction (f) sol-structure	Boden-Bauwerk Interaktion (f)
spring	ressort (m)	Feder (f)
spring-dashpot-mass model	modèle (m) de ressort, amortisseur et masse	Feder-Daempfer-Massen Modell (n)
static stiffness	rigidité (f) statique	statische Steifigkeit (f)
steady-state response	réponse (f) harmonique	harmonische Antwort (f)
strength of materials	resistance (f) des matériaux	Festigkeit (f)
structure-soil interface	interface (f) structure-sol	Struktur-Boden Kontaktflaeche (f)
suction caisson	caisson (m) foncé par dépression d'air	Saug-Senkkasten (m)
substructure	sous-structure (f)	Substruktur (f)
termination criterion	critère (m) de termination	Abbruchkriterium (n)
thin-layer method	méthode (f) des couches minces	Methode (f) der duennen Schichten
trapped mass	masse (f) enfermée	eingeschlossene Masse (f)
truncated cone	cone (m) tronqué	Kegelstumpf (m)
unbounded	infini	unendlich

Index

Arbitrary excitation, 152–3
Aspect ratio:
  computer procedure, 181
  definition, 13
  horizontal:
    full-space calibration, 53
    half-space calibration, 51
  rocking:
    full-space calibration, 53
    half-space calibration, 40, 51
    incompressible, 44, 51
  torsional, 38, 51
  vertical:
    full-space calibration, 26
    half-space calibration, 15, 33, 51
    incompressible, 44, 51

Binary tree, 76–7
Block on layered half-space, 204–6
Boundary element method, 11, 105

Calibration:
  full-space, 26, 53, 54, 119
  half-space, 15, 33, 34, 38, 40, 44, 51, 54, 119
Complex amplitude, 146
Computer implementation:
  embedded disk, 93–7
  reflection–refraction, 72–80
CONAN, 170–6
Cone:
  double, 24–6, 52–4, 70
  features, 7, 28
  initial, 6, 29–54
  radiating, 18, 22
  rotational, 35–40
    incompressible, 44–6

    reflection–refraction, 60–2
  translational, 30–5
    incompressible, 44–6
    reflection–refraction, 55–9
Constrained modulus:
  definition, 12
  incompressible, 30
Control point:
  cylinder embedded in half-space, 98
  load in equation of motion, 160–1
  location, 157
  structure embedded in layered half-space, 131
Correspondence principle, 29, 35
Crank mechanism, 121–3
Cutoff frequency:
  layer fixed at base, 69
  multi-layered half-space, 113–14
Cylinder embedded in half-space:
  dynamic stiffness, 98–101
  effective foundation input motion, 100–2
  incompressible, 101–3
  static stiffness, 98–9
Cylinder embedded in layer overlying half-space:
  dynamic stiffness, 101,103–5
  modelling, 101
Cylinder embedded in multi-layered half-space, 108–12
  definition of site, 107
  dynamic stiffness, 109–12
  incompressible, 113–14

Damping coefficient, 5
Design criterion, 129
Dilatational wave:
  definition, 12
  velocity, 12
Dimensionless frequency (*see* Frequency, dimensionless)
Dirichlet-to-Neumann method, 11

216 Index

Disk embedded in full-space:
    dynamic stiffness:
        rocking, 54
        vertical, 54
Disk embedded in half-space:
    dynamic stiffness:
        torsional, 72–3
        vertical, 72–3
    wave pattern, 72
Disk on half-space, 46–52
    dynamic stiffness:
        horizontal, 47
        rocking, 49
        torsional, 48
        vertical, 48
Disk on inhomogeneous half-space, 112–13
    dynamic stiffness, 112–13
    modelling, 112
Disk on layer fixed at base, 19–22
Disk on layer overlying half-space:
    dynamic stiffness:
        rotation, 68, 70
        translation, 68–9
Disk on multi-layered half-space:
    definition of site, 107
    dynamic stiffness, 82–3, 107–9
Double cone, 24–6, 52–4
Doubly-asymptotic approximation, 17
Driving force, 156
Dynamic flexibility:
    free-field, 86–7
    single-degree-of-freedom system, 149
    two-degree-of-freedom system, 150
Dynamic load:
    reciprocating machine, 123, 125–6
    wind turbine tower, 138
Dynamic soil-structure interaction (*see* Soil-structure interaction)
Dynamic stiffness:
    computer procedure, 188
    cone rotation:
        rocking, 40
        torsional, 37–8
    cone translation, 32–3
    cylinder embedded in half-space, 98–103
    cylinder embedded in layer overlying half-space, 101, 102–5
    cylinder embedded in multi-layered half-space, 109–12
    definition, 4
    disk embedded in full-space, 54
    disk embedded in half-space, 72–3
    disk on half-space:
        horizontal, 46–7
        rocking, 47, 49
        torsional, 47–8
        vertical, 46–8
    disk on inhomogeneous half-space, 112–13
    disk on multi-layered half-space, 82–83, 107–9
    formulation, 87–91
        coupling, 90
        horizontal, 90
        rocking, 90
        torsional, 89
        vertical, 89
    free field, 87
        computer procedure, 187
    hemi-ellipsoid embedded in half-space, 118
    incompressible, 47–9, 102–3, 115
    magnitude, 6, 54, 108, 109
    physical interpretation, 5, 127
    reflection coefficient:
        rotational cone, 62
        translational cone, 58
    single-degree-of-freedom system, 148
    soil column, 159–60
    soil system ground with excavation, 155–56
    sphere embedded in full-space, 119
    structure, 155
    suction caisson, 139
    two-degree-of-freedom system, 150

Earthquake:
    horizontal:
        effective foundation input motion, 6, 91
        free field, 157–61
        Newcastle earthquake, 132
    vertical:
        effective foundation input motion, 6, 89
        free field, 157–61
        Newcastle earthquake, 204
Echo constant:
    flexibility, 200–1, 208
    stiffness, 202–4
Effective foundation input motion:
    cylinder embedded in half-space, 100–2
    definition, 6, 155
    equation of motion, 156
    horizontal earthquake:
        computer procedure, 192
        formulation, 91
    structure embedded in layered half-space, 131–2
    vertical earthquake:
        computer procedure, 191
        formulation, 89
Embedded:
    fully, 3
    partially, 3
Engineering accuracy, 7, 28, 108
Equation of motion:
    block on layered half-space, 203

cone:
    coupled horizontal-rocking, 42
    rotational, 36, 206
    translational, 14, 31
    unified, 43
free field, 160–1
machine foundation, 127–9
prismatic bar, 162
soil-structure interaction, 156
structure embedded in layered half-space, 133
wind turbine tower with suction caisson, 140

Frequency-domain response analysis, 146–53
Frequency-response function, 148–50
Foundation input motion, effective (*see* Effective foundation input motion)
Fourier integral, 153
Fourier series:
    block on layered half-space, 204
    definition, 151
    structure embedded in layered half-space, 131
    wind turbine tower with suction caisson, 137
Fourier transform pair, 153
Free field:
    analysis, 157–61
    computer procedure, 190
    definition, 1
    dynamic flexibility, 86–7
    dynamic stiffness, 87
    half-space, 100
Frequency, dimensionless:
    definition, 5, 33, 37, 148
    hysteretic damping, 35

Generating function, 19, 63, 197
Ground (*see* System ground with excavation)

Hemi-ellipsoid embedded in half-space, 115–18
    dynamic stiffness, 118
    modelling, 117
Hysteretic:
    correspondence principle, 35
    damping ratio, 1

Impedance, 59, 166–7
Incident wave:
    concept, 18
    definition, 7
    layer fixed at base, 21
    layer overlying half-space, 63
    prismatic bar, 163
    rotational cone, 60
    time domain, 199–207
    translational cone, 56

Incompressible:
    dynamic stiffness:
        cylinder embedded in half-space, 102–3
        cylinder embedded in multi-layered half-space, 114–15
        disk embedded in half-space, 72–3
        disk on half-space, 44–9
    embedded, 91
    rocking cone, 51
        aspect ratio, 44
        dynamic stiffness, 46
        trapped mass moment of inertia, 44
        wave velocity, 44
    vertical cone, 51
        aspect ratio, 44
        dynamic stiffness, 46
        trapped mass, 44
        wave velocity, 44
Interaction force-displacement relationship:
    rotational cone, 37, 43
    soil system ground with excavation, 156, 202
    translational cone, 16, 32, 43

Lamé constants, 12
Lumped-parameter model:
    definition, 11
    rotational cone, 47, 49–50, 51
    translational cone, 13, 16, 51

Machine foundation, 121–9
Magnitude, phase angle, 146

Nearly-incompressible (*see* incompressible)
Newcastle earthquake, 131–2, 204

Opening angle (*see* Aspect ratio)
Outward wave propagation:
    prismatic bar, 164
    rotational cone, 35–40
        time domain, 207
    translational cone, 30–5
        time domain, 197–99

Periodic excitation, 151–2
Primary system, 27, 85

Radiation condition:
    cone, 16
    prismatic bar, 166
Radiation damping:
    cone, 16
    cutoff frequency, 68–9, 113–14
    definition, 4, 14
    layer fixed at base, 22
    multi-layered half-space fixed at base, 107
    prismatic bar, 166

Radiation dashpot, 17, 32, 39
Reciprocating machine, 121, 124
Recursion, 77
Redundants, 27, 85
Reflected wave:
  computer procedure, 183
  concept, 18
  definition, 7
  layer fixed at base, 21
  prismatic bar, 167
  rotational cone, 61
  translational cone, 57
    time domain, 199–201
Reflection coefficient:
  computer procedure, 185
  high frequency limit, 59, 200
  jth impingement, 66
  low frequency limit, 59, 199–200
  prismatic bar, 167
  rotational cone, 60–2
  translational cone, 55–9
    time domain, 199–200
Refracted wave:
  computer procedure, 183
  concept, 18
  definitioin, 7
  prismatic bar, 167
  rotational cone, 61
  translational cone, 57
    time domain, 199
Response spectrum, 6, 132
Rigorous method, 11
Rock outcrop, 157, 160

Shear wave:
  definition, 13
  velocity, 13
Soil-structure interaction:
  block on layered half-space, 204–5
  definition, 3–4
  equation of motion, 154–7
  machine foundation, 121–9
  structure embedded in layered half-space, 129–35
  wind turbine tower with suction caisson, 135–41
Sphere embedded in full-space, 117–20
  dynamic stiffness, 119
  modelling, 119
Spring-dashpot-mass model (*see* Lumped-parameter model)

Static stiffness:
  cone:
    rotation, 37, 39
    translation, 15, 26
  cylinder embedded in half-space, 98–9
  decomposition, 5
  disk embedded in full-space:
    horizontal, 53
    rocking, 53
    torsional, 53
    vertical, 26, 53
  disk on half-space:
    coupling, 109
    horizontal, 34
    rocking, 40
    torsional, 38
    vertical, 15
  hemisphere embedded in half-space, torsional, 117
  sphere embedded in full-space:
    rectilinear, 117
    torsional, 120
Structure embedded in layered half-space, 129–35
Substructure
  soil, 3
  structure, 3
  system ground with excavation, 154–5
Suction caisson, 135–6
System ground with excavation, 2, 6, 89, 91, 155–7

Thin-layer method, 11, 83
Termination criterion, 81–3
  depth:
    definition, 82
    multi-layered half-space, 107, 109
  magnitude, 81
Trapped mass and trapped mass moment of inertia:
  computer procedure, 182
  embedment, 88, 90
  incompressible, 44–6
Traversal, 77
Truncated semi-infinite cone (*see* Cone)

Wave velocity:
  computer procedure, 180
  dilatational, 12
  shear, 13
Wind turbine tower with suction caisson, 135–41